Technology of Proximal Probe Lithography

SPIE Institutes for Advanced Optical Technologies series

Transformations in Optical Signal Processing, William T. Rhodes, James R. Fienup, Bahaa E. A. Saleh, Editors, 1984, SPIE Volume 373 (Out of print)

Optical and Hybrid Computing, Harold H. Szu, Editor, 1987, SPIE Volume 634

Photonics: High Bandwidth Analog Applications, James Chang, Editor, 1987, SPIE Volume 648

Large-Area Chromogenics: Materials and Devices for Transmittance Control, Carl M. Lampert, Claes G. Granqvist, Editors, 1990, Volume IS 4

Dosimetry of Laser Radiation in Medicine and Biology, Gerhard J. Müller, David H. Sliney, Editors, 1989, Volume IS 5

Future Directions and Applications in Photodynamic Therapy, Charles J. Gomer, Editor, 1990, Volume IS 6

Automatic Object Recognition, Hatem N. Nasr, Editor, 1991, Volume IS 7

Holography, Pál Greguss, Tung H. Jeong, Editors, 1991, Volume IS 8

Invisible Connections: Instruments, Institutions, and Science, Robert Bud, Susan E. Cozzens, Editors, 1992, Volume IS 9

Technology of Proximal Probe Lithography, Christie R. K. Marrian, Editor, 1993, Volume IS 10

Medical Optical Tomography, Gerhard J. Müller et al., Editors, 1993, Volume IS 11

Cover:

Feature patterned on H-passivated, n-Si (111) $\rho = 10$ Ωcm, using a STM operating in air. From J. A. Dagata and J. Schneir, "Scanning tunneling microscope-based fabrication and characterization on passivated semiconductor surfaces," p. 104.

Technology of Proximal Probe Lithography

Volume IS 10

Institutes for
Advanced Optical
Technologies

Christie R. K. Marrian
Naval Research Laboratory
Editor

Roy F. Potter
Series Editor

SPIE OPTICAL ENGINEERING PRESS

A publication of SPIE—The International Society for Optical Engineering
Bellingham, Washington USA

Library of Congress Cataloging-in-Publication Data

Technology of proximal probe lithography /
    Christie Marrian, editor.
        p. cm. — (SPIE institutes for advanced optical technologies ; v. IS 10)
        Includes bibliographical references.
        ISBN 0-8194-1232-5 (hardcover). — ISBN 0-8194-1233-3 (softcover)
        1. Molecular electronics. 2. Lithography, Electron beam. 3. Probes
(Electronic instruments) 4. Nanotechnology. I. Marrian, Christie R. K.
II. Series.
TK7874.8.T43 1993
621.3815'31—dc20   93-8684
                                                                                                           CIP

Published by
SPIE—The International Society for Optical Engineering
P.O. Box 10
Bellingham, Washington 98227-0010

Copyright © 1993 The Society of Photo-Optical Instrumentation Engineers

All rights reserved. No part of this publication may be reproduced or distributed in any form or by any means without written permission of the publisher.

Printed in the United States of America

Introduction to the Series

The Institute Series of the SPIE Press, of which this is the 10th volume, has gone through some evolutionary changes over the years. While changes such as format, book size, etc., are readily apparent, the other changes in author relations, editing procedures, and timeliness are more subtle in nature; they were undertaken to improve SPIE's performance in reaching the primary goal for the series. That goal is to provide, in a timely manner, authoritative overall introductions and reviews of an emerging technology based on or related to optics and optoelectronics.

The subject of "probe lithography" lies on the frontier for those technologies emphasizing nanodimensions, i.e., technologies based on a regime of physical dimensions of less than 100 nanometers. In addition, the papers collected in this volume reflect the extremely rapid progress from the announcement of the tunneling microscope to the potential applications described here. The papers in this volume demonstrate that fundamental basic research is not far ahead (in time) of technology applications. Indeed, the synergism existing between research and applications in optoelectronic fields can be ignored only at peril to a healthy technology.

Although the emphasis of this book is on lithography, it could prove valuable to device engineers, especially those working in nanotechnologies. It is evident that the field of metrology will develop new concepts and language along with the new limits for precision and accuracy for nanotechnologies. This new metrology will have a large influence on the devices and instruments of the future.

Roy F. Potter
General Editor
SPIE Institutes for Advanced Optical Technologies

Table of Contents

ix   *List of Contributors*

INTRODUCTION

3   **Technology of proximal probe lithography: an overview**
    J. A. Dagata, National Institute of Standards and Technology;
    C. R. K. Marrian, Naval Research Lab.

PART I.   NANOLITHOGRAPHY

16   **Principles and techniques of STM lithography**
     M. A. McCord, IBM Corp.; R. F. W. Pease, Stanford Univ.

33   **Role of scanning probe microscopes in the development of nanoelectric devices**
     A. Majumdar, S. M. Lindsay, Arizona State Univ.

58   **Low voltage e-beam lithography with the STM**
     C. R. K. Marrian, E. A. Dobisz, Naval Research Lab.; J. A. Dagata, National Institute of Standards and Technology

74   **Nanolithography and atomic manipulation on silicon surfaces by STM**
     F. Grey, A. Kobayashi, H. Uchida, D. H. Huang, M. Aono, Aono Atomcraft Project (Japan)

100 **Scanning tunneling microscope–based fabrication and characterization on passivated semiconductor surfaces**
     J. A. Dagata, J. Schneir, National Institute of Standards and Technology

111 **Scanning tunneling microscope–based nanolithography for electronic device fabrication**
     J. W. Lyding, R. T. Brockenbrough, P. Fay, J. R. Tucker, K. Hess, Univ. of Illinois; T. K. Higman, Univ. of Minnesota

127 **Arrayed lithography using STM-based microcolumns**
     T. H. P. Chang, IBM T.J. Watson Research Ctr.; L. P. Muray, Cornell Univ.; U. Staufer, Univ. of Basel (Switzerland); M. A. McCord, D. P. Kern, IBM T.J. Watson Research Ctr.

PART II.   FABRICATION

162 **Nanofabrication by scanning probe instruments: methods, potential applications, and key issues**
     R. Wiesendanger, Univ. of Basel (Switzerland)

188 **Direct writing of metallic nanostructures with the scanning tunneling microscope**
     A. L. de Lozanne, W. F. Smith, E. E. Ehrichs, Univ. of Texas/Austin

200 **Fabrication of nanometer-scale structures**
M. H. Nayfeh, Univ. of Illinois/Urbana-Champaign

218 **Modification and manipulation of layered materials using scanned probe microscopies**
C. M. Lieber, Harvard Univ.

234 **Atomic force microscopy experimentation at surfaces: hardness, wear, and lithographic applications**
T. A. Jung, Univ. of Basel (Switzerland) and Paul Scherrer Institute (Switzerland); A. Moser, Univ. of Basel (Switzerland); M. T. Gale, Paul Scherrer Institute (Switzerland); H. J. Hug, U. D. Schwarz, Univ. of Basel (Switzerland)

268 **Fabrication and characterization using scanned nanoprobes**
G. C. Wetsel Jr., S. E. McBride, H. M. Marchman, Univ. of Texas/Dallas

289 **Ballistic electron emission microscopy: from electron transport physics to nanoscale materials science**
H. D. Hallen, AT&T Bell Labs.

PART III. METROLOGY

322 **Generating and measuring displacements up to 0.1 m to an accuracy of 0.1 nm: Is it possible?**
E. C. Teague, National Institute of Standards and Technology

364 **Metrology with scanning probe microscopes**
J. E. Griffith, AT&T Bell Labs.; D. A. Grigg, Digital Instruments; G. P. Kochanski, M. J. Vasile, AT&T Bell Labs.; P. E. Russell, North Carolina State Univ.

390 **Metrology applications of scanning probe microscopes**
L. A. Files-Sesler, J. N. Randall, F. G. Celii, Texas Instruments Inc.

List of Contributors

M. Aono
Aono Atomcraft Project
Exploratory Research for Advanced Technology
Japan Research Development Corp.
Tokodai 5-9-9
Tsukuba, Ibaraki 300-26, Japan

Roger T. Brockenbrough
Univ. of Illnois
Dept. of Electrical and Computer Engineering
Urbana, IL 61801, USA

Francis G. Celii
Texas Instruments Inc.
MS 147
P.O. Box 655936
Dallas, TX 75265-5936, USA

T. H. Chang
IBM T.J. Watson Research Ctr.
Yorktown Heights, NY 10598, USA

John A. Dagata
National Institute of Standards and Technology
Bldg. 220, Rm. A107
Gaithersburg, MD 20899-0001, USA

A. L. de Lozanne
Univ. of Texas/Austin
Physics Dept.
Austin, TX 78712-1020, USA

Elizabeth A. Dobisz
Naval Research Lab.
Electronics Science & Technology Div.
Code 6864
4555 Overlook Dr.
Washington, DC 20375-5347, USA

E. E. Ehrichs
Univ. of Texas
Dept. of Physics
Austin, TX 78712-1081, USA

Patrick Fay
Univ. of Illinois
Dept. of Electrical and Computer Engineering
Urbana, IL 61801, USA

Leigh A. Files-Sesler
Texas Instruments Inc.
Central Research Labs.
MS 134
13588 N. Central Expwy.
Dallas, TX 75243-1108, USA

Michael T. Gale
Paul Scherrer Institute
Optics Group
Badenerstrasse 569
CH-8048 Zürich, Switzerland

François Grey
Aono Atomcraft Project
Exploratory Research for Advanced Technology
Japan Research Development Corp.
Tokodai 5-9-9
Tsukuba, Ibaraki 300-26, Japan

Joseph E. Griffith
AT&T Bell Labs.
Rm. 6F-225
P.O. Box 636
Murray Hill, NJ 07974-0636, USA

David A. Grigg
Digital Instruments
520 E. Montecito
Santa Barbara, CA 93103-3252

Hans D. Hallen
North Carolina State Univ.
Dept. of Physics
P.O. Box 8202
Raleigh, NC 27695-8202, USA

Karl Hess
Univ. of Illinois
Beckman Institute
405 North Mathews Avenue
Urbana, IL 61801-2325, USA

T. K. Higman
Univ. of Minnesota
Dept. of Electrical Engineering
200 Union St. SE
Minneapolis, MN 55455-0160

D. H. Huang
Aono Atomcraft Project
Exploratory Research for Advanced Technology
Japan Research Development Corp.
Tokodai 5-9-9
Tsukuba, Ibaraki 300-26, Japan

H. J. Hug
Univ. of Basel
Institute of Physics
Klingelbergstrassse 82
CH-4056 Basel, Switzerland

T. A. Jung
Univ. of Basel
Institute of Physics
Klingelbergstrassse 82
CH-4056 Basel, Switzerland

Dieter P. Kern
IBM Thomas J. Watson Research Ctr.
MS 17-207
P.O. Box 218
Yorktown Heights, NY 10598-0218, USA

A. Kobayashi
Aono Atomcraft Project
Exploratory Research for Advanced Technology
Japan Research Development Corp.
Tokodai 5-9-9
Tsukuba, Ibaraki 300-26, Japan

G. P. Kochanski
AT&T Bell Labs.
Murray Hill, NJ 07974, USA

Charles Lieber
Harvard Univ.
Applied Sciences
Dept. of Chemistry
Cambridge, MA 02138, USA

Stuart M. Lindsay
Arizona State Univ.
Dept. of Physics
Tempe, AZ 85287-0001, USA

Joseph W. Lyding
Univ. of Illinois
Dept. of Electrical and Computer Engineering
Urbana, IL 61801, USA

Arun Majumdar
Univ. of California/Santa Barbara
Dept. of Mechanical & Environmental Engineering
Santa Barbara, CA 93106-5070, USA

Herschel M. Marchman
AT&T Bell Labs.
Rm. 7E-411
600 Mountain Ave.
Murray Hill, NJ 07974-2008, USA

Christie R. K. Marrian
Naval Research Lab.
Electronics Science & Technology Div.
Code 6864
4555 Overlook Dr.
Washington, DC 20375-5347, USA

Sterling E. McBride
Univ. of Texas/ Dallas
Erik Jonsson School of Engineering and Computer Sciences
MP 33
P.O. Box 830688
Richardson, TX 75083-0688, USA

M. A. McCord
IBM Corp.
Semiconductor Research and Development Ctr.
P.O. Box 218
Yorktown Heights, NY 10598, USA

A. Moser
Univ. of Basel
Institute of Physics
Klingelbergstrasse 82
CH-4056 Basel, Switzerland

L. P. Muray
Cornell Univ.
Knight Lab.
National Nanofabrication Facility
Ithaca, NY 14853, USA

Munir Nayfeh
Univ. of Illinois/Urbana-Champaign
Dept. of Physics
1110 W. Green St.
Urbana, IL 61801, USA

R. F. W. Pease
Stanford Univ.
Solid State Electronics Lab.
MC 4055, Electrical Engineering Dept.
McCullough 204
Stanford, CA 94305-4055, USA

John N. Randall
Texas Instruments Inc.
Central Research Labs.
MS 134
13588 N. Central Expressway
Dallas, TX 75243-1108, USA

Phillip E. Russell
North Carolina State Univ.
Materials Science Dept.
P.O. Box 7916
Raleigh, NC 27695-0001, USA

Jason Schneir
National Institute of Standards and Technology
Building 220, Room A107
Gaithersburg, MD 20899-0001, USA

U. D. Schwarz
Univ. of Hamburg
Institute of Applied Physics
Jurginsstrasse 11
20355 Hamburg, Germany

W. F. Smith
Univ. of Texas/Austin
Physics Dept.
Austin, TX 78712-1020, USA

U. Staufer
Univ. of Basel
Institute of Physics
Kingelbergstr. 82
CH-4056 Basel, Switzerland

E. C. Teague
National Institute of Standards and Technology
Precision Engineering Div.
Building 220, Room A117
Gaithersburg, MD 20899, USA

John R. Tucker
Univ. of Illinois
Dept. of Electrical and Computer Engineering
Urbana, IL 61801, USA

H. Uchida
Aono Atomcraft Project
Exploratory Research for Advanced Technology
Japan Research Development Corp.
Tokodai 5-9-9
Tsukuba, Ibaraki 300-26, Japan

Michael J. Vasile
AT&T Bell Labs.
MS 2C-105
600 Mountain Avenue
New Providence, NJ 07974-2008, USA

Grover C. Wetsel
Univ. of Texas/Dallas
Erik Jonsson School of Engineering and Computer Sciences
MP 33
P.O. Box 830688
Richardson, TX 75083-0688, USA

Roland Wiesendanger
Univ. of Basel
Dept. of Physics
Klingelbergstrasse 82
CH-4056 Basel, Switzerland

Introduction

THE TECHNOLOGY OF PROXIMAL PROBE LITHOGRAPHY: AN OVERVIEW

John A. Dagata
Precision Engineering Division
National Institute of Standards and Technology
Gaithersburg MD 20899

and

Christie R. K. Marrian
Electronics Science and Technology Division
Naval Research Laboratory
Washington DC 20375

ABSTRACT

The papers presented in this volume illustrate the tremendous diversity of proximal probe lithography (PPL) which we define as the modification and manipulation of material with nanometer scale control. Indeed, the papers highlight examples where the precision achieved with a proximal probe is not attainable with conventional tools. The field has matured following the excitement generated by the first dramatic demonstrations of material manipulation with proximal probes, such as the scanning tunneling microscope. Research has been directed to uncover the fundamental mechanisms responsible and to put the field on a sound scientific basis. The success of these efforts can be gauged by the present interest in technological applications based on the unique capabilities of PPL. This overview discusses the present status and viability of PPL in general terms with specific issues being considered in the rest of the volume. The need to distinguish between research directed towards the development of a viable technology and that required for basic science studies is highlighted. Further, it is argued that to fully take advantage of the unique capabilities of PPL, a co-evolutionary development with other emerging technologies, such as nanoelectronics, nanometrology and molecular nanotechnology, will be needed.

1. INTRODUCTION

This monograph is concerned with the technology of proximal probe lithography (PPL). The present interest in PPL research is an outcome of the tremendous success of scanned probe microscopes, such as the scanning tunneling microscope (STM), at providing real space structural and electronic information about surfaces with atomic resolution.[1,2] These new imaging and modification capabilities have appeared in research laboratories at a time when nanometer-scale measurement and control of surface properties have become significant issues in production environments.[3,4] For example, the microelectronics industry will soon be faced with the lithography and characterization of circuits with critical dimensions approaching the 100 nm range. Similarly, precision optical components require fabrication and metrology with nanometer scale precision.

It is essential to recognize that there are fundamental differences between the forces that sustain scientific research and those that drive technological progress. These differences must be appreciated before the technological potential of proximal probe instrumentation can be properly addressed, a point that is not often considered in presentations in either the research literature or the popular press. The enthusiasm for potential applications of scanned probe instruments is, unfortunately, often untempered by a realistic assessment of their practicality. However, given the resolution capabilities of scanned probe instruments, it is natural to ask if the proximal probe concept can be implemented into an instrument of use in a manufacturing environment and, if so, how can the implementation be best brought about?[5,6] In the following chapters of this volume, these questions are considered in the context of the specific interests of the individual authors. In this overview we examine the more general aspects of the questions by highlighting the differences between science and technology on the nanometer scale and focusing on three points. First, nanometer science and technology is cross disciplinary in nature and has already spawned new technologies where the control and properties of material on the nanometer scale are the major focus. Those relevant to this volume are introduced in section 2.

The second focus is the identification of the most promising PPL applications and the research directions needed to implement them. These issues are discussed in section 3. Nanolithography and metrology are at the leading edge of PPL applications and will provide the first opportunities for dealing with the challenges of the nanometer regime in a production environment. It makes sense, therefore, to build on the advances in instrumentation, processing, and methodology achieved by the contributors to this volume and others worldwide, as the basis of the forthcoming efforts required to develop viable applications. Furthermore, doing so highlights that PPL, and nanotechnology as a whole, involves an interrelationship between processing and instrumentation that differs significantly from that inherent in extensions of existing manufacturing technologies.

The third focus is the potential for PPL in the emergence of nanoelectronics which is discussed in section 4. There is an obvious synergy between the capabilities of proximal probes and the electronic properties of nanometer scale structures and devices. As illustrated in the chapters of this volume, proximal probes can be used for both the fabrication and characterization of individual structures. The need to make a continued impact on the development of nanoelectronics represents a critical challenge for PPL researchers.

2. SCIENCE VERSUS TECHNOLOGY ON THE NANOMETER SCALE

The scanning tunneling microscope (STM), the first of the scanned proximal probe instruments, was a product of years of effort by numerous researchers to overcome the difficulties of combining ultrahigh vacuum (UHV) surface preparation with the demands of precision instrument design.[7] The understanding of the critical factors in the design of the mechanical and electronic components of proximal probes has evolved rapidly since then. The STM and AFM are now indispensable analytical tools for atomic-level surface studies.

It was realized during those early days that proximal probes offered a novel approach to the study of nanoscale electronic structures and devices with potential applications in the nanoelectronics area. The reduction of device size is of vital importance to the computer and microelectronics industries, and as a consequence STM development received a high level of support from major corporate research labs and funding agencies in the US. However, basic research into the fabrication and characterization of quantum effect-based electron device structures is exceedingly difficult.[8,9,10] While the physics of choice for future devices is uncertain, it is likely that progress in this technology will depend on the creation and use of a new set of fabrication tools. In section 4, we shall return to the opportunities for proximal probe lithography in this field.

The advent of nanoelectronics is, in fact, only part of a much broader, cross-disciplinary interest in fabrication with molecular-level control. A second example can be found in the new field molecular nanotechnology[11] where the focus is structures based on biomolecules such as proteins and DNA. Several research laboratories have reported striking demonstrations of rudimentary fragments of this technology and the associated speculations have become part of the discourse in the scientific [12] and popular press.

The term nanotechnology has been adopted by a third group largely based within the precision engineering community. The focus here is the extension of machining and measurement into the nanometer regime and is referred to as ultra-precision machining and nano-metrology.[3,4,13] The chief concerns of this field are mechanical positioning and metrology with the atomic level precision available with proximal probes. Here, the understanding of the interaction of the instrumentation with the test sample becomes a central issue. We shall return to this issue from the perspective of PPL applications in section 3.1.

The instrumental and process control requirements at the nanometer scale cut across many scientific and engineering disciplines. The technological challenges are very different from those faced in the scientific field by early STM workers who carried out their investigations using well-characterized surfaces with atomically ordered regions. In the next section, these challenges are considered in the context of promising applications for PPL.

3. THE FUTURE OF PROXIMAL PROBE LITHOGRAPHY

Much of the research published to date on material modification with proximal probes has been analytic in nature. The observation of a particular phenomenon has been correlated with certain actions (e.g. applying a voltage pulse or increasing the tip loading) with the probe. It is only recently that serious efforts to uncover the basic mechanisms at work have been made. The chapters of this volume illustrate the success of this activity and show that the field now has a sound scientific footing. The promise of material modification on the near atomic scale inevitably led to early speculation about technological applications. However, the development of a robust and reproducible technique for nanolithography or metrology (for example) needs a radically different focus to the scientific studies published here and elsewhere. Some specific suggestions are made by the authors of the following chapters as to future directions for PPL. In this section, we examine in general terms the problems that need to be overcome to develop viable PPL based applications. This is illustrated by considering what we see as the first application areas: nanolithography, fabrication, data storage and metrology.

In identifying PPL applications, it is important to note that the development of the appropriate instrumentation is an integral part of the needed technology. The work reported in the following pages was carried out, with few exceptions, using instruments virtually identical to the microscopes presently used for surface science studies. However, a production-level tool for nanolithography will need to incorporate a number of functional requirements[14] along with the proximal probe concept. For example, the lithographic technique must have a sufficient throughput and process latitude. The instrumentation must be robust against environmental variation, must include the appropriate levels of metrology and must be user friendly.

3.1 Compatibility With A Processing Environment

How do we exploit the extreme surface sensitivity of proximal probe techniques reliably and reproducibly on samples which must pass through environments and processes which will not preserve the structural perfection of the surface? This is a crucial issue, because, without a detailed understanding of surface properties during a process step, atomic-scale control will be impossible. This is further complicated at the nanometer scale as the information available from analytic tools such as proximal probes is also critically dependent on the physical and chemical state of the surface. Variation of the contrast mechanism on the scale of features to be fabricated or characterized implies that the very notion of measurement artifacts must be re-thought. For example, the analysis of STM constant current images requires knowledge of surface work function variation. Similarly, monolayer phase changes can provide non-topographic variations in contrast and severely limit our ability to calibrate nanometer-scale artifacts.

We want to examine this point a little more deeply since it is key to the development of nanotechnology. If we adopt the language of traditional metrology we would tend to use the terms size, form, and texture to describe the physical aspects of machined structures.[4] The physical and chemical nature described by crystal lattice structure, electronic band structure, alloy phases, and chemical bonding and their consequences are traditionally not considered. For nanoscale structures, the macroscopic concepts of size, form, and texture merge into one another and the physical and chemical origins of these properties become significant. For example, if we wish to carry out "simple" dimensional measurements of structures with an AFM, it must be first demonstrated that local variations in the surface forces (van der Waals, capillary, electrostatic, etc.) will not induce distortions in an intended geometric measurement.[15] Again, we emphasize that the physical and chemical properties of the surface must be understood and controlled on the nanoscale whatever the research goals.

Modifying and monitoring the properties of a surface during a nanometer scale lithographic process is even more complex. In this case, many variables must be controlled within a localized volume encompassing the bulk and surface properties of the probe and sample as well as the air, liquid, or vacuum interface. A judicious choice of molecular interactions, electrochemical forces, and instrumental operation during the modification process must be made. The full range of strategies for proximal probe lithography has not yet been explored, and even the rich possibilities suggested by the work presented in this volume only demonstrate the first efforts toward discovering processes in which the unique instrumental advantages of proximal probes can be realized.

In this context we return again to the importance of surface preparation as an essential feature of these studies. Whether one chooses to exploit the particular advantages of resist based nanolithography or to emphasize direct-write schemes in one's research, the maximum thickness of the patterned features will be on the order of a few nanometers. In the case of resist exposure, for example, there are enormous problems to be overcome in the deposition of ultrathin uniform films which adhere properly to the substrate to allow high resolution pattern definition.[5] Similarly for direct-write processes, the nanometer scale topographic and chemical uniformity of the substrate will directly affect the minimum feature size that can be obtained.[6] It is to be hoped that a consensus among applications researchers will eventually emerge as to the most promising techniques for PPL based nanolithography. At this point the choices are not at all clear, but it is not too early to begin the selection process. As with other forms of advanced lithography, certain functional requirements must be met. For example, the ancillary technologies for instrumentation and sample preparation must exist, feature placement accuracy must match lithographic resolution and a degree of compatibility with other fabrication practices must be demonstrated. The PPL community can learn much from discussion of these issues by proponents of other advanced technologies, such as the recent review of x-ray lithography by Peckerar and Maldonado.[16]

3.2 Instrumental Requirements for Proximal Probe Lithography

Up to now we have not addressed the instrumental evolution which will be required for a realization of practical proximal probe based technology. The process development described in this monograph is occurring in parallel with the design and fabrication of high throughput proximal probe tools.[17] We now describe the unique properties of scanned probe instruments which make them attractive for use in advanced manufacturing environments and the challenges that must be met in the near future. We consider first the nanolithography area which represents a significant technological challenge because both new instrumentation and new ancillary technologies must be developed. In contrast, the areas of data storage and nanometrology will probably require only the development of novel instrumentation which can be used with refinements of existing techniques.

A. Nanolithography and Fabrication

Scanned probes have several advantages over high energy electron or ion beam microprobes for nanolithography. They can operate over a wide range of pressure and temperature conditions, they utilize low energy particles thus reducing sample damage and the probe-surface interaction is localized thereby minimizing the possibility that unwanted reactions occur in other parts of the system. The interplay between PPL and nanoelectronics is considered in the next section. Here, we consider the types of instrumentation development needed for nanolithography in general.

We envision two roles for PPL for nanolithography; the fabrication of individual structures in a research and development (R & D) environment and high throughput lithography on a production scale. A fast (~50 µm/s tip velocity) single tip proximal probe instrument with independent tip position measurement would have numerous advantages for R & D based nanolithography. At present, converted scanning electron microscopes or expensive e-beam lithography systems are used. A proximal probe based system would give superior lithographic resolution and be significantly easier to build and maintain. Examples of such custom built instrumentation are described in this monograph.

The second role of PPL would exploit the advantages of the proximal probe concept for a production environment.[17] Multiple tip arrays will be needed to carry out lithographic exposures at a throughput to make mask making or direct-write realistic.[5] This evolution of PPL might influence future electronics circuit design strategies to take advantage of the dense, parallel features that would be easily patterned by multiple tip arrays. The technology exists currently to microfabricate large-scale arrays of STM elements, either in the microlens configuration proposed by Chang et al.[18] or the independently servo-control multiple AFM elements being considered by Quate.[17] As these and other efforts to develop the multiple tip technology proceed worldwide, it is important to keep in mind that such tip densities are likely to present new challenges for those currently working in the process area of PPL. Lithographic processes are needed which are viable under low energy irradiation from 10^{2-3} tips/cm^2. Specifically, processes that minimize the diffusion of activated chemical species and the loss of material during exposure are needed. The particular process (e.g. direct write or resist based) and the relevant mechanisms (e.g. resist exposure, surface oxidation or precursor decomposition) are, at present, completely open questions.

B. Data Storage

Data storage techniques, e.g. optical, electrical, mechanical or magnetic, appear to be largely compatible with proximal probe capabilities. There is a close coupling between proximal probe-based approaches to high-throughput nanolithography and rapid reading and writing of mass data storage. The instrumental stability of the proximal probe control loop during high frequency operation is, as with nanolithography, a key concern. Thus refinement of an existing data storage method is a good strategy as this would allow researchers to focus on optimizing proximal probe instrumentation. Approaches based on existing magnetic and magneto-optical technologies, such as the use of proximal optical probes by Betzig et al.[19], are promising for this reason. However, the thermal assisted mechanical indentation technique developed by Mamin and Rugar[20] and the system based on charge storage in insulators Barrett and Quate[21] may prove to have other advantages.

C. Metrology

Dimensional metrology, such as the accurate measurement of nanometer sized structures and the roughness of optical, mechanical, and electronic surfaces, is an extremely important component of PPL and nanotechnology in general. At this length scale, it is no longer possible to ignore the basic mechanisms of the interactions between the individual constituents of the instrument and sample.[13] It has been argued that manufacturing on the nanoscale will almost certainly require closed-loop control of critical steps during every stage of production.[4] Thus in addition to the problems associated with nanometer scale metrology, an implementation compatible with manufacturing processing will be required. Issues which must be addressed include: (1) quantifying the probe-sample interaction for a specific surface; (2) determining the errors due to mediating layers of contamination, variations in surface properties, and mechanical deformations of the tip and sample; (3) developing methodologies for deconvoluting effects of tip shape and; (4) implementing the instrumental design for rapid and repeatable measurements, i.e. developing a robust and user friendly metrology tool.

4. NANOELECTRONICS: OPPORTUNITIES FOR PROXIMAL PROBE TECHNOLOGY

The evolution of PPL will be critically dependent on its success in the nanoelectronics field. Viable manufacturing applications will be determined largely by the success of instrumentation researchers in integrating proximal probe arrays with a practical metrology. However, an impact on nanoelectronics research and development is possible with single tip probes. Three particular directions are envisioned. First, advances are expected from the use of proximal probes to fabricate and characterize novel electronic device structures. Research programs aimed at introducing proximal probe techniques for dimensional and electrical metrology of such structures are have already been implemented. Indeed, the success of some these first efforts is described in this volume. Valuable experience is being gained in overcoming the practical difficulties associated with the development of a viable proximal probe technology. Surface preparation, the physical mechanisms of the sample-tip interaction during the measurements, and the question of methodology and standardization of measurement are examples of these challenges.

The second direction will focus on the fabrication of individual devices. It is important, however, to distinguish between research on device physics for precision electrical measurements (such as single-electron tunneling (SET) and quantum dots) and research concerned with devices for advanced integrated circuit architectures for information processing.[8,9] The device and fabrication requirements for each are quite different, and are likely to encompass very different functional requirements. For example, whereas low temperature (a few degrees K) operation may be acceptable for precision electrical measurements using SET devices, room temperature operation is essential for consumer electronics applications. Circuit architectures have more stringent alignment requirements than individual devices so greater lithographic precision is necessary. The fabrication of individual SET structures would appear to be an early opportunity for PPL because of the simplicity and the greater flexibility in materials and structures possible with low temperature devices.

The third challenge for PPL is the demonstration of lithography of devices and circuits for room temperature operation. Whereas resolution in the 100 nm range is needed for cryogenic operation, feature sizes of an order of magnitude smaller are going to be needed for higher temperature devices. PPL has the potential to break the ~20 nm resolution limit of extensions of 'conventional' lithography with high energy electron and ion microprobes. PPL also promises improved feature placement (e.g. how close individual features can be defined) due to the absence of proximity effects which plague high energy e-beam lithography.[22] The properties of the actual devices will be quite different from the existing silicon technology. Heterojunction technology,[23] made possible by advances in epitaxial growth techniques, may overcome the problems encountered in scaling the current silicon based device technology. PPL coupled with epitaxial growth techniques offers a radically different approach for the development of nanoelectronics devices and circuit architectures.[9] Indeed, considerable research is now underway on room temperature device concepts for both III-V and IV-IV semiconductor materials.

PPL researchers will need to gain considerable expertise in fabricating individual device structures as well as the means to integrate these techniques with multiple tip tools before a viable PPL production technology can be said to exist. Reaching this stage will require the sustained interaction of scanned probe and device researchers so that the two communities will have a mutual understanding of the nature of nanometer scale device physics and fabrication. It is crucial for the sustained development of PPL that the desirability and compatibility of this new field be demonstrated to and embraced by device researchers. This activity will provide the first test of the viability of PPL, and it is for this reason it is vitally important that research be directed toward nanoelectronics applications.

5. SUMMARY

This monograph brings together a broad, and hopefully, representative, sample of the current efforts and aspirations of proximal probe lithography. As such, these reviews reflect what may at first seem to be a surprising variety in the main concerns of this emerging field. The authors also demonstrate significant differences in the articulation of the challenges and the emphasis of their work. This variety is representative of the newness of the field itself and reflects the excitement encompassing all aspects of nanotechnology. Our hope is that this volume will initiate further discussion of the issues facing proximal probe research at the present time.

At this time, the most direct applications of PPL, namely nanolithography for nanoelectronics, metrology and mass data storage, represent what may be the first opportunities to realize the potential and to confront the challenges of a broader range of nanotechnology applications. It is crucial to foster the cross disciplinary interactions needed to facilitate the successful transfer of nanotechnology out of the research laboratory. An important concern for researchers and program managers must be the organization of the national (and, increasingly, the international) research enterprise so that this transfer can proceed as rapidly as possible. The potential impact for nanometer scale control of surface structure is exceedingly rich in both scientific and technological terms. The development of PPL is vital to the success of new technologies based on the unique properties of material on the nanometer scale.

ACKNOWLEDGMENTS

The authors thank E. C. Teague (NIST), E. A. Dobisz (NRL), F. K. Perkins (NRC/NRL), M.C. Peckerar (NRL) and J. S. Murday (NRL) for reading this manuscript and offering many helpful suggestions.

REFERENCES

1. J. S. Murday and R. J. Colton, "Proximal Probes: Techniques for Measuring at the Nanometer Scale", in Chemistry and Physics of Solids VIII, (Springer NY 1990) p. 347.

2. C. F. Quate, "Manipulation and Modification of Nanometer Scale Objects with the STM", in Highlights of the Eighties and Future Prospects in Condensed Matter Physics, L. Esaki, ed., NATO ASI (Plenum NY 1991).

3. N. Taniguchi and T. Miyazaki, "Background and Development of Nanotechnology on Advance Intelligent Industry- Current Status of Ultraprecision and Ultrafine Materials Processing", Advances of Nanotechnology, K. Kawata and D. J. Whitehouse eds., Science University of Tokyo, Noda, 1992), p. 1.

4. D. J. Whitehouse, "Present and Future Problems in Nanotechnology", Advances of Nanotechnology, K. Kawata and D. J. Whitehouse eds., Science University of Tokyo, Noda, 1992), p. 32.

5. C. R. K. Marrian, E. A. Dobisz, and J. A. Dagata, "Scanning Tunneling Microscope Lithography: A Viable Lithographic Technology?", SPIE 1671 166 (1992); C. R. K. Marrian, E. A. Dobisz, and J. A. Dagata, "e-Beam Lithography with the Scanning Tunneling Microscope", J. Vac. Sci. Technol. B 10 2877 (1992).

6. J. A. Dagata, W. Tseng, J. Bennett, E. A. Dobisz, J. Schneir, and H. H. Harary, "Integration of STM Nanolithography and Electronics Device Processing", J. Vac. Sci. Technol. A 10 2105 (1992).

7. G. Binnig and H. Rhorer, "Nobel Prize Address", Rev. Mod. Phys. 59 615 (1987).

8. R. W. Keyes, "The Future of Solid-state Electronics", Physics Today, August 1992, p. 44; R. W. Keyes, "What Makes a Good Computer Device?", Science 238 138 (1985).

9. J. N. Randall, M. A. Reed, and G. A. Frazier, "Nanoelectronics: Fanciful Physics or Real Devices?", J. Vac. Sci. Technol. B 7 1398 (1989).

10. K. Hess and G. Iafrate, "Approaching the Quantum Limit", IEEE Spectrum, July 1992, p. 44.

11. K. E. Drexler, "Molecular Directions in Nanotechnology", Nanotechnology $\underline{2}$ 113 (1992).

12. P. Ball and L. Garwin, "Science at the Atomic scale", Nature $\underline{355}$ 761 (1992).

13. E. C. Teague, "Nanometrology", <u>Scanned Probe Microscopy: STM and Beyond</u>, H. K. Wickramasinghe ed., (AIP NY 1992).

14. N. P. Suh, <u>The Principles of Design</u>, (Oxford University Pr. NY), chap. 1.

15. T. P. Weihs, Z. Nawaz, S. P. Jarvis, and J. B. Pethica, "Limits of Imaging Resolution for Atomic Force Microscopy of Molecules", Appl. Phys. Lett. $\underline{59}$ 3536 (1991).

16. "X-ray Lithography - An Overview", M.C. Peckerar and J.R. Maldonado, to be published in the Proceedings of the IEEE, September 1993.

17. C.F. Quate, "A Kinder, Gentler Chip Inspection", Science $\underline{258}$ 1575 (1992).

18. T. H. P. Chang, L. P. Muray, U. Staufer, M. A. McCord, and D. P. Kern, "Arrayed Lithography Using STM-based Microcolumns", this volume.

19. E. Betzig, J. K. Trautman, R. Wolfe, E. M. Gyorgy, P. L. Finn, M. H. Kryder, and C. H. Chang, "Near-field Magneto-optics and High Density Data Storage", Appl. Phys. Lett. $\underline{61}$ 142 (1992).

20. H. J. Mamin and D. Rugar, "Thermo-mechanical Indentation with an Atomic Force Microscope", 39th American Vacuum Society Symposium, November 9-13 1992, Chicago IL.

21. R. C. Barrett and C. F. Quate, "Large-scale Charge Storage by Scanning Capacitance Microscopy", Ultramicroscopy $\underline{42}$ - $\underline{44}$ 262 (1992).

22. Proximity effects and their correction are described in more detail in the paper by Marrian, Dobisz and Dagata in this volume.

23. F. Capasso, "Resonant Tunneling and Superlattice Devices: Physics and Circuits", in <u>Physics of Quantum Electron Devices</u>, F. Capasso, ed., (Springer-Verlag Berlin 1990), chap. 8.

Part I
Nanolithography

NANOLITHOGRAPHY

This section is concerned with applications of scanned probe instruments which are closest to conventional lithographic technologies such as e-beam, optical or x-ray lithographies. A pattern is defined in some medium or on a surface directly and is then replicated into the underlying substrate or is used to build up patterned layers on that substrate. These steps form the basis of the fabrication technology used in microelectronics based industries. Minimum feature sizes are constantly shrinking and by the end of the decade integrated circuits with design rules approaching 100 nm will be in production. As one looks further ahead to even smaller feature sizes, it becomes clear that many existing lithographic technologies will no longer be tenable. A paradigm shift in lithographic technology is necessary for fabrication with sub 100 nm feature sizes. Proximal probe lithography has the potential to meet this need at a significantly lower cost than present day lithographic systems and their extensions.

The first paper reviews some of the earliest STM lithographic studies performed by Mark McCord (now at IBM) and Fabian Pease at Stanford University. They describe studies of a number of different electron sensitive materials and impressive results with the mainstay of high resolution lithography, PMMA. Christie Marrian and Elizabeth Dobisz (both NRL) and John Dagata (NIST) demonstrate that there are significant advantages using low voltage electrons for lithography in resists. Enhanced resolution, superior process latitude and an absence of proximity effects were observed when STM lithography was compared to lithography with a tightly focused 50 kV e-beam. The authors also discuss the lithographic speed of multiple proximal tips writing in parallel. Arun Majumdar and Stuart Lindsay from Arizona State University describes lithographic resist exposure using an AFM based instrument. This demonstration has interesting implications for a future lithography tool as multiple AFM tips in contact with a resist covered surface may well prove easier to implement than the corresponding array of STMs. The second half of this paper considers the vexing question of how electron transport occurs through organic and biological materials which are ordinarily viewed as insulators.

The next three papers are concerned with the selective modification of a silicon surface under the action of voltage pulses from the STM. This can be thought of as a 'resistless' lithography. Francois Grey and co-workers at RIKEN describe the STM induced modification of the atomic arrangement of a silicon surface in ultra high vacuum. A mechanism based on field ionization is discussed. When similar experiments are performed in ambient conditions, as described in the next two papers, a thin surface oxide is formed. This is described in detail by John Dagata and Jason Schneir (NIST) who performed studies to elucidate the oxidation mechanism and kinetics. An intriguing attempt to integrate this technique into the fabrication of a novel electronic device is presented by Joe Lyding and his co-workers at the Universities of Illinois at Urbana-Champaign and Minnesota. The selective oxidation method has been used to modulate the thickness of the gate oxide of a MOSFET type device.

The development of arrays of probes is vital to the success of proximal probe lithography as a viable production technology. Philip Chang and his colleagues at IBM, Cornell and Basel are pursuing the integration of proximal probe techniques and miniaturized electron optical elements. They describe the design and first results from such a system and point out that the design is compatible with batch processing and can be integrated into to arrays for direct-write applications.

Principles and Techniques of STM Lithography

M. A. McCord
IBM Semiconductor Research and Development Center, P. O. Box 218, Yorktown Heights, NY 10598

R. F. W. Pease
Solid State Electronics Laboratory, Stanford University, Stanford, CA 94305

Abstract
Conventional electron beam lithography suffers from limitations set by source brightness, lens aberrations and space charge effects. The scanning tunneling microscope is an entirely new configuration that avoids these limits and so offers the possibility of higher throughput and greater choice of patterning processes. Here we review early experiments aimed at demonstrating these advantages and to determine the practicality of this form of patterning. Both conventional resist exposure and resistless processes were demonstrated and showed sub-100nm resolution and pattern transfer was also demonstrated. Achieving high throughput appears to be very challenging because the small separation (<100nm) needed for adequate resolution makes high speed scanning impractical. One possible way around this problem is to micromachine arrays of STM tips so that the exposure can be carried out in parallel.

Introduction

Almost since its invention in 1981, researchers have been considering ways of using the scanning tunneling microscope (STM)[1] for lithography.[2,3] The combination of a probe with potentially atomic resolution and extremely high current densities make it a natural application at first glance. Also, the ability to image what it writes would add another dimension to the fabrication process. Finally, the low cost and ease of operation make the STM available to a much larger pool of researchers than would have access to complex and expensive electron beam lithography tools.

The ability to image the structures that it writes allows the user to obtain rapid feedback on the results of the fabrication process. However, this feature is severely limited in practice for a variety of reasons. First, the STM requires conducting substrates for good imaging, which eliminates a lot of interesting lithographic systems, most importantly, the polymeric resists. Also, most resist systems require a separate development step which involves removing the sample from the STM; locating the exposed area after removing the sample is a formidable task for most STMs. Also, the STM is very poor at imaging rough structures; rather, it is likely that the sample images the tip rather than the other way around. Indeed, it is only on samples that are nearly atomically flat that atomic resolution is achieved. Thus, imaging the tall structures that can be obtained by direct deposition, for example, would give a poor representation of the actual structure that was fabricated. In general, since the resolution of STM lithography is in the several nanometer to several tens of nanometer range, more conventional techniques, most importantly the SEM, are sufficient and frequently preferable means of imaging lithographic results. However, for certain cases, especially atomic and molecular manipulation, the ability to image the sample with atomic resolution is indispensable.

The variety of lithographic techniques demonstrated with the STM is truly remarkable, ranging from crude scratching of the surface to direct atomic manipulation. Other techniques include exposure of conventional and novel resist systems, direct deposition, local heating, and electrochemical etching. No other technique is capable of fabricating structures with such a variety of mechanisms. This paper will cover a variety of these techniques, giving examples as well as discussing some of the limitations.

In addition, some of the factors governing resolution and writing speed will be discussed, as well as possible means of overcoming some of these limitations. Only the authors' work will be discussed in detail but brief mention will be made to the experiments of a number of other researchers where appropriate. References are included when they are relevant to the discussion at hand, but they should not be regarded as a complete bibliography of the field.

The experiments described in this paper were performed with custom-built STMs[4,5] for a variety of reasons. Perhaps most important was the need for a scan field larger than available on most commercial STMs, with a minimum of 10x10 um, while 100x100 um was preferred. This allows a variety of exposure trials to be performed on a single sample, and also facilitates alignment of the STM to pre-existing structures on the sample. Another reason is that the creation of arbitrary patterns requires integrating a digital pattern generator into the STM control electronics. Today, however, while an

STM specifically designed for lithography may not be available, most of the hardware required can be purchased off the shelf. Apart from a large scan range, other features that are desirable include a long focal length optical microscope along with a coarse x-y sample stage to roughly align the STM tip to features on the sample; some form of pattern generation capability generally consisting of a computer connected to a pair of digital-to-analog converters with appropriate software; and gas handling and vacuum equipment for working with gases for direct deposition.

Lithographic techniques

Physical modification of surfaces

One of the most common methods used for lithography is to take a flat, homogeneous sample and either pulse the tip voltage or drive the tip into the surface to create a small structure under the tip.[6,7] Usually the structure is shallow depression about 5 nm in diameter, although occasionally a bump is reported. Either silicon or gold are common substrates. The techniques of pulsing the tip voltage or extending the tip are lumped together since it is not always clear in either case whether the mechanism is due to the physical touching of the tip or due to local heating from very high current densities. When pulsing the tip voltage, it may be possible that a destabilization of the feedback loop causes the tip to touch the sample. Similarly, extending the tip towards the surface results in a very high current pulse that may be responsible for the patterning.

The main drawback of these techniques is that a homogeneous sample is technologically uninteresting for the most part since it doesn't create any devices or useful structures, nor is there any easy way to transfer such patterns into a film or substrate that would have useful properties. In this respect, it is little better than scratching the sample with a piece of sandpaper, which would in all probability make some very fine lines, albeit of little practical value. However, such experiments have value in terms of studying patterning processes, energy dissipation and atomic diffusion in very small sample volumes, and for looking into techniques for high density storage. There is also the advantage of being able to use the STM to image the results immediately.

An interesting variation of this brute force technique is to deposit a thin insulating layer on the surface of a conducting sample. The STM is then used to micromachine the insulating film from the surface.[8,9] This technique has the advantage of being able to create or transfer patterns into potentially useful material systems. Two different metal halide films, calcium fluoride and aluminum fluoride, as well as a polymer film of PMMA, were experimentally studied using this technique.

In operation, the STM tip is brought towards the sample until a tunneling current is established. Presumably, the tip has then penetrated the insulating film until it is a few angstroms above the conducting/insulating interface. As the tip is scanned in x and y, the feedback loop maintains the tunneling current and tip height so that the tip machines away the insulating film. Shown in figure 1 are some lines 0.36 µm in width machined into a 20 nm thick film of calcium fluoride on a silicon substrate. The calcium fluoride was thermally deposited as polycrystalline film with a grain size of 5-10 nm. The calcium fluoride appears to have been completely removed by the tip, leaving behind apparently undamaged bare silicon. As the tip cuts through the insulating layer, it leaves small shavings along the sides of the lines, as well as small piles of crystallites at the end of each line segment. The machinings could mostly be removed by ultrasonic agitation in water for 60 seconds. Before-and-after micrographs of the tip show that it suffered no appreciable damage during the machining process. The radius of the tip, estimated to be 0.19 μm, implies a linewidth of 0.17 μm for a 20 nm thick film. A likely explanation for the wider lines observed experimentally is that debris builds up around the tip, effectively increasing its radius.

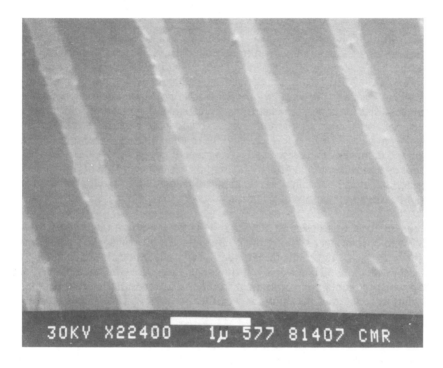

Figure 1. 0.36 μm lines machined with the STM in a 20 nm thick layer of polycrystalline calcium fluoride thermally evaporated on a silicon substrate.

The nature of the machining is strongly influenced by the structure of the insulating film. Lines similar to those shown in figure 1 were also machined in a 20 nm film of amorphous aluminum fluoride. While the shavings from the calcium fluoride consisted of small piles of crystallites, the aluminum fluoride was removed in long continuous strips that spiraled around like machinings from a soft metal. Attempts to machine fine features in a film of PMMA were less successful. The film appeared instead to tear away from the substrate in strips about 1 μm wide, and the width was apparently determined more by the film properties than the tip shape. Such behavior is indicative of the fact that the long, intertwined polymer chains adhere better to themselves than to the silicon substrate.

The nature of the machining is thus strongly determined by both the film characteristics and the tip shape. The smallest features obtained so far for calcium fluoride are approximately 100 nm wide, however, it may be possible to extend this technique to smaller dimensions by using thinner films and extremely sharp tips. Attempts to machine less than the full thickness of the insulating film were unsuccessful; this is consistent with the poor adhesion such films typically have for the substrate compared to their self-adhesion and with the difficulty of maintaining a large but stable tip-to-substrate distance under the conditions encountered during machining. In addition to possible lithographic applications, this technique may also prove useful in identifying mechanical properties of thin films, particularly in conjunction with an AFM designed to measure sideways as well as vertical forces.[10]

Chemical modification of resist films

In microlithography we normally pattern a radiation-sensitive resist film and subsequently transfer the pattern into the underlying functional material. Generally, resist films are composed of polymers, although a variety of inorganic compounds have also been used. Most of these materials are insulators, which pose a problem for the STM with its inability to work with thick insulating films. Fortunately, many of the techniques used for applying resist films are capable of producing film thickness of a few tens of nanometers or less, a thickness that can be penetrated by electrons field emitted from an STM with a few tens of volts of energy. These techniques include spin casting and thermal evaporation, but the Langmuir-Blodgett (LB) technique, where films are built up a monolayer at a time by floating them off the surface of a suitable solvent, is perhaps the most interesting for this application due to the potential of obtaining flat, defect free films with very precise thicknesses.

Experimentally it was observed that 20 volts is the minimum required tip voltage for the STM to operate over a 20 nm thick film of resist; otherwise the tip penetrates into the resist (as with micromachining) with generally undesirable results. This is consistent with the field of several volts/nm required for field emission of electrons with the tip when the reduction of the field in the resist due to its relative permittivity is taken into account, and the minimum required voltage will roughly scale with resist thickness. At these voltages the STM is operating well into the field emission regime, where electrons first tunnel into the vacuum, and are then accelerated into the sample. For most of the resist experiments, a thickness of 20 nm was used as a compromise between thinner resists which would allow the use of lower voltages for higher resolution, and thicker resists, which are easier to image in an SEM and afford better pattern transfer capability.

Poly-methyl methacrylate (PMMA) has long been the resist of choice for electron beam nanolithography due to its combination of high contrast, high resolution, and ease of processing. As such, it is also a natural choice for studying STM lithography.[11] Spin-cast experiments demonstrated the capability of making usable films 20 nm thick. While thinner films could be applied, pinhole densities increased rapidly below this thickness. PMMA has the interesting property of behaving as a positive resist at low exposure doses, when chain scission of the molecules is the predominant mechanism and as a negative resist at high doses when cross-linking between polymer chains dominates.

Figure 2a shows lines exposed in a 20 nm thick layer of PMMA. The contrast has been enhanced by shadowing of gold-palladium from right to left. The exposed resist was dissolved away by a developer of 3:7 cellosolve:methanol. In figure 2b, lines of PMMA used as a negative resist are shown, where the unexposed regions were dissolved in acetone. Interestingly enough, the exposure energy is critical to the tone of PMMA. Below 25 eV, even very large doses could not bring about negative exposure. Under the right conditions, simultaneous positive and negative exposure could be observed as a set of double lines, where only the lightly exposed resist under the edges of the tip was developed away.

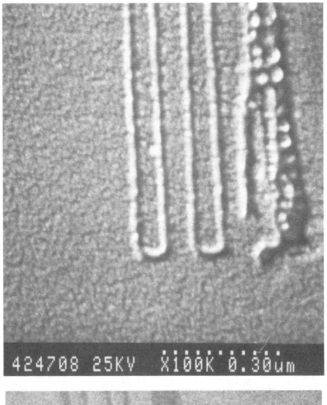

Figure 2a. 20 nm lines on a 70 nm pitch made using a 20 nm thick film of PMMA as a positive resist. The lines were exposed with a 20 volt, 10 pA beam at a writing speed of 1 μm/second. (b) 20 nm lines on a 70 nm pitch made using a 20 nm thick film of PMMA as a negative resist. The lines were exposed with a 30 volt, 30 pA beam at a writing speed of 1 micron/second.

Pattern transfer, an important capability if the resist process is to have technological applications, was demonstrated using the lift-off technique. A metal film is deposited on top of the developed and exposed resist film. The resist is then dissolved out from under the metal film, removing the metal as well. Figure 3 shows a line of aluminum on silicon patterned by lift-off using PMMA resist exposed with the STM.

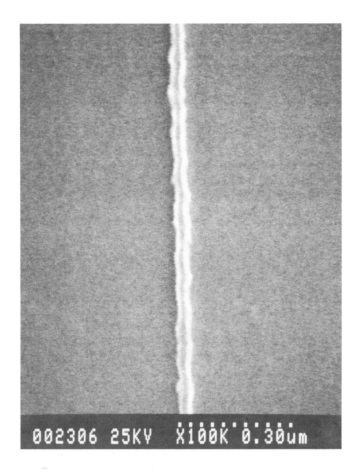

Figure 3. A 30 nm aluminum line fabricated using the lift-off process. First a 30 nm thick film of PMMA was exposed with the STM; then, a 26 nm thick film of aluminum was deposited and the underlying resist was subsequently dissolved away.

Exposure of a 10 nm thick film of docosenoic acid deposited by the Langmuir-Blodgett technique was also demonstrated.[12] This particular film was chosen for its known sensitivity to electron irradiation. Unfortunately, a good technique for imaging or transferring the patterns was not developed, making interpretation of the results rather difficult. Exposures were also made on a conductive polymer film in order to see if this would allow exposure of thicker films or at lower voltages. Although some preliminary results were obtained, they were not reproducible.

The highest resolution lithography has been achieved using inorganic resists[13] since the large size of polymer molecules, 5-10 nm, ultimately limits the performance of most

Figure 4. Lines exposed in a 20 nm thick film of CaF$_2$ on silicon using the STM and developed in water for 2 minutes. (a) 0.23 μm lines exposed with a 60 v, 1 nA beam and a writing speed of 0.5 μm/s. (b) 20 nm lines exposed with a 20 v, 2 nA beam at a writing speed of 0.85 μm/s.

organic resists. Of many types of inorganic resists, the best known are the metal halide family. Under intense irradiation, the halides desorb, leaving behind the metal. In some cases, the metal diffuses away from the exposed area, resulting in a self-developing resist. Several different resists were tried with the STM, including calcium fluoride, lithium fluoride, and aluminum fluoride.[14] Lithium fluoride was chosen as one of the most sensitive of the metal halides, which require doses orders of magnitude higher than most polymer resists. Aluminum fluoride was picked because it can be deposited as an amorphous film, eliminating any effects from the polycrystalline grain structure present in most of these films. Unfortunately, neither aluminum fluoride nor lithium fluoride behaved as a self developing resist with the STM, and no further development step was found that could elucidate the patterns that were faintly evident in the STM. Calcium, however, has the advantage that water can be used as a developer to remove it after the fluoride has been dissociated. Shown in figure 4 are lines of calcium fluoride on a silicon substrate. Lines as small as 20 nm were made, unfortunately at this point the grain structure evident in the film appears to be limiting the resolution of the resist. Attempts were also made at patterning germanium selenide and arsenic trisulphide; however, the results were either irreproducible, unsatisfactory, or subject to ambiguous interpretation.

Direct deposition

In its most primitive form, direct deposition makes use of organic molecules naturally present in any moderate vacuum to form a carbonaceous deposit on the surface. Such deposits are well known to electron microscopists as generally unwanted and unavoidable carbon contamination, although it has been used as a very high resolution resist in electron beam lithography.[15] This was the first technique attempted with the STM.[12] Shown in figure 5 are lines of contamination written with a 10 volt, 60 nA beam at a writing speed of 0.25 um/sec. The sample consisted of a 100 nm gold film on a silicon substrate. The dark rectangles are additional contamination deposited by the SEM used to take the micrograph. Figure 5b shows the identical pattern after the gold was sputter etched away, demonstrating transfer of the resist pattern into the gold film. Enhanced deposition can be obtained by deliberately introducing an organic compound into the vacuum system. This was tried with propylene, ethylene, and acetylene, but with only limited success.[16] The difficulty seems to due to the fact that the deposits are insulating, which makes the deposition a self-limiting process, since once the gap between the tip and sample is filled, no more deposition can take place. Worse, in some instances the deposit takes the form of a sticky substance, which then gets smeared over the sample by the tip.

By using an organometallic gas as a precurser, it is possible to to form conductive deposits, a technique long practiced using lasers and focused ion beams. In order to achieve this, the STM was placed in a special subchamber into which various gases could be introduced.[17, 18] Figure 6 shows a 10 nm wide line made using tungsten carbonyl, and an array of dots deposited from iron carbonyl. Direct deposition turns out to be a complex process, with a number of complications caused by the STM. First, there is the fact that a conductive deposit growing on the surface alters the electric field between the tip and the sample, and thus affects the emission current and in turn, the way the deposit grows. For instance, a large deposit tends to steal current and growth from a neighboring smaller deposit, which puts limits on how close two structures can be placed, the distance being determined by a combination of the deposit height and the tip radius.

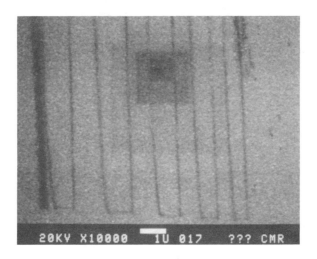

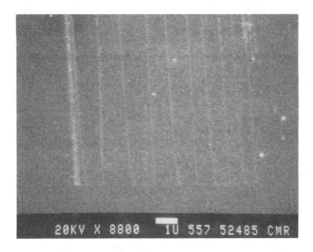

Figure 5. (a) Lines of contamination formed by a single pass of the STM with a voltage of 10 V, a current of 60 nA, and a writing speed of 0.25 μm/s. The sample is 100 nm of gold on a silicon substrate. (b) Same field of view as (a) showing gold lines formed by sputter etching through the resist pattern of (a).

 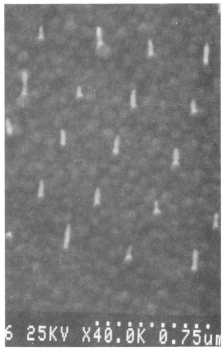

Figure 6. (a) A 10 nm metallic line deposited from W(CO)$_6$ using the STM. Writing was done with a 30V, 10 nA beam at a speed of 0.25 μm/s and a gas pressure of 16 mtorr. (b) Array of 20 nm diameter dots deposited from Fe(CO)$_6$ that stand approximately 150 nm tall. The dots were deposited with a 40 V, 300 pA beam at a pressure of 60 mtorr; each dot took 0.8 seconds to deposit.

It also makes it difficult to grow continuous lines, possible restricting the technique to making isolated dots. Another difficulty is that some deposition occurs on the tip as well as the sample. This may imply a limited tip life as well as difficulty in controlling the shape of deposits to the extent that they are controlled by the tip shape, although the dependence is weak. In addition, a significant amount of carbon impurities tend to be incorporated into the deposits. Auger analysis indicated a composition for the tungsten deposits of roughly 50% W, 40% C, and 10% O.

An interesting application of the direct deposition technique uses iron carbonyl as the precurser gas to form deposits in order to study the properties of nanometer-scale magnets.[19, 20] In particular, the experiment was designed to look for evidence of macroscopic quantum magnetic tunneling, where the magnetic field of a small enough magnet spontaneously reverses its polarity without the application of thermal energy or an external force. Arrays of 100 dots were deposited from the iron carbonyl directly into the pickup coil of a planar SQUID magnetic susceptometer. The dots were typically 20 nm in diameter by 80 nm tall. Compared to the tungsten carbonyl, higher voltages (> 27) were needed to initiate deposition, although lower currents (< 1 nA) were found to give the best results. Cryogenic measurements of the particles revealed frequency dependent peaks in the susceptability, although a disagreement with the theory of quantum magnetic tunneling remains to be resolved.

Resolution Limits

Since the STM is capable of atomic resolution when used as an imaging tool, it might be expected that such resolution would be obtainable for writing as well. Unfortunately, there are several factors that make this goal difficult if not impossible. One of the great attractions of the STM is its ability to image samples with essentially no damage, due to the very low energy of the tunneling electrons that limits their effect either through altering chemical bonds or by simple thermal effects. In order to achieve any lithographic result, it is generally necessary to increase the energy of the electrons to around 4 eV, and often much higher than that. At these energies the STM is operating in the field emission regime,[21, 22] where electrons tunnel from the tip into the vacuum and are then accelerated into the sample. Although the physics of the electron transfer may change, in practice it makes little difference to the operation of the instrument. Field emission of electrons as described by the Fowler-Nordheim equation varies exponentially with the electric field; for most reasonable currents this field is in the range of a few volts/nm. By calculating the field, and thus the current, in the vicinity of the tip and sample, the beam diameter of an STM tip operating in the field emission mode can be determined,[3] as shown in figure 7. For a constant value of the peak electric field, the minimum beam diameter varies linearly with the tip voltage. An interesting result is that for any given tip voltage, there is an optimum tip radius required to give the minimum beam diameter. Blunter tips degrade the beam diameter by increasing the emission area, but sharper tips also reduce the beam diameter since the separation between the tip and sample must be increased in order to maintain a constant field and emission current. Other factors that limit resolution for writing include the fact that lithography generally involves working with a non-planar surface, while the STM can only achieve atomic resolution when the surface is atomically smooth; otherwise the resolution is limited by the macroscopic shape of the tip, which is far from atomically sharp.

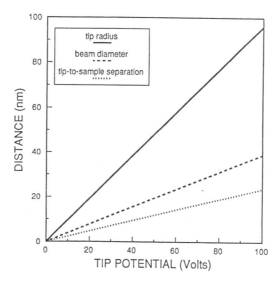

Figure 7. Minimum beam diameter as a function of tip voltage along with the optimum tip radius at a maximum field of 5 V/nm for an STM operating in the field emission mode.

Another factor limiting the writing resolution compared to imaging resolution may be the effect of backscattered and secondary electrons that result when the STM is operated in the field emission mode.[22, 16] While there are only limited data on the number and energies of electrons that are backscattered from a surface at low primary energies[30,31,32], it appears to be a reasonable assumption that 30% of the electrons are backscattered from a sample regardless of its chemical composition when the primary energy is less than 100 volts, and that these electrons are reflected without loss of energy. Using a finite element model to calculate the fields in the vicinity of the tip and sample, trajectories of reflected electrons from the sample were simulated. The majority of reflected electrons, in cases including various tip diameters and voltages, were found to be forced back to the sample by the intense electric field between the tip and sample as shown in figure 8a. A Monte Carlo approach simulating the trajectories of thousands of electrons, was used to estimate the total electron distribution at the sample. As shown by figure 8b, the net effect is that of a double gaussian distribution, with the reflected electrons spread out over an area several times that of the primary electrons. This is similar to exposure profiles found in conventional electron beam lithography, where backscattered electrons from deep within the sample also result in an exposure distribution wider than the primary beam. The result is the well known proximity effect, which can degrade resolution and linewidth control in electron beam lithography. The trajectories of secondary electrons were also studied, but were found to have little impact on the overall beam profile since they return to the surface after traveling at most a few nanometers due to their low energy (typically 2 eV).

An interesting variation on electron beam lithography that can be implemented in an STM is to reverse the polarity of the tip so that electrons are field emitted from the sample up through the resist.[11] The field strength in the resist being about 1 V/nm, an electron can quickly gain enough energy to cause exposure to all but the bottom few nm of the resist film. This technique has several advantages over a more conventional approach. First, problems that might be caused by reflected electrons are avoided, since the electric field now pulls stray electrons away from the resist. In addition, the electrons have no chance to diverge before interacting with the resist, as they would if they had to first travel the gap between the tip and sample. At voltages above 20V, this results in a noticeable improvement in experimental linewidths[32]. Finally, the tunneling current is observed to be much more stable when the polarity is reversed, presumably because the tip is a much poorer source for ions that can be emitted under electron bombardment and result in electrical breakdown across the gap between the tip and sample.

Writing speed
All the patterns shown above were written at a speed that is orders of magnitude too slow for manufacturing; the scan speed of the tip was only about 1μm/s. Assuming a 100 nm linewidth this translates to a throughput of 10^{-9} cm^2/s; for wafer exposure we require a throughput of 1cm^2/s and for mask making a throughput of 1mm^2/s might suffice. However there is potential for much faster writing because the current from the tip is limited neither by aberrations nor by space charge (as it is in a conventional, focused electron beam) but only by heating or other catastrophic effects. Indeed in the experiments reported we went to some lengths to avoid overexposing the resist. Some improvement would result if we increased the scan speed but the difficulty is coming up with a control system that would allow the tip to follow the terrain of the workpiece.

From fig. 7 we can see that to obtain 100 nm resolution we need to have the tip-to-target distance only about 50 nm. This is considerably less than the fly-height of a magnetic disk head (greater than 100 nm) and so it is unlikely that we could scan the tip with respect to the target at the speed of a hard disk (several m/s) and maintain control of a 50 nm fly-height.

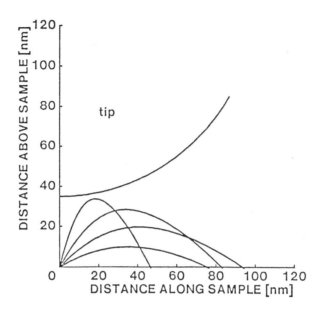

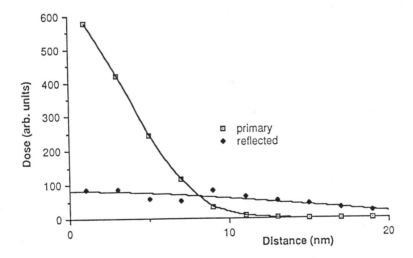

Figure 8. Monte Carlo simulations of (a) sample trajectories of electrons elastically reflected from the surface under an STM tip, and (b) distribution of primary and reflected electrons for a tip voltage of 20 V.

There are three possible ways around this problem:
1. Find a way to stop the beam spreading so fast after it leaves the tip.
2. Operate with the tip in contact with the workpiece.
3. Use many tips so the scan speed can be reduced without sacrificing throughput.

The first way amounts to putting a lens between tip and target.[23-26] Both magnetic and electrostatic focusing ('microlenses') have been tried. Neither has yet performed well (200nm diameter beam at 1KV) about this approach although some ingenious configurations have been described [27, 28].

The second approach suffers from obvious unanswered questions about tip reliability but should not yet be ruled out.

The third approach may be the most practical in light of the recent acceleration of effort in the area of micro-machining single crystal silicon using orientation-dependent etching. A micromachined 'STM-on-chip' has been described[29] and we would need to build an array of these, say, on 20μm centers over a length 2cm long to allow a total of 1000 tips so that a mechanical scan speed of 1 cm/s would be needed to give a throughput of $1mm^2$/s (sufficient for 1 X mask patterning). Obviously we could multiplex such arrays and increase the speed still further to make the process more attractive for direct (maskless) writing. At the slower scan speeds it might be more practical to use also the second approach and have the tip in contact with the workpiece.

Conclusion

The STM has been shown to be capable of a wide variety of lithographic techniques, including exposure of various types of resist, direct deposition from organometallic gases, and mechanical micromachining. The resolution for most of these techniques seems to be typically 20 nm, roughly equivalent to the best resolution that can be achieved using sophisticated electron beam lithography tools.

The STM has already proven useful in the laboratory for demonstrations of the technological limits of lithography, data storage, and even chemical synthesis. This provides motivation and inspiration for improving more conventional technologies. Speeding up the process by several orders of magnitude appears at first daunting; but using micromachining to engineer arrays of STM tips, each individually blankable, appears one possible approach.

From the point of view of achieving accurate overlay, a lithography system based on such an array would encounter the same advantages (and difficulties) as does any form of direct-write lithography system.

Acknowledgements

The authors appreciate the help provided by numerous people over the years including L. S. Hordon, T. H. Newman, A. Bryant, D. Smith, E. Crabbe, P. Maccagno, K. E. Williams, K. J. Polasko, S. Eisensee, C. F. Quate, D. P. Kern, T. H. P. Chang, K. Lee, M. Angelopoulolis, D. Awscholom, J. Smythe, and S. Rishton. The preparation of this paper was partly supported by NSF grant ECS 8920652.

References

1. G. Binnig, H. Rohrer, C.h. Gerber, and W. Weibel, Appl. Phys. Lett. **40**, 178 (1982).

2. M. Ringger, H. R. Hidber, R. Schlogl, P. Oelhafen, and H. J. Guntherodt, Appl. Phys. Lett. **46**, 832 (1985).

3. M. A. McCord and R. F. W. Pease, J. Vac. Sci. Technol. **B3**, 198 (1985).

4. M. A. McCord and R. F. W. Pease, J. Physique **47**, c2-485 (1986).

5. M. A. McCord, Rev. Sci. Instrum. **62** (2), 530 (1991).

6. U. Staufer, R. Wiesendager, L Eng, L. Rosenthaler, H.-R. Hidber, H.-J. Guntherodt, and N. Garcia, J. Vac. Sci. Technol. **A 6**, 537 (1988).

7. D. W. Abraham, H. J. Mamin, E. Ganz, and J. Clarke, IBM J. Res. Dev. **30**, 492 (1986).

8. M. A. McCord and R. F. W. Pease, Appl. Phys. Lett. **50**, 569 (1987).

9. J. Gobrecht and J. B. Pethica, Microciruit Eng. **5**, 471 (1986).

10. C. M. Mati, R. Erlandsson, G. M. McClelland, and S. Chiang, J. Vac. Sci. Technol. **A6,** 575 (1988).

11. M. A. McCord and R. F. W. Pease, J. Vac. Sci. Technol. **B6**, 293 (1988).

12. M. A. McCord and R. F. W. Pease, J. Vac. Sci. Technol. **B4**, 86 (1986).

13. M. Isaacson and A. Murray, J. Vac. Sci Technol. **19**, 1117 (1981).

14. M. A. McCord and R. F. W. Pease, J. Vac. Sci. Technol. **B5**, 430 (1987).

15. A. N. Broers, W. W. Molzen, J. J. Cuomo, and N. D. Wittels, Appl. Phys. Lett. **29**, 596 (1976).

16. M. A. McCord and R. F. W. Pease, Surf. Sci. **181**, 278 (1987).

17. E. E. Ehrichs, R. M. Silver, and A. L de Lozanne, J. Vac. Sci. Technol. **A6**, 540 (1988).

18. M. A. McCord, D. P. Kern, and T. H. P. Chang, J. Vac. Sci. Technol. **B6**, 1877 (1988).

19. D. D. Awscholom, M. A. McCord, and G. Grinstein, Phys. Rev. Lett. **65**, 783 (1990).

20. M. A. McCord and D. D. Awschalom, Appl. Phys. Lett. **57**, 2153 (1990).

21. R. Gomer, *Field Emission and Field Ionization* (Harvard University, Cambridge, MA, 1961).

22. R. Young, J. Ward, and F. Scire, Rev. Sci. Instrum **43**, 999 (1972).

23. M. A. McCord, T. H. P Chang, D. P. Kern, and J. L. Spiedell, J. Vac. Sci. Technol.**B7**, 1851 (1989).

24. T. H. P. Chang, D. P. Kern, and M. A. McCord, J. Vac. Sci. Technol. **B7**, 1855 (1989).

25. L. S. Hordon and R. F. W. Pease, J. Vac. Sci. Technol. **B8**, 1686 (1990).

26. C. A. Spindt, I. Brodie, L. Humphrey, and E. R. Westerberg, J. Appl. Phys. **47**, 5248 (1976).

27. L. P. Muray, U. Staufer, D. P. Kern, and T. H. P. Chang, J. Vac. Sci. Technol. **B10**, 2749(1992).

28. T. H. P. Chang, D. P. Kern, and L. P. Muray, J. Vac. Sci. Technol. **B10**, 2743 (1992).

29. T. R. Albrecht, S. Akamine, M. J. Zdeblick, and C. F. Quate, J. Vac. Sci. Technol **A8**, 317 (1990).

30. Z. Tollkamp, L. Reimer, Scanning **3**, 35, (1980)

31. G. A. Harrower, Phys. Rev., **104**, 52, (1956).

32. R. Browning J. Vac. Sci. Technol. **B10**, 2453 (1992).

33. L. S. Hordon, H. Zhang, S. W. J. Kuan, P. Maccagno, R. F. W. Pease, J. Vac. Sci. Technol. **B7**, 170 (1989).

Role of Scanning Probe Microscopes in the Development of Nanoelectronic Devices

Arun Majumdar[†]
Department of Mechanical and Aerospace Engineering
Arizona State University
Tempe, AZ 85287

Stuart M. Lindsay
Department of Physics
Arizona State University
Tempe, AZ 85287

ABSTRACT

The last six years have witnessed the rapid development of several techniques of nanometer and atomic scale material manipulation using scanning probe microscopes (SPM). A stage has now been reached where we can explore the possibilities of using this technology to make practical nanometer-scale electronic devices. We feel that there are two key elements to this issue - (i) a nano-device must be physically accessible to the macro-world; and (ii) a nano-device that exploits quantum phenomena should be electronically accessible and should operate under ambient conditions. In view of this, the first part of the paper is devoted to discussing the different SPM lithographic techniques that will form the physical interface between the micrometer and the nanometer scales. The second part critically explores the possibilities of using SPM for developing molecular electronic devices, arguing that their small size may allow single-electron effects to be exploited at ambient temperatures.

1. INTRODUCTION

Current trends in miniaturization indicate that by the year 2000, nanometer-scale electronic, sensing and actuating devices will play a significant role in engineering and biomedical fields. Science and engineering at atomic scales is at present in its infancy and there is serious concern about whether nanotechnology[1] currently exists. Even though nanotechnology may not be available now, the ability to access nanometer scales controllably is definitely possible. Therefore, it is fair to expect that for nanotechnology to be fully developed, the next decade must witness tremendous growth in the science and engineering at nanometer scales. The foundations of engineering at these scales will be developed only by the ability to: (1) observe features at nanometer scales; (2) fabricate nanoscale structures; (3) measure physical properties at nanoscales; (4) develop theories

---

[†] Current address: Department of Mechanical and Environmental Engineering, University of California, Santa Barbara, CA 93106-5070.

for phenomena unique to nanoscales; and (5) exploit these phenomena in the design of suitable devices.

Until the recent inventions of the scanning probe microscopes (SPM) - in particular, scanning tunneling microscope[2] (STM) and the atomic force microscope[3] (AFM) - experimental research in nanoengineering was extremely difficult and expensive. The unprecedented resolution, the versatility and the ease of use of the STM and AFM have now provided new and multi-faceted opportunities to conduct experimental research at nanometer scales. Although the STM and the AFM were initially invented to study surface morphology, they have unexpectedly found applications in an area that is of critical importance for the development of nanoengineering - material manipulation at nanometer and atomic scales. The most striking evidence of this was found in the ability to manipulate individual atoms of xenon[4] and silicon[5]. Several other techniques have demonstrated material deposition or removal at nanometer scales by field-induced evaporation[6], field-induced chemical reaction[7] as well as electron-energy induced chemical reaction[8-10]. Excellent reviews of the state-of-the-art material manipulation techniques by SPM are now available[11,12]. These techniques leave no doubt that material manipulation at nanometer scales using SPM is possible. The burning question that must now be addressed is - how can these material manipulation techniques be used for the development of nanotechnology?

The most obvious application of material manipulation using SPM is to fabricate electronic devices. Devices on the nanometer scales must operate on fundamentally different physics where transport of charge carriers are wave-like and sometimes ballistic in nature. Besides electronic devices, chemical and biological sensors can be made by manipulating single or a cluster of molecules. No matter what kind of device one fabricates, it should be kept in mind that for practical operation of such devices, one must be able to *physically* and *electronically* interface the nanometer-scale device with the macro world. The problem of interfacing is non-trivial and seems to be the biggest concern at present[1]. The goal of this paper is to explore the possibilities of developing nanometer-scale electronic devices based on the current state of knowledge.

Modern microelectronic devices are physically interfaced with the macro-world by using photolithographic techniques to fabricate structures that are in the micrometer range. For nanotechnology to be successful, we must take this hierarchy a step further and now develop techniques to link the micrometer to the nanometer scales. The first part of this paper will discuss the several SPM lithographic techniques that we have developed at Arizona State University for this purpose. These include the etching of semiconductors using the STM under solution, deposition of metals on semiconductors under solution and field-induced deposition of metals in air. Since most microfabrication techniques use polymeric resists to produce the desired patterns, we have used the AFM for the first time to chemically modify positive and negative polymeric resists. These methods are described in detail in section 2.

The ability to fabricate nanometer structures opens the world of *nanoelectronics*, one fundamentally different from that built around familiar semiconductor technology. Conventional technologies fail when a single defect dominates the behavior of a device or the transition of a single electron causes significant shifts in energy levels. This world is dominated by the quantum aspects of electron transport and it has been the subject of intense theoretical study and some experimental exploration. An excellent review is given by Averin and Likharev[13]. One of the quantum phenomena which has recently attracted a lot of attention for developing logic and memory devices, is *single electron tunneling*. Much of the effort to date has focused on building very small (mesoscopic) structures which, when operated at low temperatures, serve as reasonable realizations of solvable quantum mechanical models. The STM has played a significant role in realizing a "working model" of some of these devices[14-18].

In section 3 of this article, we will explore a quite different approach to the construction of such single electron devices by the use of organic molecules. This is motivated by our long-standing interest in STM of biomolecules[19] rather than by any firm theoretical or experimental basis for a technology based on 'molecular electronics'. From the preliminary studies, it appears to be that molecular single electron devices can not only operate at room temperatures but also be electronically accessible by voltages of the order of 1V. Despite the promising prospects, it appears to us that the theoretical challenges in this field may be somewhat greater than encountered in the approach based on building 'mesoscopic devices'. Nonetheless, several STM experiments have produced tantalizing results and we will review some aspects of tunneling in adsorbed organic molecules in section 3 of this paper and speculate about some mechanisms for practical devices. Section 4 contains the summary and some speculation regarding the future of nanotechnology.

2. SPM LITHOGRAPHY AT NANOMETER SCALES

The success of modern microelectronic devices lies mainly in the ability to fabricate structures in the micrometer range by optical lithography[20]. Typically, a film of photoresist, that would chemically modify under exposure to photon radiation, is deposited on a surface. The exposed or chemically modified regions would etch either faster or slower than the unexposed regions, thus leading to positive or negative resists, respectively. To generate a certain architectural pattern, the photoresist is exposed in that pattern and developed in the etchant to form the desired trenches and hillocks. The spatial resolution of optical lithography is limited by diffraction and is therefore of the order of the wavelength of visible light (~ 0.5 μm).

To fabricate devices in the nanometer range and to physically interface them to the micrometer scales, it is necessary to develop lithographic techniques with nanometer scale resolution. One way to achieve this is to reduce the wavelength of the incident radiation by using either X-rays or high-energy electron beams. An excellent review on this topic has been compiled by Chang et al.[21]. Although X-ray lithography is a faster process and suitable for mass production by direct contact pattern transfer, its resolution (~ 50 nm) is currently limited by fabrication of patterned masks. Scanning electron-beam lithography does not have high throughput although single devices have been fabricated with resolution in the range of 10-20 nm. Its resolution is limited by scattering of incident high-energy electrons and by the production of secondary electrons that would expose resist films in undesired regions. A different approach is to use SPMs for nanolithography. Over the past six years, several studies have shown that materials can be deposited, removed and arranged on the surface at nanometer and atomic scales. The resolution achieved by this technique is comparable or, in several occasions, higher than other lithographic techniques. Although the scanning process makes it unsuitable for high throughput, it is too early to decide whether that will be a limitation. Therefore, it is instructive to study the different techniques for SPM lithography.

This paper will discuss in detail the different SPM lithographic techniques developed at ASU. Since there are excellent reviews available in the literature[11,12], there will be no attempt to discuss all the different SPM techniques that have so far been developed. The first part will discuss lithography using the AFM and the second part, that using STM.

2.1 Lithography Using the AFM

Although the STM has been successfully used for lithography (see discussion in section 2.2), one has to keep in mind that it can be used only on conducting or semiconducting materials. However, most of the resist materials used for fabrication of electronic devices and sensors, are polymeric materials which are insulators. One of the most common one used is poly(methylmethacrylate) (PMMA).

Despite the inability to operate the STM under tunneling mode on an insulating surface, it has been used in a field-emission mode to chemically modify thin insulating polymeric resist films[22-25]. Under a ultra-high vacuum environment, the Stanford group[22] observed chemical modification of a 20 nm thick PMMA film when they bombarded the film with electron energies above 25 eV provided by an STM tip in field emission mode. The doses they used were larger than 10^{-2} C/cm^2. They observed that below 25 eV, even large doses could not bring about negative exposure. They obtained a lateral resolution of 20 nm and found that their resolution decreased for increasing tip voltage when the tip was negatively biased. Subsequently, the group in Naval Research Laboratories have used the STM to chemically modify several types of polymer resists with nanometer-scale spatial resolution[23-25]. Although the field-emission technique is successful, there are two inherent obstacles. (1) There is no control on the force between the STM tip and the polymer film. Surface irregularities or slight variations in the film thickness can therefore push the tip into the film thereby damaging it. (2) During imaging in a field-emission mode, it is possible to chemically modify the film which makes it unsuitable for preferential exposure. To circumvent these problems, we have developed[26] a lithographic technique using the AFM.

We used the commercially available AFM, Nanoscope II from Digital Instruments. The AFM consists of a sharp tip mounted on a cantilever which is brought very close to or in contact with a substrate surface by piezoelectric actuators. The deflections of the cantilever due to tip-substrate interatomic forces are optically detected[27] by reflecting a laser beam from the cantilever on to a photodetector. With the spring constant of the cantilever known, the tip-substrate force can then be obtained. While maintaining a constant tip-substrate force of 10 - 20 nN by a feedback control of the piezoelectric actuator, the tip is scanned over the substrate surface to obtain topographical images with atomic resolution capability. To chemically modify a material at nanometer scales using the AFM, it is necessary to provide enough energy either by mechanical forces, by electron or by photon bombardment. The commercially available tips are made of silicon nitride which cannot be used to provide energetic electrons or photons. We, therefore, deposited a 500 Å gold film by thermal evaporation on the silicon nitride tip

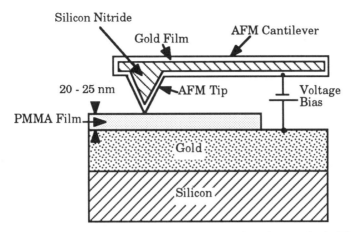

Fig. 1 Schematic diagram the experimental set-up for chemical modification of a poly(methylmethacrylate) (PMMA) film using a gold-coated, silicon nitride AFM cantilever and tip. A voltage bias is applied across the 20 - 25 nm PMMA film between the metallized tip and the underlying gold film.

and cantilever[†]. The gold-coated AFM tip was then used as a highly-localized electron source under a voltage bias as shown in Fig. 1. The PMMA film of molecular weight 950,000 was spun on to a 2000 Å film of gold that was evaporated on to a silicon substrate. Starting with an initial PMMA concentration of 0.4 percent in a chlorobenzene solution and spinning at a speed of 2640 rpm for 30 seconds, it was estimated from a spin curve that the film thickness was between 20 - 25 nm. After spinning, the film was baked at 170 °C in ambient atmosphere for 60 seconds before conducting our experiment.

Our experiments started by first scanning the PMMA film surface under a constant force of 10 - 20 nN with the gold-coated AFM tip at zero bias. Upon finding the film quality to be satisfactory (that is without any pin holes), the tip was withdrawn about 50 µm away from the surface. A DC voltage bias in the range of 0-18 V (see discussion below) was then applied to the tip and the tip was again advanced towards the surface while monitoring the force. Once a tip-substrate repulsive contact force of 10 - 20 nN was reached, the tip was scanned over the PMMA surface at the lowest allowable scan rate of 0.13 Hz per line under a constant force. The idea was to have maximum electron exposure of the PMMA film so that the dosage was well above the critical threshold[21] of 100 µC/cm^2. During the exposure, the current fluctuated between 1- 10 pA. After the desired exposure, the AFM tip was withdrawn and the voltage bias set to zero. To check for any mechanical damage, the tip was engaged again at zero bias and scanned over the exposed area. We did not observe any trace of mechanical damage of the film with tip-substrate forces in the range of 10-20 nN. The sample was then removed and developed for 60 seconds in a solution of methyl-isobutyle-ketone (MIBK) and isopropyl alcohol (IPA) in a 1:1 ratio. The sample was rinsed in methanol and blow dried to remove any

---

[†] We found that the adhesion between gold and silicon nitride is very poor. In addition, stresses in the gold film would often warp the cantilever. We later started by depositing a 40 Å layer of chrome on silicon nitride followed by 300 Å of gold on the chrome using electron beam evaporation. The chrome acts as a good buffer layer for adhesion and stress relief.

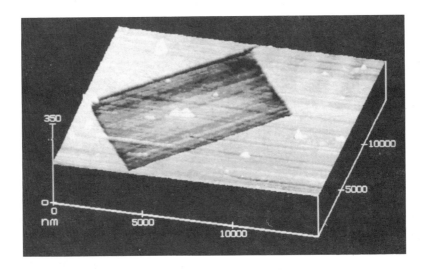

Fig. 2 An AFM image of a 7 μm x 10 μm box created on a PMMA film by first scanning the gold-coated AFM tip under a -7V bias.

moisture. A silicon nitride tip was then used to image the PMMA film to look for any surface features in the exposed regions.

We found the first evidence of lithography when we observed the 10 μm x 7 μm box on the PMMA film shown in Fig. 2. For this case the PMMA film was exposed under a tip voltage bias of - 7 V. Since smaller polymer chains dissolve faster than longer ones in the developing solution, the experimental results suggest that the energy of the 7 eV electrons emitted from the gold-coated AFM tip was sufficient to break the polymer chains to smaller lengths thus utilizing PMMA as a positive resist. We scanned another exposed region to find a grating pattern of lines with 68 nm periodicity and line widths of about 34 nm as shown in Fig. 3. This again was an evidence of positive resist phenomenon under a tip bias of -7 V. To further improve the resolution of this technique, it is necessary to produce sharper AFM tips. The line width of 34 nm possibly corresponded to the radius of curvature of the gold-coated AFM tip. Using the exposed area of a single line to be 34 nm x 10 μm, the measured exposure current (1-10 pA) and the known scan rate (0.13 Hz), the dosage was calculated to be in the range of 2-20 mC/cm^2, which is well above the critical threshold for PMMA exposure.

When the bias was increased to -18 V, we found protruding lines in the exposed regions as shown in Fig. 4. This gave us evidence of a negative resist phenomenon due to bond formation to build longer-chain polymer molecules which dissolve more slowly

Fig. 3 An AFM image showing a grating line pattern created by scanning the gold-coated AFM tip under a -7V bias. In the confined regions where the tip was in contact with the PMMA film, the emitted electrons broke the chemical bonds creating shorter polymer chains which etched out faster in the developing solution. The line periodicity is 68 nm and the line widths are about 34 nm.

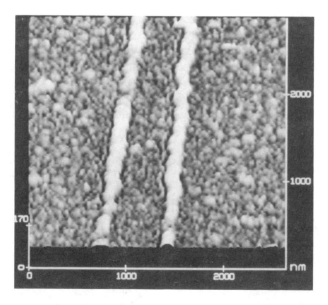

Fig. 4 An AFM image of a two protruding lines created by scanning the gold-coasted AFM tip under a -18V bias. The higher electron energy was responsible for forming chemical bonds that created longer polymer chains in the confined regions under the AFM tip which etched away slower than the unexposed regions. The width of the lines is about 100 nm.

than the unexposed molecules. The observations of protruding lines also eliminated the possibility of mechanical damage under repulsive forces. The observation of bond breakage at -7V and bond formation at -18V suggests that these two competing effects are strongly energy-dependent with bond formation preferred at higher energy. The bias-dependent results of positive (bond breakage) and negative (bond formation) exposure are in qualitative agreement with those of McCord and Pease[22] who used the STM under field-emission mode to expose a PMMA film.

The main idea of using the AFM, as opposed to STM, for PMMA lithography is to decouple the method to maintain a constant tip-surface distance from the method to expose the polymer film. There are two clear advantages: (1) mechanical damage can be eliminated by maintaining a constant force of 10 - 20 nN during exposure; (ii) the surface can be imaged under zero tip bias, thus eliminating the exposure of PMMA film during imaging.

The successful demonstration of using AFM for lithography on PMMA opens the road to using other insulating resists on which the STM cannot operate. The resolution of PMMA is limited by size of the polymers which is typically about 10 nm. For lithography of higher resolution, one must use inorganic resists. Utsugi[7] used the STM on a resist of film of Ag-Se to write "NTT" with 2 nm resolution. The reason STM could be used on Ag-Se is because this material is both an ionic and an electronic conductor. However, there are several other chalcogenide materials such as As-S and As-Se which are electrically insulating but have been used as resist in electron-beam lithography[28]. It would be interesting to use the AFM with a metallic tip to expose such inorganic resists.

The AFM lithography described so far involves the conventional method of patterning resists and then transferring the patterns to make the final devices. The transfer process usually leads to a loss of resolution. Therefore the direct fabrication of devices is preferred. This is described in detail in the following section on STM lithography.

2.2 Lithography Using the STM

This section is divided into two subsections on etching and deposition of materials that were conducted at ASU.

2.2.1 Etching of Semiconductors by STM. Semiconducting materials such as Si and GaAs form the majority of current electronics devices. Such devices are made by several processes that involve optical lithography as a crucial step for obtaining the desired pattern[20]. To fabricate devices smaller than this scale, one must be able to make patterns on semiconductors with higher spatial resolution. With this motivation, a new technique of etching semiconductors with 20 nm resolution using the STM was developed at ASU. This is described as follows.

The experiments conducted by Nagahara et al.[29] used an STM on a semiconductor under dilute solution (0.05 %) of hydrofluoric acid as shown in Fig. 5. The STM tips were electrochemically etched from $Pt_{0.8}Ir_{0.2}$ wires. To minimize unwanted Faradaic current through the solution, the tips were coated with Apiezon wax up to their extreme ends[30]. Only tips with an interface capacitance less than 10 pF and a leakage current of less than 100 pA (with 1 V applied in a 1 M NaOH solution) were

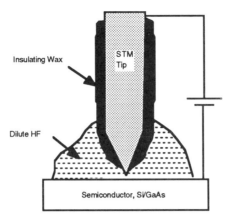

Fig. 5 Schematic diagram of localized etching of semiconductor under dilute HF solution using STM.

selected for the experiments. The samples used were n-Si (100) (10^{18} cm^{-3}) and p-GaAs(100) (10^{19} cm^{-3}) wafers that were chemically and mechanically cleaned to remove organic grease and oxide films. Ohmic contacts were made using a Ga-In eutectic alloy for Si and electrochemically deposited gold for GaAs.

The etching of Si and GaAs samples was performed by scanning a surface area while maintaining the tip bias at +1.4 V and a tunnel current at about 1 nA. It was observed that when the surface was scanned under these conditions, a hole developed in the scanned area. The depth of etching increased as the area of the scan decreased. This is shown quantitatively in terms of current dosage in Fig. 6. It can be seen that below a dose of about 4 A/cm^2, etching does not take place. To generate a desired pattern, this technique was used to write the letters "A S U" on a GaAs (100) surface as shown in Fig. 7. The size of each letter is about 500 nm x 500 nm with the line widths of less than 100 nm. In further experiments the feature size was reduced to about 20 nm.

The mechanism of the STM etching technique is still not clear, although most of the evidence suggest that it is due to chemical etching of a field-induced oxide layer formation under the STM tip. The electric field generated between the STM tip and the semiconductor surface is of the order of 10^7-10^8 V/cm. Such high fields can induce oxide growth which can be etched in the presence of dilute HF solution. Mechanical etching was eliminating by demonstrating that etching did not occur under distilled water or dilute H_2SO_4.

The ability to make patterns on semiconductors immediately opens roads to making nanometer-scale electronic devices. In this case, the lithography is directly used for device fabrication as opposed to the sole purpose of interfacing between nanometer and micrometer scales

2.2.2 Deposition of Metals on Semiconductors: Metals are often deposited on semiconductors for devices involving potential barriers such as the Schottky barrier. To fabricate nanometer-scale devices such as a MESFET, it is necessary to deposit a gate with line width in the nanometer range. Conventionally, this is done by exposing a

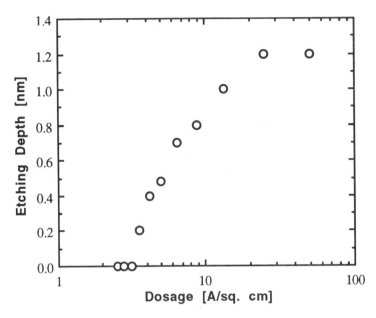

Fig. 6 Depth of etching as a function of current dosage. The data was obtained from the scan rate, tunnel current and area scanned from the experiments of Nagahara et al.[29].

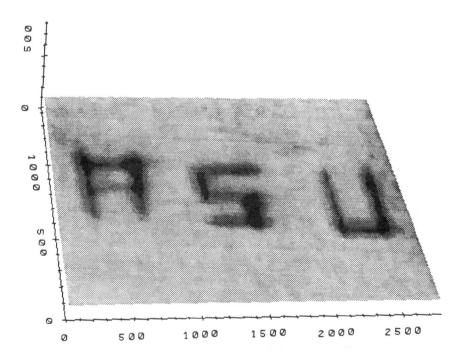

Fig. 7 The letters "A S U" written on p-GaAs(100) using an STM under a dilute HF solution. The line widths are typically 50 nm.

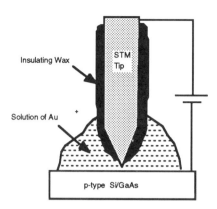

Fig. 8 Schematic diagram of metal deposition under solution by STM

positive resist film between the source and drain by photolithography, etching the exposed are and then depositing metal on the exposed semiconductor to get a gate. In contrast with such an indirect process, the STM can be used to directly deposit metal on semiconductor with nanometer-scale resolution. This was done[31] at ASU and is described as follows.

A p-type GaAs was chosen as the sample. A droplet of $KAu(CN)_2$ was placed on the sample and a STM tip was immersed into the droplet. The STM tip was coated with Apeizon wax such that only the very end was exposed and other regions were insulated to reduce Faradaic currents. When a tip bias of +2V was applied to the STM, it was observed that a Au was deposited on the surface. When the bias was kept for a longer period of time, the mound grew in size. Figure 9 shows two gold dots deposited in this manner. The dots are about 100 nm at the base and are about 2 nm in height. Although the mechanism for gold deposition is not fully understood, one possible mechanism is as follows. When positive bias is applied to the STM in a tunnel junction formed by metal-insulator-p:semiconductor, the conduction and valence band in the semiconductor would bends as shown in Fig. 10. An incident photon on the semiconductor excites an electron from the valence to the conduction band. In the presence of band bending, the electrons migrate towards the surface creating a negative surface charge. The Au^+ ions move towards the semiconductor surface to neutralize the charge thereby leading to gold deposition.

Although the size of the Au dots are 100 nm at the base, the resolution can be further improved by increasing the doping concentration of the semiconductor. This would increase the carrier concentration in the semiconductor which would eventually confine the electric field lines in the semiconductor to be localized under the STM tip. This would result in band bending in a more localized area that would lead to higher resolution. After the metal is deposited on the p-type GaAs, the top layer can be oxidized to get a MOSFET.

Other STM lithographic techniques are described in this book as well as in other reviews[11-12] of the field.

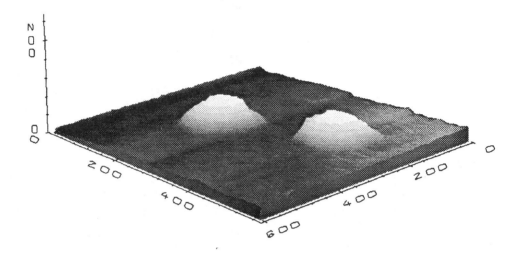

Fig. 9 Two gold dots of 100 nm base deposited on p-GaAs by STM under a Au⁺ solution.

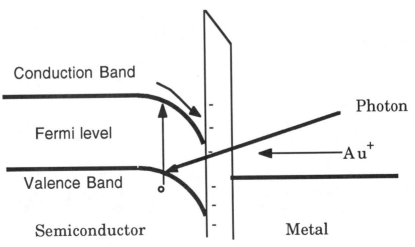

Fig. 10 Energy diagram showing the physics of photon-induced electrochemical deposition of metal on semiconductor by STM.

2.3 Application of SPM Lithography for Nanoelectronic Devices

The lithographic techniques that have been discussed so far can be utilized in two different ways for the development of nanoelectronic devices. (1) They can be used to directly fabricate conventional devices such as FETs. (2) They can be used to physically access devices fabricated by other techniques.

Although there are several techniques of SPM lithography, there has not been much progress in using them to fabricate nano-devices. The feasibility of fabricating a device like a FET can be demonstrated by writing a metal line on a semiconductor for the gate. The source and drain can be fabricated by photolithography leaving a region of about 1 µm between them for the gate. The main difference between SPM and photolithography conventionally used is that, the former requires a conducting or a semiconducting substrate whereas the latter can be performed on insulating substrates too. In light of this, there must be some changes made in the process of device fabrication although such problems are easy to overcome.

Although the fabrication of conventional devices such as FETs seems possible, the main contribution of SPM lithography might come in the development of non-conventional devices that exploit quantum transport phenomena. SPM lithography can be used to either fabricate them or to provide the interface for physical access to the macroworld. Non-conventional devices that have recently attracted a lot of attention are based on single electron tunneling. The next section describes how molecules can be used for such devices.

3. MOLECULAR ELECTRONIC DEVICES AND THE STM

Although the physics of electron tunneling has been understood for several decades, the phenomena of *single electron tunneling* (SET) has recently attracted a lot of attention. This is mainly due to the promising prospects of making nanometer-scale electronic logic and memory devices. When electrons tunnel via a small particle, the tunneling can become highly correlated in time, the current containing regular spaced bursts of charge corresponding to the transition of a single electron. This is because the energy of the intermediate particle changes as it is charged by tunneling which prevents further tunneling until the particle is discharged. This modulation of SET is the basis of various devices that have been proposed[13,32]. The conventional approach has been to fabricate small semiconductor or metal structures which can operate only at low temperatures. This is because the energy required for a single electron to tunnel through is $e^2/2C$ where e is the charge of the electron and C is the capacitance of the tunnel junction. The capacitance of conventional tunnel junctions made of oxides or semiconductors is quite high so that the thermal energy fluctuations of the order of kT has to be very small to observe SET effects. However, organic molecules have capacitance low enough such that $e^2/2C$ can be of the order of 1 - 10 eV which is much above the thermal energy at room temperature. This opens the possibility of fabricating nano-scale devices using organic molecules which can be operated at room temperature. In this section we introduce some elementary aspects of tunneling through molecular monolayer.

3.1 Tunneling in Organic Monolayers

The development of oxide tunnel junctions[33] in the 1960's resulted in many studies of organic monolayers, although most of the work focused on inelastic processes owing to phonons of meV energies[34]. Indeed, it appears that convincing evidence for effects involving *electronic states* has been obtained by these methods only recently[35,36].

The field has been given added impetus by the large number of STM and AFM studies of molecular adsorbates (recently reviewed by Frommer[37]). On the whole, the goal of these studies has been limited to obtaining reliable images. More extensive studies are few in number owing to the difficulty of obtaining reproducible electrical characteristics for a single molecule. Aviram *et al.*[38] report switching of conductance through application of a voltage pulse to a surface covered with hemiquinones. Pomerantz *et al.*[39] have described rectification at a graphite surface covered with Phthalocyanine molecules. Negative differential resistance has been observed on the surface of boron doped silicon[40] and asymmetric current-voltage characteristics have been recorded over DNA adsorbates[41].

There is, as yet, no general, computationally tractable, theoretical approach for describing the electronic properties of molecular adsorbates. In the next section we will review some aspects of the problem in a semiqualitative manner.

3.2 Electron Transfer and Molecules on Metals

The problem of charge transfer in adsorbed layers has been the subject of a great deal of study because of its importance in electrochemistry[42] and heterogeneous catalysis[43]. It has received some attention from the physics community in connection with enhanced field emission[44]. In complex molecules, the theoretical approach is usually based on hopping conductivity[45]. The transfer event is considered to be essentially instantaneous on the timescale of molecular vibration (Franck-Condon principle) so that the transfer rate is given, in a 'Golden Rule' approach by the product of a matrix element and a density of states which takes account of the electron-phonon interaction (for reviews see Devault[46] and Boxer[47]). In the electrochemical environment, the contribution from solvent motion must also be considered[48]. For a good intuitive description of molecular adsorption (it combines band structure theory with traditional chemical arguments) see the review by Hoffmann[43].

We will attempt to make a connection with this complex problem by extending a simple model of a molecule in an STM tunnel gap. In doing so we shall see *that there may be important effects from the time-evolution of the charge distribution over timescales where molecular motion occurs*, leading to effects that would not be anticipated on the basis of the simple hopping approach[45].

3.3 A simple model for molecules in an STM gap and resonant tunneling.

Chen[49] has emphasized the profound connection between tunneling and the formation of chemical bonds, a concept familiar in Pauling's notion of 'resonance' energy. Working with Otto Sankey, we have pursued a similar line of reasoning in using a tight binding model for a molecule in a tunnel gap. We explored the quantum transmission of electrons through a 'bound' molecule[50]. The simplest version of that model is a one electron, one dimensional, conduction band described by only two

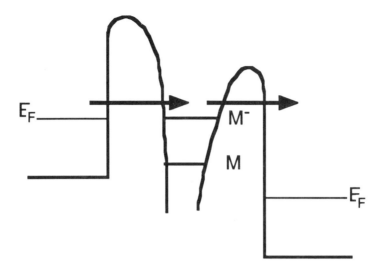

Fig. 11 Energy Diagram of the tunnel junction formed by the molecule between two metallic electrodes. The energy state of the molecules is shown as the two horizontal lines.

parameters[51]. The first is the 'on site energy', ε_0 which describes the matrix element for a valance state at a particular atomic site evaluated from the Hamiltonian for the whole crystal (if interactions between atoms were vanishingly small, this would be the atomic orbital energy). The second parameter, which describes the interaction between nearest neighbors only, is the matrix element taken between adjacent sites. It is called the hopping matrix element, τ. It is straightforward to show that a 1D row of atoms (lattice constant, a) has a conduction band of energies $E_k = \varepsilon_0 + 2\tau \cos ka$ where k is the wave vector of a Bloch wave, e^{ika}. More realistic versions of this empirical tight binding model have proved very useful in describing the electronic properties of solids[52].

The extension to a molecule in a tunnel gap is straightforward. The energy diagram of the tunnel junction is shown in Fig. 11 where the molecule is assumed to have two energy states near the Fermi level of the metals. At some point in our infinite row of 'metal atoms', we change on-site energies from ε_0 to $\varepsilon_{1,2...n}$ and hopping matrix elements from τ to $\tau_{1,2,...n-1}$. In this way we make a 'molecule' of n atoms, sandwiched between two metal electrodes. We call the hopping matrix element that connects the 'molecule' and 'STM tip' τ_R and the hopping matrix element that connects the 'molecule' and the substrate, τ_L. The states in the metals to the left and right are still Bloch waves with energies in the conduction band, so the problem reduces to a matrix of order ~ n^2 and it can be solved for the transmission of electrons as a function of energy[50]. An example is given in Figure 12. Here, the 'metal' has $\varepsilon_0 = 1$ and $\tau = 1$ while the molecule has two atoms of on-site energy 0 (open circles) and two of on-site energy -4. The hopping matrix elements have been kept equal to one, with the exception of those between 'molecule' and 'metal'. τ_R has been kept small (0.1) while τ_L has been varied from 0.1 to

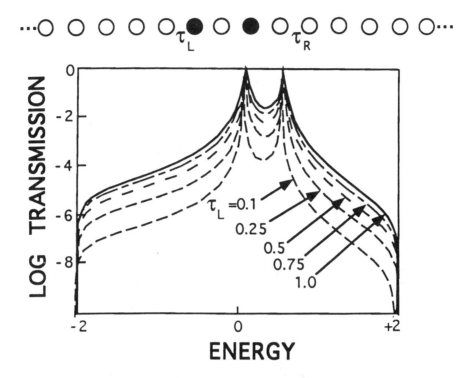

Fig. 12 Transmission vs. energy for an electron incident on a molecule in a tunnel gap. The model is illustrated above the plot. The open circles to the left and right represent semi-infinite chains corresponding to the STM substrate and tip. The molecule (2 open circles, 2 filled circles) is coupled to the substrate via the hopping matrix element τ_L and to the tip via the hopping matrix element, τ_R. The filled circles are atoms with $\varepsilon_0 = -4$ and the open circles have $\varepsilon_0 = -0$, giving two states in the conduction band. The plots are shown for $\tau_R = 0.1$ and for values of τ_L between 0.1 and 1.0 (hopping matrix element in the crystal is to be taken to be 1). There are two resonances in transmission, corresponding to the two molecular eigenstates in the conduction band. Increased coupling broadens the maxima, but does not diminish their value. This no longer holds if coupling is such as to cause charge transfer to the substrate (see discussion in Lindsay and Sankey[55]).

1.0 (simulating ever stronger coupling between 'molecule' and 'substrate'). The molecule has two states that lie in the conduction band (±2 energy units in width) and, when the incident electron energy is equal to one of the `molecular' eigenenergies, *the transmission becomes unity.* The high transmission is a consequence of electron delocalization, and the effect is called *resonant tunneling*. Whenever a localized state in a double well is degenerate with eigenstates outside the well, an incident electron can penetrate the well with a transmission *on the order of unity, independent of how `opaque' the barriers that* *define the well are!*[53]. This amazing phenomenon was exploited by Esaki in proposing (the now familiar) quantum-well devices[54].

3.4 Resonant tunneling and limitations of the simple model.

Fabricated quantum wells are relatively large and rigid when compared to molecules, so the simplest tight binding model[50] has flaws if applied to molecules.

We need to consider the following added complications: 1) The time evolution of the resonant state is important in that significant local charging might occur (our Hamiltonian contained no explicit Coulomb term to take account of such effects); 2) On timescales longer than phonon periods ($\sim 10^{-13}$ s) relaxation will play a role (our model nuclei are rigid) and 3) many-electron interactions are important in molecular interactions with metal surfaces (e.g., Newns[56], Hoffmann[43]). We will discuss the first two points below, abandoning the latter as beyond the scope of our present treatment. Unfortunately, many-electron effects may well be important. Indeed, the current oscillations that result from the highly correlated nature of the electron tunneling events are a fundamental manifestation of single electron tunneling[13,57]. (We have discussed some many electron effects in a qualitative way elsewhere[51]).

We first discuss the time evolution of the resonant state. Working with Kevin Schmitt, we have computed numerical solutions of the time dependent Schrödinger equation for a tight binding model of a single atom in a tunnel gap[58]. The sequence of events that lead to resonant tunneling is as follows (see Figure 13). The resonant state in the gap is initially uncharged. Electrons incident on the barrier (from the left) cause a slow charging of the resonant state. Amplitude builds up in the resonant state and begins to 'leak' out of the barrier on the right. The `leaking' rates are set by the hopping matrix elements on the left and right of the local state, τ_L and τ_R. The degree of excess charge accumulated on the local state is controlled by the inverse of these parameters. An optical analog is the Fabry-Perot interferometer where amplitude builds up in a resonant cavity to just the amount required to give unit transmission at resonance. Mirrors with low reflectance (the analog of bigger $\tau_{L,R}$) still yield unity transmission but give a broader response (as seen in the quantum case when τ_L is increased from 0.1 to 1.0; Fig. 12.). Key parameters in distinguishing between resonant tunneling (as a simple 'bond forming' process between tip, molecule and substrate[49]) and more complex charge transfer processes, are the *time for charging the intermediate state* and the *amount of excess charge accumulated in the intermediate state*. The hopping matrix element, τ has a simple interpretation in terms of the uncertainty principle and an effective 'time for charge transfer' between sites. In crystals where $\tau \sim$ eV, this time is $\sim 10^{-16}$ s, growing roughly exponentially with increasing gap between sites[58] (of course, the charging process itself is unobservable within this timescale). Thus, for roughly 7 lattice constants (say 20 Å gap) the 'charging time' approaches phonon frequencies, and nuclear motion can begin to play a role in lowering the increased energy due to charging (i.e. in forming polarons; stronger coupling can lead to trapped charge[59]).

The effects may not be dramatic at these timescales. A simple model shows that coupling to a single phonon introduces a sideband in the transmission spectrum but otherwise leave it unaffected[60]). Stockman *et al.*[61] have recently considered coupling to a thermal bath of phonons in a multiple quantum well structure, showing that quantum interference effects can become washed out by dephasing of the wave function. In any case, it is clear that these effects will be important on much longer timescales. At nA currents, the increase in charge (over a period of $\sim 10^{-13}$ s at the molecule is very small indeed $\sim 10^{-6}e$. If, however, the molecule where to become fully charged (hopping times of $\sim$ nS) the transmission would appear to be *qualitatively* different.

At the molecular level, this charging of the (previously neutral) molecular state, M, is the energy required to form the molecular ion, M^- (if a bound state exists). These energies can be large, and it is instructive to consider them in the context of a well-studied phenomenon in charge transfer through small particles, the *Coulomb blockade*; see for example, Mullen et al.[16] and references therein. If a particle (molecule) of 3 Å diameter (area, $A \sim 10^{-19} m^2$) is separated from a surface by a 3 Å vacuum gap (x), then the equivalent 'parallel plate capacitor' would have a capacitance ($C = \varepsilon_0 A/x$) of 10^{-20} F. The energy of the charged state ($e/2C$) would be raised by more than 10 eV! This charging could not be achieved without applying an electric field that raised the electronic energy appropriately, and this is the origin of the `gap' in current-voltage curves attributed to the Coulomb blockade.

The energy required to overcome the Coulomb blockade could also be obtained from an electrochemically applied field; in this case the 'blockade energy' is simply the potential change required to form the molecular ion on the electrode. *Thus, there is a very direct connection between electrochemical charge transfer, `tunneling' via small insulated*

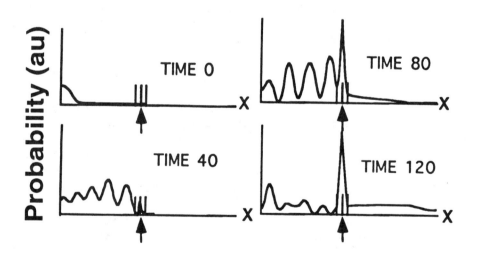

Fig. 13 Showing the time evolution of a resonant state. The arrow points to a resonant atom in a gap between two metal lattices. Charge is incident from the left. The series shows how probability builds up on the resonant state until unity transmission is achieved. The fine structure is due to interference with the reflected wave. For details of the calculation see Lindsay et al[58].

particles (Coulomb blockade) and the problem of charge transfer via charging of intermediate molecular states. We are exploring this connection with experiments in which STM contrast is measured as a function of electrochemical conditions[62].

For mesoscopic metal particles the small blockade energy requires cryogenic experiments[16] but the small capacitance of an individual atom or molecule raises the Coulomb energy to many times k_BT so that, in principle, these effects might be observed at room temperature. There is some experimental evidence for this. The Coulomb blockade has been observed at room temperature by Venkateswaran et al[63] who attribute large charging energies to single electron traps in a silicon dioxide film. Nejoh[64] has reported a 'Coulomb staircase' structure in current-voltage curves obtained over liquid crystal molecules at room temperature. He has carried out calculations of the ionization potential for these molecules, concluding that the observed energy step (~ 100 mV) is much smaller than expected. One explanation of this discrepancy is the neglect of structural relaxation effects in the calculation.

The transition from simple resonance (in effect, chemical bonding) to discrete charge transfer is illustrated beautifully by recent experiments with a fabricated double-well device. Bo et al.[65] have built a device with very asymmetric barriers so that the coupling between the local states in the double well and an emitter is weak in one bias direction and strong in the other. Current flowing in the direction that gives strong coupling shows distinct sharp peaks due to resonances with states in the well. Tunneling in the other (weak coupling) direction shows the classic 'staircase' form of current vs. voltage associated with the Coulomb blockade. Theory for the transition between the two limiting forms of transmission is given by Korotkov et al.[66].

The much greater Coulomb energies obtained by charging molecules provide a strong driving force for structural relaxation. Indeed, this is one of the complications that makes the calculation of the electronic states of an adsorbate on a metal such a formidable problem. An extreme case of relaxation occurs in heterogeneous catalysis, where bonds are broken. An interesting case arises when relaxation can lower the energy of the excited state enough to trap the ionizing charge[59]. Since the relaxed molecule will have considerably altered electronic states, this phenomenon could form the basis of a very simple molecular memory. The charge trapping state is used to 'switch' the molecule and some other state is used to 'read' it via its quantum transmission, as altered by structural relaxation. These problems have not been addressed from a first-principles approach in the past. Clearly, solution of the time-dependent Schrödinger equation is needed to describe the charging and relaxation process in full. However, the recent development of quantum molecular dynamics (which combines the most useful aspects of quantum and classical mechanics) gives grounds for optimism that these problems can, at last, be attacked with currently available computers[67,68].

3.6 Experimental studies on molecular monolayers.

Some examples of STM studies of ordered molecular monolayers are Sleator and Tucko[69], Ohtani et al.[70], Foster and Frommer[71], Smith et al.[72], Spong et al.[73] and Heckl et al.[74]. Systematic studies of the contrast variation with the tip to substrate voltage, V_{ts}, have been reported for liquid crystals[75,76] and decanol on graphite[77]. At low bias, the underlying graphite lattice is seen, the adsorbate becoming visible as the

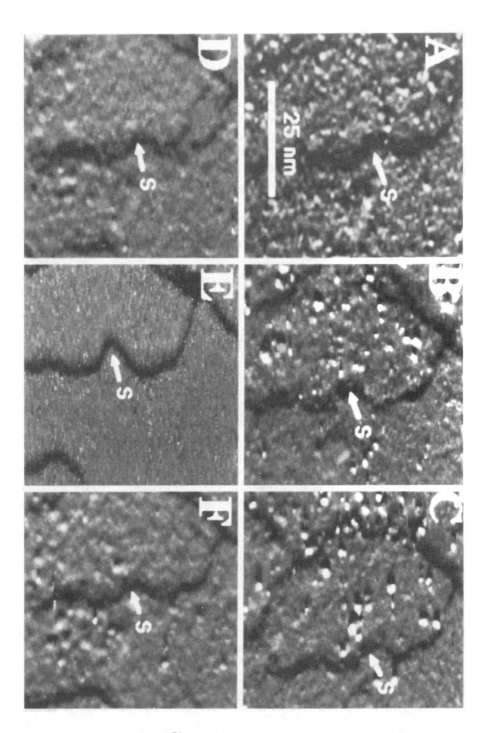

Fig. 14 Porphyrin on Au(111)[79]: V_{ts} is A = -2 V; B = -1.5 V; C = -1.2 V; D = -0.4 V; E = -0.05V. Contrast is restored on returning to -1.2 V while scanning the same area in F. The response of contrast to changes in V_{ts} is *not* immediate, suggesting the slow charging of an intermediate involved. A step on gold is marked "S".

bias is raised. Mizutani et al.[75,76] attributed this effect (in liquid crystals) to resonant tunneling. We[62] have studied all four of the DNA bases adsorbed onto an Au (111) surface under electrochemical potential control. We find that thymine has an anomalously small contrast when imaged at an electrochemical potential where it undergoes an electron transfer reaction with the substrate. We have observed[78] similar behavior in the case of iodine adsorbed on Au (111) an effect we tentatively attribute to delocalization of the atomic or molecular state nearest the Fermi energy.

Porphyrin is a particularly interesting case of an adsorbate that is 'transparent' at low V_{ts}[79]. We have studied films in which the porphyrin rings were attached by a pendant isocyanate group which reacted strongly with the gold. This system was developed so that the electron transfer involved in binding did not disrupt the states in the organic ring (expected to be important in tunneling). Preliminary optical studies showed that the optical transitions in the porphyrin ring appeared to be little disrupted by this binding scheme[80]. We show some STM results below in Figure 14.

These changes in contrast *were not instantaneous* but took about 10 s after a change in V_{ts} for the equilibrium value of contrast to be reached. This strange behavior might be a consequence of the unusual binding scheme, chosen to minimize interaction between states associated with the porphyrin ring and the gold. If these states are well enough isolated from the tip or substrate, they may be able to trap charge (over long timescales) and the subsequent relaxation would be noticed by consequent changes in the transmission of the adsorbate.

4. FUTURE POSSIBILITIES

In the first part of this paper we reviewed the different techniques of nanofabrication using SPM. The question that must now be asked is how do we use this technology. We feel that there are two possible options: (1) to use the nanofabricated structures to make conventional devices such as FETs; (2) to use the structures made by SPM lithography to physically access devices made by other techniques. Although the first option is the first step towards using this technology, the more long-term impact may actually come from the second option. This is because once we access the nanometer scales, electronics devices based on electron tunneling and, in particular, single electron tunneling are attractive. To electronically access single-electron-tunneling devices at room temperatures with voltages in the range of 1-10 V, it is necessary to use organic molecules with very low capacitance. Therefore the development of the field of molecular electronics must go parallel with the advances in nanolithography using SPM. In view of this, the second part of this paper discusses some of the recent developments of STM studies on molecules.

The number of straightforward STM experiments that show interesting electronic effects is small. Undoubtedly, one complication lies in controlling the atomic nature of the tip, and this illustrates the technological problems that might be encountered in adapting some of these phenomena to practical devices. Have we come full circle to be limited by macroscopic aspects of film preparation, as with the oxide tunnel junctions of the 1960's? We doubt it, because the STM and AFM, even with the vagaries of tip-contamination, permit useful microscopic data to be obtained on many occasions. In contrast, macroscopic film methods must rely on a complete absence of pinhole shorts etc. Thus, systems can be examined to search for material combinations such that the desired conformation is also the lowest conductance conformation (out of the many possible, perhaps uncontrollable, conformations). Given that a single correctly oriented molecule could do the job, the techniques of nanofabrication could now be used to build devices with enough molecules in the gap to ensure function, but still small enough to avoid fatal defects. This is a difficult task, but not as difficult as it was prior to the introduction of scanning probe microscopy.

5. ACKNOWLEDGEMENTS

Our sincere thanks to Patrick Oden, Larry Nagahara, Juan Carrejo, John Graham and John Alexander, Otto Sankey and Kevin Schmidt for their help. Stuart Lindsay was supported by NSF (DIR 89-20053) and ONR (N0014-90-J-M55). Arun Majumdar gratefully appreciates the support of NSF through the Young Investigator Award.

6. REFERENCES

1. P. Ball and L. Garwin, "Science at the atomic scale,"*Nature* 355, 761-766 (1992).
2. G. Binnig, H. Rohrer, Ch. Gerber and E. Weibel, "Surface Studies by Scanning Tunneling Microscopy," *Physical Review Letters* 49, 57-61 (1982).
3. G. Binnig, C. F. Quate and Ch. Gerber, "Atomic Force Microscope," *Physical Review Letters* 56, 930-933(1986).
4. D. M. Eigler and E. K. Schweizer, "Positioning Single Atoms with a Scanning Tunneling Microscope," *Nature* 344, 544-546 (1990)
5. I. W. Lyo and Ph. Avouris, "Field-Induced Nanomter to Atomic Scale Manipulation of Silicon Surfaces with the STM," *Scienc* 253, 173-176 (1991).
6. H. J. Mamin, P. H. Guenther and D. Rugar, "Atomic Emission from a Gold Scanning Tunneling Microscope Tip," *Physical Review Letters* 65, 2418-2421 (1990).
7. Y. Utsugi, "Nanometer-scale Chemical Modification Using a Scanning Tunneling Microscope," *Nature* 347, 747-749 (1990).
8. T. R. Albrecht, M. M. Dovek, M. D. Kirk, C. A. Lang, C. F. Quate and D. P. E. Smith, *Applied Physics Letters* 55, 1727-1729 (1989).
9. W. Mizutani, J. Inukai and M. Ono, "Making a monolayer hole in a graphite surface by means of a scanning tunneling microscope," *Japanese Journal of Applied Physics* 29, L815-L817 (1990).
10. J. P. Rabe, S. Buchholz and A. M. Ritcey, "Reactive graphite etch and the structure of an adsorbed organic monolayer-a scanning tunneling microscopy study," *J. Vac. Sci. Technol. A* 8, 679-683 (1990).
11. C. F. Quate, "Manipulation and Modification of Nanometer Scale Objects with the STM," presented at the NATO Science Forum '90, *Highlights of the Eighties and Future Prospects in Condensed Matter Physics*, Biarritz, France, Sept. 16-21, 1990; to be published in a NATO Series, Plenum Press.
12. J. A. Stroscio and D. M. Eigler, "Atomic and Molecular Manipulation with the Scanning Tunneling Microscope," *Science* 254, 1319-1326 (1991).

13. D. V. Averin and K. K. Likharev, in *Mesoscopic Phenomena in Solids*, Altshuler, B.L., Lee, P.A. and Webb, R.A. Eds., North Holland, Amsterdam, 1991, chapter 6.
14. P. J. van Bentum, H. van Kempen, L. E. van de Leemput, and P. A. Teunissen, *Physical Review Letters* 60, 369 (1988).
15. P. J. van Bentum, R. T. Smokers and H. van Kampen, *Physical Review Letters* 60, 2543 (1988).
16. K. Mullen, E. Ben-Jacob, R. C. Jaklevic and Z. Schuss, *Physical Review B* 37, 9810 (1988).
17. R. Wilkins, E. Ben-Jacob and R. C. Jaklevic, *Physical Review Letters* 63, 801 (1989).
18. K. A. McGreer, J-C. Wan, N. Anand and A. M. Goldman, *Physical Review B* 39, 12260 (1989).
19. S. M. Lindsay and B. Barris, J. Vac. Sci., Technol. A6, 544 (1988).
20. L. F. Thompson, C. G. Willson and M. J. Bowden, *Introduction to Microlithography*, ACS Symposium Series 219, American Chemical Society, Washington D.C. (1983)
21. T. H. P. Chang, D. P. Kern, E. Kratschmer, K. Y. Lee, H. E. Luhn, M. A. McCord, S. A. Rishton and Y. Vladimirsky, "Nanostructure Technology," *IBM J. of Res. Develop.*32, 462-493 (1988).
22. M. A. McCord and R. F. W. Pease, "Lift-off metallization using poly(methylemthacrylate) exposed with a scanning tunnelin microscope," *J. Vac. Sci. Technol. B* 6, 293-296 (1988).
23. C. R. K. Marrian and R. J. Colton, "Low-voltage electron beam lithography using scanning tunneling microscope," *J. Vac. Science Technology B* 6, 293-296 (1990).
24. C. R. K. Marrian, E. A. Dobisz and R. J. Colton, "Investigation of undeveloped e-beam resist with a scanning tunneling microscope," *J. Vac. Science Technology B* 9, 1367-1370 (1991).
25. E. A. Dobisz and C. R. K. Marrian, "Sub-30 nm lithography in a negative electron beam resist with a vacuum scanning tunneling microscope," *Applied Physics Letters* 58, 2526-2528 (1991).
26. A. Majumdar, P. I. Oden, J. P. Carrejo, L. A. Nagahara, J. J. Graham and J. Alexander, "Nanometer-scale lithography using the atomic force microscope," *Applied Physics Letters* 61, 2293-2295 (1992).
27. O. Marti, B. Drake and P. K. Hansma, *Applied Physics Letters* 51, 484 (1987).
28. G. H. Bernstein, W. P. Liu, Y. N. Khawaja, M. N. Kozicki and D. K. Ferry, "High-resolution electron beam lithography with negative organic and inorganic resists," *J. Vac. Sci. Technol. B.* 6, 2298-2302 (1988).
29. L. A. Nagahara, T. Thundat and S. M. Lindsay, "Nanolithography on Semiconductor Surfaces Under an Etching Solution," *Applied Physics Letters* 57, 270-272 (1990).
30. L. A. Nagahara, T. Thundat and S. M. Lindsay, *Reviews of Scientific Instrument* 60, 3128 (1989).
31. T. Thundat, L. A. Nagahara and S. M. Lindsay, "Scanning tunneling microscopy studies of semiconductor electrochemistry," *Journal of Vacuum Science Technology* A8, 539-543 (1990).
32. T. A. Fulton and G. J. Dolan, *Physical Review Letters* 59, 109 (1987).
33. I. Giaever, *Physical Review Letters* 5, 147 (1960).
34. E. L. Wolf, *Principles of electron tunneling spectroscopy* Oxford University Press, New York (1988).

35. K. W. Hipps and U. Mazur, *J. Phys. Chem.* 91, 5218 (1987).
36. K. W. Hipps, and J. J. Hoagland, *Langmuir* 7, 2180 (1991).
37. J. Frommer, *Angewandte Chemie*, in press (1992).
38. A. Aviram, C. Joachim and M. Pomerantz, *Chem. Phys. Lett.* 146, 490 (1988).
39. M. Pomerantz, A. Aviram, R. A. McCorkle, L. Li and A. G. Schrott, *Science* 255, 1115 (1992).
40. P. Bedrossian, D. M. Chen, K. Mortensen and J. A. Golovchenko, *Nature* 342, 258 (1989).
41. S. M. Lindsay, Y. Li, J. Pan, T. Thundat, L. A. Nagahara, P. I. Oden, J. A. DeRose, U. Knipping and J.W White, "Studies of the electrical properties of large molecular adsorbates," *J. Vac. Sci. Technol.* B9, 1096 (1991).
42. J. O. M. Bockris and A. K. N. Reddy, *Modern Electrochemistry*, chap. 8, Plenum, NY(1970)
43. R. Hoffmann, *Rev. Mod. Phys.* 60, 601 (1988).
44. J. W. Gadzuk, *Phys. Rev. B* 1, 2110 (1970).
45. J. J. Hopfield, *Proc. Natl. Acad. Sci.* (USA) 71, 3640 (1974).
46. D. D. DeVault, *Quantum mechanical tunneling in biological systems* Oxford University Press (1981).
47. S. G. Boxer, *Ann. Rev. Biophys. Chem.* 19, 267 (1990).
48. V. G. Levich, *Advances in electrochemistry and electrochemical engineering* (Delahay, P., and Tobias, C.W., eds) (Interscience, N.Y.) 4, 249-372 (1965).
49. C. J. Chen, *Introduction to Scanning Tunneling Microscopy*, Oxford University Press, New York (1992)
50. S. M. Lindsay, O. F. Sankey, Y. Li., C. Herbst and A. Rupprecht, *J. Phys. Chem.* 94, 4655 (1990).
51. S. M. Lindsay, in *Scanning Tunneling Micxroscopy, Theory, Techniques and Applications* Bonnell, D. (ed.) VCH Press, New York, 1992, in press.
52. W. A. Harrison, *Electronic Structure and the Properties of Solids* (W.H. Freemen, San Francisco) (1980).
53. D. Bohm, *Quantum Theory* (Prentice Hall, Englewood Cliffs, N.J.) (1951).
54. L. Esaki, *IEEE J. Quant. Electronics* QE22, 1611 (1986).
55. S. M. Lindsay and O. F. Sankey, "Contrast and conduction in STM images of biopolymers" in *Scanned Probe Microscopies, STM and Beyond,* ed. H.K. Wickramasinghe, AIP, NY, 125-135 (1992).
56. D. M. Newns, *Phys. Rev.* 178, 1123 (1969).
57. K. K. Likharev and T. Claeson, *Scientific American,* June, 80-85 (1992).
58. S. M. Lindsay, O. F. Sankey and K. E. Schmidt, "How does the scanning tunneling microscope image biopolymers?" *Comments on Molecular and Cellular Biophysics* 7, 109 (1991).
59. A. S. Davydov, *Biology and Quantum Mechanics* (Pergamon Press, N.Y.) (1982).
60. P. Hyldgaard and A. P. Jauho, *J. Phys: Condens. Matter* 2, 8725 (1990).
61. M. I. Stockman, L. S. Muratov, L. N. Pandey and T. F. George, *Phys. Rev. B* in press (1992).
62. N. J. Tao, J. A. DeRose and S. M. Lindsay, ``Self assembly of molecular monolayers: The DNA bases on Au(111) studied by in situ STM" *J. of Phys. Chem.* 97, 910-919 (1992).
63. N. Venkateswaran, K. Sattler, U. Muller, B. Kaiser, G. Raina and J. Xhie, *J. Vac. Sci. Technol.* B9, 1052 (1991).
64. H. Nejoh, *Nature* 353, 640 (1991).

65. Bo, Su, V. J. Goldman and J. E. Cunningham, *Science* 255, 313 (1992).
66. A. N. Korotkov, D. V. Averin and K. K. Likharev, *Physica B* 165 & 166, 927 (1990).
67. O. Sankey, G. B. Adams, X. Weng, J. Dow, Y-M. Huang, J. C. H. Spence, D. A. Drabold, W-M. Hu, R. P. Wang, S. Klemm and P. A. Fedders, *Superlattices and Microstructures* 110, 407 (1991)
68. G. B. Adams, PhD Thesis, Arizona State University, 1992 (unpublished).
69. T. Sleator and R. Tucko, *Phys. Rev. Lett.* 60, 1418 (1988).
70. H. Ohtani, R. J. Wilson, S. Chaing and C. M. Mate, *Phys. Rev. Lett.* 60, 2398 (1988).
71. J. S. Foster and J. E. Frommer, *Nature* 333, 542 (1988).
72. D. P. E. Smith, H. Horber, Ch. Gerber and G. Binnig, *Science* 245, 43 (1989).
73. J. K. Spong, H. A. Mizes, L. J. LaComb, Jr., M. M. Dovek, J. E. Frommer and J. S. Foster, *Nature* 338, 137 (1989).
74. W. Heckl, D. P. E. Smith, G. Binnig, H. H. Klagges, T. W. Hansch and J. Maddocks, *Proc. Natl. Acad. Sci.* (USA) 88, 8003 (1991).
75. W. Mizutani, M. Shigeno, M. Ono and K. Kajimura, *Appl. Phys. Lett.* 56, 1974 (1989).
76. W. Mizutani, M. Shigeno, M. Ohmi, M. Suginoya, K. Kajimura and M. Ono, "Observation and control of adsorbed molecules," *J. Vac. Sci. Technol.* B9, 1102 (1991).
77. G. C. McGonigal, R. H. Bernhardt, Y. H. Yeo and D. J. Thompsom, "STM Imaging of Physisorbed Molecules at the Liquid-Solid Interface" in *Scanned Probe Microscopies, STM and Beyond* (H.K. Wickramasinghe, ed, AIP, NY) in press (1992).
78. N. J. Tao and S. M. Lindsay, ``In situ scanning tunneling microscopy study of bromine and iodine adsorption on Au(111) under potential control" *J. Phys. Chem* 96, 5213-5217 (1992).
79. D. K. Luttrull, J. Graham, D. Gust, T. A. Moore and S. M. Lindsay, *Langmuir* 8, 965-968 (1992).
80. D. K. Luttrull, Ph.D. Thesis, Arizona State University, unpublished (1992).

LOW VOLTAGE E-BEAM LITHOGRAPHY WITH THE STM

Christie R.K. Marrian and Elizabeth A. Dobisz
Nanoelectronics Physics Section
Surface and Interface Sciences Branch
Electronics Science and Technology Division
Naval Research Laboratory
Washington DC 20375

and John A. Dagata
Precision Engineering Division
National Institute of Standards and Technology
Gaithersburg MD 20899

ABSTRACT

Low voltage e-beam based techniques have significant advantages for nanometer scale fabrication. Energies of only a few eV are required to induce the chemical change, such as bond scission, associated with the exposure of radiation sensitive materials (resists). The spatial scattering of electrons with such low energy is so small that the proximity effects associated with high energy e-beam lithography are absent. A scanning tunneling microscope (STM) has been operated in the field emission mode to produce a spatially confined, low energy (5-50 eV) electron beam for studies of the exposure of e-beam resist materials. STM lithography results have been compared to identically processed samples exposed with a tightly focussed 50 kV e-beam. In a commercially available polymeric resist, improved resolution (~25 nm) and the absence of proximity effects have been observed. STM patterning of ultra thin resist layers, especially suited to low voltage lithography, is described. The development of a proximal probe based lithographic technology is discussed highlighting methods to increase lithographic speed.

1. INTRODUCTION

The physical phenomena apparent at sub-100 nm dimensions have led to numerous concepts for nanoelectronics devices, such as resonant tunneling structures, quantum confinement and interference based devices, single electron transistors and Coulomb blockade switches. In a few cases working prototype devices have been fabricated with 100 nm scale geometries, but operation at liquid He temperatures is required for device demonstration. Higher temperature operation should be possible for smaller scale geometries. However, fabrication procedures for 100 nm geometries are presently extremely difficult and time consuming and operational devices are only produced with very low yields. Radically different approaches to nanofabrication are needed to achieve the necessary dimensional control, reliability and reproducibility.

Fabrication in manufacturing production based on masked UV and/or X-ray lithographic exposure of a suitable radiation sensitive material (resist) is expected to suffice for structures down to 100 nm, a limit which should be reached shortly after the turn of the century. For smaller dimensions the lithography of choice is unclear. However, a steered beam lithography, such as e-beam lithography, will be an essential component of this future fabrication technology either for direct write of electronic circuits or for mask making. However, high voltage beam lithographies are extremely expensive and have demonstrated problems (beam induced damage, sensitivity to thermal drift and magnetic fields and proximity effects) for dimensions of the order of 100 nm. Indeed, the understanding and correction of proximity effects are the major obstacles to increased resolution with e-beam lithography. Proximity effects are the result of resist exposure away from the point of impact of the primary electron beam. This exposure is caused by primary electrons forward scattered in the resist layer, electrons backscattered from the substrate and secondary electrons generated by the primary and scattered electrons. Even in current e-beam lithography systems, resolution is limited by electron scattering and proximity effects as opposed to the size of the focussed beam. The conventional approach to overcoming proximity effects is to increase the primary beam energy and create a more diffuse 'fog' of backscattered electrons for which correction can be made. This technique has worked in some applications but it is not clear that correction can be made for a broad range of resists and substrates at a resolution of the order of 100 nm.

An alternative strategy is to decrease the primary beam energy so the scattered electrons responsible for proximity effects are spatially localized. The logical extension of this approach is to reduce the electron energy to the threshold required to induce the resist chemistry. However, creation of such a focussed low energy electron beam is difficult with conventional electron-optical techniques. The scanning tunneling microscope (STM) provides a technique for spatially confining the electron beam.

In the STM, a sharp tip is maintained very close to a surface by controlling the tip-sample separation with a simple servo loop. The servo controls the position of the tip to maintain a constant tip-sample current. In most applications, the STM is operated with a small (between ±4 V) bias between the tip and sample, so the tip-sample current is determined by electron tunneling through the electronic barrier between tip and sample. As the tip is scanned laterally, the servo loop adjusts the tip height to track variations in the surface electronic structure. With suitable sample preparation, atomic resolution imaging can be achieved. At higher tip-sample biases, the tip-sample current is determined by field emission. For a given tip-sample current, the tip-sample separation is greater in the field emission mode than in the tunneling mode. However, the servo loop is still able track variations in the surface topography for imaging (with nanometer resolution) and lithography.

As the other chapters of this volume indicate, a wide range of material modification and manipulation techniques have been demonstrated. Our work has concentrated on the study of radiation sensitive materials used in various kinds of lithography. In particular, we have concentrated on high resolution e-beam resists as part of our investigations into the resolution determining mechanisms of nanofabrication processing. This paper summarizes these recent studies at the Naval Research Laboratory of the exposure of resist materials with low energy electrons in the STM. The various advantages of this low voltage approach to lithography are discussed along with some key issues that need to be addressed to develop a proximal probe based low voltage lithography system.

2. EXPERIMENTATION

The STM head was obtained commercially from W.A. Technology, and is mounted in a stainless steel chamber with ion and turbo pumps. The STM is driven by custom built

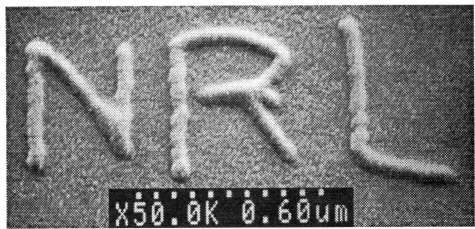

Fig. 1. SEM micrograph of non-contiguous pattern written with the STM at a tip-sample voltage of -25 V in SAL-601 and transferred into GaAs with a BCl_3 reactive ion etch.

electronics and in-house developed software. Lithography is performed at pressures of about 10^{-7} torr. Operation in vacuum is required because of the high tip-sample biases used. Samples are introduced under dry N_2 and the chamber can be evacuated in about 10 min with the turbo pump which is then valved off and turned off for the lithographic patterning.

Lithographic exposure is performed with an elevated bias between tip and sample (typically -10 to -50 V) and moving the tip laterally (at a constant speed). The STM servo loop is operated to maintain a constant tip-sample current. By appropriate movement of the tip with the STM scanner patterns can be defined in the resist material. As operation is in the constant current mode, the exposure dose applied to the resist can be easily monitored and varied. In the results described here, the STM tip is biased negatively with respect to the sample so electrons pass through the resist from the tip and into the substrate.

Operation of the STM requires a conductive substrate. Care must be taken to remove non-conductive oxides from the semiconductor surface. Si was etched in HF diluted with ethanol in a 10:1 ratio to remove the native oxide and passivate the surface with hydrogen. GaAs was etched in 7:1:1 $H_2SO_4 : H_2O_2 : H_2O$ solution, rinsed in water and coated with 10 nm of silicon which stabilizes the surface and improves resist adhesion. Films of resist were spin coated onto the prepared substrates. Resist thickness (typically between 30 and 70 nm) was verified with a surface profilometer. The GaAs samples were etched in a small custom built reactive ion etch (RIE) system with BCl_3. Etching was performed at an RF power of 80 W, a pressure of ~1 mtorr and a flow rate of 6 sccm. Samples were coated with 10 nm of Au or Au-Pd for inspection in a scanning electron microscope (SEM).

The results described here are predominantly with the resist SAL-601 from Shipley. It is a novolak based negative acting resist which uses chemical amplification to enhance its sensitivity. E-beam exposure releases acid which catalyses crosslinking during a post exposure bake. The resist was processed in the standard fashion as recommended by the manufacturers.[1] A post exposure bake at 107 °C for 7 min was followed by development in MF-322 for 12 min. Some results are also presented for a negative acting polydiacetylene resist P4BCMU developed at NRL.[2] Thin films were deposited by spin

Fig. 2. SEM micrograph of an 2 rows of 5 dot exposures written in the STM by raising the tip-sample voltage to -35 V over each dot.

coating a dilute solution of the resist in chloroform onto freshly etched substrates. Development of the lithographic patterns was by rinsing in chloroform for 60 s.

High voltage e-beam exposure was performed with a JEOL JBX-5DII. Typically it was operated at 50 kV with a 10 nm (17 nm 1/e diameter) focussed e-beam at a current of about 30 pA. Parallel studies of resist resolution and proximity effect reduction have been carried out with the JEOL.[3,4]

3. RESULTS
3.1 Low voltage lithography

Successful resist exposure and patterning of GaAs has been achieved at tip-sample voltages up to -35 V. The smallest voltage at which resist exposure has been observed is -12 V. With our present experimental set-up, 3 pA is the minimum tip-sample current possible and 1 μm/s the maximum lateral tip velocity. Even this minimum line dose, 30 nC/cm, is sufficient to expose the resist. An example of STM lithography of SAL-601 is shown in figure 1, which is an SEM micrograph of an etched GaAs sample at an tilt of 45 °. The micrograph has not been expanded vertically to correct for the tilt. The developed resist was used as a mask for an approximate 100 nm RIE of the substrate. The pattern was written at -25 V with a line dose of 200 nC/cm. The beam was 'blanked' between the non-contiguous parts of the pattern by retracting the tip before moving it laterally to the start of the next element of the pattern (i.e. letter). It would clearly be faster if blanking could be achieved without retracting the tip by adjusting the operation of the STM so that the tip could be moved laterally without exposing the resist. Figure 2 shows a series of dots that have been exposed in a film of SAL-601 on Si. Blanking of the exposure was achieved by reducing the tip-sample current and voltage as the tip is moved between dots.[5] This has not proved a reliable technique (as can be seen by the

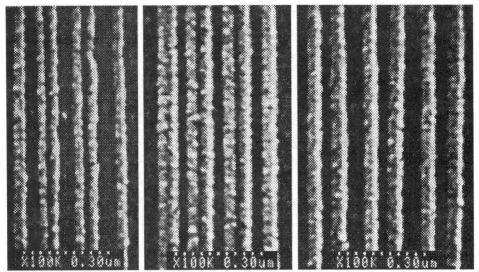

Fig. 3. SEM micrographs of STM lithography in 30 nm of SAL-601 on Si. The patterns were written at -15 V (left), -25 V (center) and -35 V (right) with a line dose of 2 µC/cm.

evidence of resist exposure between the dots) due to limitations in the speed of our STM. As mentioned above, the exposure sensitivity of SAL-601 is such that it is exposed even at the lowest line doses. Figure 2 was only possible with some resist with a lower than usual sensitivity caused, we believe, by a prolonged exposure to the atmosphere. However, with a suitable match between resist sensitivity and STM performance, it will be possible to reliably define non-contiguous patterns, clearly essential for any practical lithography.

The finest resolution STM lithography has been performed in a 30 nm film of resist on Si as shown in figure 3. Here 23 nm wide lines of resist written at -15 V are shown. This is a few nm less than have been written in a 50 nm film. The issue of resolution is discussed further in the next section. Also shown in the figure are lines written at -25 and -35 V illustrating the ability to control the feature size with the tip-sample bias. Results from a series of lithographic exposures are at different writing voltages are summarized in figure 4. Feature size in the developed resist patterns is plotted as a function of tip-sample voltage for lithography performed at an exposure of 200 nC/cm. The points lie close to a line passing through the origin which can be explained by considering STM operation in the field emission mode. As the tip-sample bias is increased the action of the STM servo is to retract the tip to maintain the constant tip-sample current. Field emission from the tip is a function of the electric field at the tip as given by the Fowler-Nordheim equation. To first order the field at the tip is proportional to the field between tip and sample. As a result, the tip-sample separation will be proportional to tip-sample bias for a given tip-sample current. As the electrons flowing from the tip diverge, the diameter of the beam impinging on the sample will be proportional to the tip-sample separation and hence bias.

3.2 Comparison To High Voltage Lithography

Comparisons have been made between the STM lithography and identically processed resist films exposed with a tightly focussed 10 nm 50 kV e-beam. We have shown that there is a significant advantage in going to a low voltage for e-beam lithography. Smaller

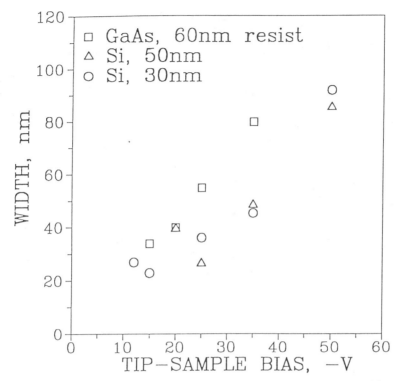

Fig. 4. Measurements of feature size in developed SAL-601 versus STM writing voltage.

feature sizes can be written with the STM than with the 50 kV e-beam. In a 50 nm film of SAL-601 on silicon 27 nm lines have been written at -15 V. This is three times smaller than has been defined with a 10 nm, 50 kV e-beam in an identically prepared and developed 50 nm resist film. Most importantly, proximity effects are eliminated in STM lithography. Figure 5 compares identically prepared and processed samples of SAL-601 on silicon. On the left, STM written 40 nm lines on 55 nm spacing are clearly resolved. Isolated lines written under the same conditions (-25 V, 200 nC/cm) are of the same size. In contrast, on the right 'lines' written on 100 nm centers with the 10 nm, 50 kV are not resolved. In the absence of proximity effects, such a grating should be resolved as isolated 75 nm lines can be written.

A further attractive feature of low voltage e-beam lithography is the wide process latitude. The variation of feature size with dose as shown in figure 6 is significantly less than that observed in lithography with a tightly focussed 50 kV e-beam on identically prepared samples of SAL-601 on GaAs or Si. A thirty times increase in line dose results in only a 40% increase in feature size. The lowest dose points shown in figure 6 correspond to the lowest tip-sample current (3 pA) possible. Even at this current, no evidence of underexposure was observed. The data for figure 6 was recorded from SEM micrographs of developed resist patterns similar to those shown in figure 3. Some variation from sample to sample (±15%) which we attribute to variation in the sensitivity of SAL-601 and possibly variations in surface morphology.

Similar minimum feature sizes have been achieved with proximal probe lithography in PMMA using both STM [6] and AFM [7] based instruments. However, the same resolution in PMMA can be achieved with a tightly focussed high voltage e-beam. Thus,

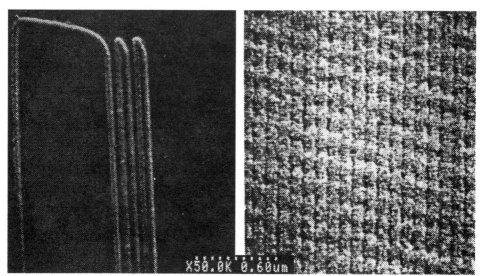

Fig. 5. Comparison of lithography at -25 V with the STM (left) and at 50 kV (right) in identically processed 50 nm films of SAL-601 on a bulk Si substrate.

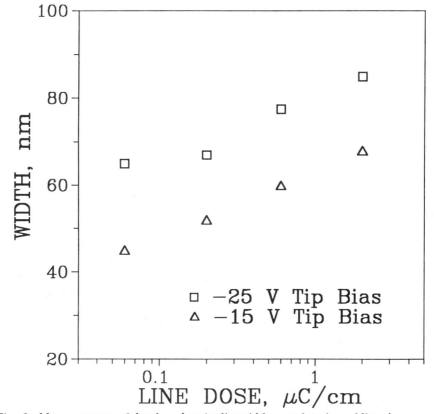

Fig. 6. Measurements of developed resist linewidth as a function of line dose.

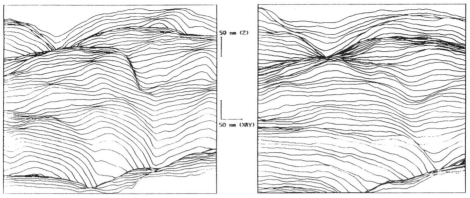

Fig. 7. The same surface imaged with the STM under tunneling (-0.6 V tip bias, left) and field emission (-35 V tip bias, right).

in contrast to SAL-601, no resolution improvement has been realized by going to ultralow exposure voltages with PMMA. Our current interpretation is that SAL-601 is more sensitive to low energy electrons than PMMA which causes the resolution to be degraded by low energy scattered and secondary electrons created by a high voltage primary e-beam. These results have been extremely valuable in our studies of proximity effect reduction in high voltage e-beam lithography.[8]

3.3 Imaging

The ability of the STM to image with atomic resolution is well known. However, the surfaces associated with microfabrication are rarely sufficiently clean and ordered to exhibit atomic scale features. The main requirement for lithography is the ability to recognize alignment and registration marks with nanometer scale precision. Surface topographs are readily obtained from the STM operating in the field emission mode. A comparison of the imaging in the tunneling mode (-0.6 V tip bias, left image) and the field emission mode (-35 V bias, right image) is shown in figure 7. The line displacement plots are of the same 500 nm by 500 nm area of a tungsten film and have the same X, Y and Z scales. The film is polycrsytalline and was deposited by CVD. Evidence of the columnar growth and faceting observed in SEM micrographs of the film is visible in the tunneling mode image. Although more detail is apparent in the tunneling image, the more pronounced feature (particularly the depressions at front and back of the images) are readily apparent in the field emission image.

Latent images in exposed but undeveloped resists can also be imaged with the STM.[9] Figure 8 is a grey scale STM image of a thin film of the experimental polydiacetylene resist P4BCMU. The latent image of a 1 μm mesh pattern written with a 50 kV e-beam is visible as depressions in the STM image. The darker lines (top left to bottom right) were written with a dose of 4.2 nC/cm, corresponding to a developed linewidth of 120 nm. The lines in the other direction (bottom left to top right) appear less dark and were written with a dose of 1.7 nC/cm, corresponding to a developed linewidth of about 80 nm. P4BCMU is comparable in resolution to SAL-601 when exposed with a 50 kV e-beam, i.e the developed feature size (~80 nm minimum) is significantly greater than the size of the 50 kV e-beam. The width of the latent images measured from the STM images (~90 nm) is close to the linewidths observed in the developed resist patterns and significantly greater than the width of the 50 kV e-beam (10 nm). This has provided

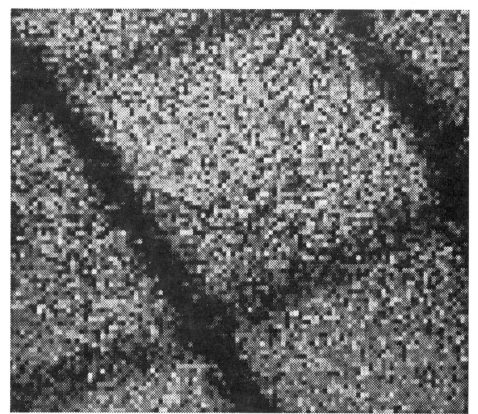

Fig. 8. STM image of the latent exposure image in an undeveloped film of P4BCMU. The 1 µm pitch mesh was written with a 50 kV e-beam at line doses of 4.2 and 1.7 nC/cm. Lighter regions are higher.

insights into the resolution determining mechanisms of the resist. It appears that it is the exposure process with the high energy e-beam as opposed to the resist development which limits the minimum feature size in P4BCMU. Essentially the same conclusion has been reached for SAL-601 as described above.

The ability to image a pattern in an undeveloped resist film has given valuable insights into the properties of the P4BCMU. It also shows that the STM would have applications in a 'mix and match' lithographic scheme where an STM type system is used to add very fine features to patterns defined with a lower resolution lithography. The same effect has been noted in PMMA where electron exposure causes shrinkage to the polymer allowing a latent image to be imaged with a proximal probe such as the atomic force microscope.

4. DISCUSSION: TOWARDS A VIABLE STM BASED LITHOGRAPHY

Our results have demonstrated that the STM can be used for lithography, with an e-beam resist. Further, the resulting patterns are sufficiently robust to act as a mask for a reactive ion etch with BCl_3. The results demonstrate that there are specific advantages in using an STM (i.e. low energy) approach to e-beam lithography in that the resist resolution is enhanced and proximity effects are absent. In terms of registration, the imaging capabilities of the STM indicate that it is possible to align STM written patterns to a previously defined pattern or registration marks on a sample. However, significant challenges must be overcome before the technique can be considered viable outside the research laboratory. Here we concentrate on two specific issues, namely resolution and speed and discuss other advantages of very low voltage e-beam lithography.

4.1 Lithographic Speed

A limitation of beam based techniques is that they are serial in nature and thus inherently slow. The issue of lithographic speed must be addressed to develop a technologically viable STM based lithography system. Speed gains can be obtained in two ways. First, the speed of the STM can be increased through improved design although new resist technology then becomes critical. Second, parallel operation of multiple tips is possible.

The STM tip can provide such a high current that the time required to provide the resist exposure dose is not the lithographic speed limiting factor. In effect, the lithographic speed of the STM is determined by the maximum lateral velocity of the tip, which is limited by the transient response of the STM servo loop. The open loop gain of the loop has contributions from the tunneling (or field emission) current, gain of electronic control circuit and the response of the piezoelectric element which adjusts the position of the STM tip. The exponential variation of tip-sample current with tip-sample separation is compensated by a logarithmic amplifier in the control circuit. An integrator is usually included in the control circuit for stability and to eliminate tip-sample offset errors. The transient response is then proportional to the inverse of the open loop gain of the servo loop. In practice the transient response is limited by the resonant frequency and/or the resonance damping of the scanner used to move the STM tip. To obtain the fastest transient response the resonance should be as high a frequency as possible and be critically damped. For example, by increasing the damping of an underdamped tube scanner, its transient response was decreased from 2 to ~0.2 ms because the servo loop could be operated at a higher loop gain.[10] Recently, with piezoelectric slabs with resonant frequencies in the MHz range, STMs have been built which can scan at close to 1 mm/s.[11] In contrast, the pattern in figure 1 was written at a speed close to 0.5 μm/s with a current of 3 pA. The same exposure dose would require 6 nA when the tip is moved laterally at 1 mm/s. Such a current is well within the capabilities of conventional STM tips.

Resist exposure time is the speed limiting factor for nanolithography with our JEOL JBX-5DII used with its smallest probe sizes. The time required to expose a 100 μm line of 30 nm width has been estimated. For the STM, a lateral tip velocity of 1 mm/s is assumed. To write the line, a single pass at a tip-sample voltage chosen to give a 30 nm feature size would take approximately 0.1 s. For the JEOL a 30 nm linewidth requires a high resolution resist such as PMMA which has a low sensitivity of 2 nC/cm. A single pass line with a beam current of 100 pA would take 0.2 s. Thus at these dimensions, an STM based lithographic tool would give a comparable writing speed.

Further increases in speed could be realized by the parallel operation of multiple STM tips, with each tip having its own piezoelectric scanner and servo loop. For example, an array of about 10 tips operating in parallel is feasible with conventional STM technology. Greater numbers of parallel STM's would require microfabrication techniques similar to that demonstrated in Japan and at Stanford University.[12] Table 1 gives an estimate of the time required to write a 'mask' requiring 30 nm minimum feature sizes over a 4 cm^2 area with a 50% fill factor. It has been assumed that 50% of the writing would be performed at -15 V (30 nm effective spot size) and 50% at -50 V (100 nm spot).

TABLE 1. 'Mask' writing times for different numbers of tips

Number of tips	1	10	1000
Time (hrs)	5000	500	5

Thus, ~1000 tips are required for this approach to full scale nanolithography. Such a number is well within the capabilities of present microfabrication technology.

4.2 Lithographic Resolution

As has been shown above, low voltage e-beam lithography in 30 to 50 nm films of polymeric resist materials has been achieved with the STM as a source of electrons. Over the range of -50 to -15 V, minimum feature size scales with tip-sample voltage. These results (summarized in figure 4) were obtained with a number of different samples and tips. The data lie close to a straight line which passes close to the origin. As the effective beam size would be expected to be proportional to the tip-sample voltage, the intrinsic resolution of the SAL-601 films is better than the size of the exposing beam. If the resist (or resist processing) was responsible for a resolution degradation, the line passing through the data in figure 4 would be offset in the vertical direction by an amount corresponding to the resolution degradation. Thus a 50 nm film of SAL-601 appears to have a very high intrinsic resolution. Interestingly, the resist was not believed to be capable of ultrahigh resolution due to problems associated with the chemical amplification. For example, thermally activated diffusion of the acid catalyst during the post exposure bake was believed to be responsible for the minimum feature sizes close to 75 nm being the smallest possible with a tightly focussed high energy e-beam. The STM lithography has provided an upper bound to the catalyst diffusion length of about 10 nm. Thus it is the exposure process rather than the intrinsic properties of the resist or resist processing which would seem to be responsible for the observed lithographic resolution with the 50 kV e-beam. A similar conclusion has been drawn for the resist P4BCMU based on measurements of the latent exposure images summarized above. The resolution of P4BCMU at 50 kV is significantly greater than expected from considerations of the probe size. As the STM measurements of the latent image show comparable widths to the developed films, the resolution degradation appears again related to the exposure process with a 50 kV e-beam rather than due to the resist development process. Thus for these two resist systems, the STM lithographic studies have demonstrated valuable insights into the properties of the resist system itself as well as demonstrating the advantages of low voltage e-beam lithography.

The mechanism by which electrons traverse the resist film is not clear. SAL-601 is non-conductive as is PMMA which has also been patterned lithographically with low voltage electrons.[6,7] However, as SAL-601 is a negative resist, the presence of developed resist features in the SEM micrographs show that the action of the STM tip has indeed exposed the resist. Other examples of STM images of non-conductive materials have been reported and models proposed.[13] Unfortunately, none of these models would appear to be applicable to the resist layers we have successfully patterned with the STM. It appears possible that the high field that is created across the resist film by the presence of the STM tip is sufficient to induce conductivity in the resist which allows the tip-sample current to flow. It appears that PMMA and SAL-601 behave differently in this respect. We have found that 15 to 20 nm is the thickest PMMA film that can be patterned with the STM. With thicker films, stable operation of the STM is not possible.[14] The need to develop a high field across the resist is also the reason that there appears to be a threshold voltage close to -12 V for lithography in 50 nm of SAL-601. Below this voltage the STM tip has to push into the resist to maintain the tip-sample current. As a result, mechanical damage to the resist occurs and less than the full thickness of the resist film is exposed and remains after development.

The ultimate resolution of a STM based lithography scheme is defined by the effective width of the beam of electrons emitted from the tip, provided that the radiation sensitive layer (i.e. resist) has a sufficient resolution capability. As has been pointed out by several researchers, there is a significant resolution barrier at about 20 nm.[15] A lithographic technology capable of 10 nm resolution on a variety of materials with arbitrary patterns does not presently exist. STM lithography would appear to be a viable candidate to reach this resolution scale. However, to achieve resolution on the 10 nm scale using the STM in the constant current mode, lithographic exposure must be performed at about 5 V. This implies that resist films thinner than 10 nm are necessary. It is not practical to use polymeric resists at these thicknesses as uniform and continuous films cannot be spin cast. An alternate approach to the formation of ultrathin resist films is needed. We are currently investigating the use of molecular self assembly (SA) to form homogeneous radiation sensitive layers for low voltage e-beam lithography.

4.2.1 Ultrathin Resist Layers

Molecular self-assembly is a chemical process in which molecules are spontaneously organized to form larger structures. Chemisorbed films are a class of self-assembling materials in which precursor molecules from solution or vapor phase react at interfaces to produce layers that are chemically bonded to those surfaces. An example of SA film precursor is an organosilane where a trichlorosilyl ($-SiCl_3$) group is bonded to an organic group. The organic group can be chosen to provide the desired chemical properties such as radiation sensitivity.

Substrates to be coated with organosilane self-assembled monolayer (SAM) films are cleaned or oxidized to produce a sufficient density of surface hydroxyl (-OH) groups. The surface is then exposed to the organosilane precursor (usually in dilute solution or vapor phase, although spin coating can be used) for several minutes to allow the chemisorption reaction (siloxane bond formation) to occur, for example:

$$R\text{-}SiCl_3 + HO\text{-}(substrate) \longrightarrow R\text{-}Si\text{-}O\text{-}(substrate) + HCl.$$

Organosilane precursors can be used to form chemisorbed SAM films on a wide variety of surfaces, unlike an approach such as organothiol adsorption which is restricted primarily to Au. Formation of SAM films on the surfaces of semiconductors (Si, poly-Si, Ge), dielectrics (SiO_2, Al_2O_3, Si_3N_4, SiC), metals (Au, Pt, W, Al, In-doped SnO_2), polymers (epoxy, novolak, polyvinylphenol), plastics (ABS, polycarbonate, polysulfone), and diamond have all been demonstrated at NRL.[16] Surfaces treated with organosilane SAM films have been patterned by exposure to deep UV, soft x-ray, and focussed ion beams.[17]

The patterned organosilane SA layers have been demonstrated to serve as a template for subsequent selective attachment and buildup of a variety of materials. For example, irradiation of films with ligating R groups with an affinity for a metal complex catalyst

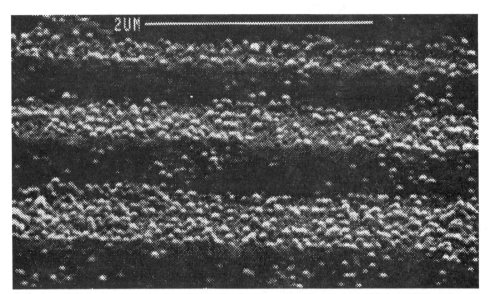

Fig. 9. SEM micrograph showing selective metallization of a STM (-20 V, 10 pA) patterned SA film; the brighter areas are the exposed Au substrate and the darker regions are the EL deposited Ni film.

damages the ligating groups and prevents a Pd-based catalyst from binding to the film. Treatment of the catalyzed surface, followed by immersion in an electroless (EL) Ni plating bath, results in selective deposition of Ni in the unexposed areas. The metal films are of high quality and can be used as conductive paths or as masks for energetic plasma etches.

An example of STM patterning of an ultrathin organosilane film and EL plating is shown in figure 9. An SEM micrograph of a Ni film (dark regions) selectively deposited on the unirradiated regions of a Au substrate is shown in the figure. The unplated (light) regions were defined by the STM with a tip-sample voltage of -20 V and current of 10 pA. Inspection of the unplated regions in the figure reveal that they are considerably rougher than the unexposed areas. In addition to destroying the ligating functionality of the R groups, the action of the STM patterning has damaged the underlying Au. Indeed this is not unexpected as -20 V is significantly higher than the threshold voltage observed by Mamin et al.[18] for the creation of Au mounds on various substrates with the STM. However, this substrate damage does not occur with Si substrates as shown in figure 10. Here the bare Si appears as the dark stripes and the EL Ni film as the brighter area corresponding to the unexposed field. The exposure was performed at -10 V with a current of 100 pA. Patterning has been achieved at tip biases down to -6 V. It is expected that the composition of these organosilane films will enable lithography at still lower voltages. Further, it will open the possibility of performing the lithography at ambient pressure, which will greatly widen the flexibility of the technique.

These results demonstrate the potential of low voltage e-beam patterning of SA films. Although the films are extremely thin, the exposed pattern an be readily replicated in a metallic film. The films are of high quality and can be used as conducting structures or as a mask resistant to plasma etching of the underlying substrate. It is worth emphasizing that the matching of such a resist technology to the capabilities of the STM is crucial for the development of an ultra high resolution low voltage lithographic tool.

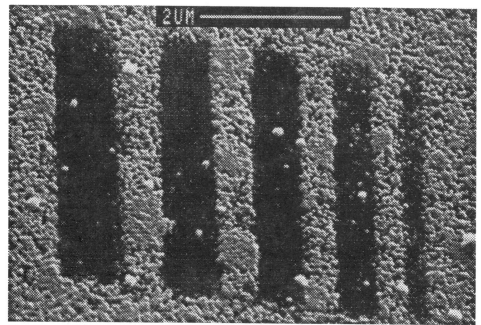

Fig. 10. SEM micrograph of a patterned SA film on Si. The darker areas are the unplated Si regions defined with the STM at -10 V and 100 pA.

4.3 Other Issues

A problem inherent in ultrahigh resolution e-beam lithography, is that shot noise in the beam current causes an unacceptable variation in exposure. At an applied dose of 100 electrons per resolved pattern element, about 10% of the pattern elements will receive less than 90 electrons, which will be insufficient to expose the resist. (A sensitivity of 2 $\mu C/cm^2$, corresponds to about 100 electrons per 800 nm^2). Our measurements indicate that SAL-601 requires about 100 times greater dose at 15 V than 50 kV. (As the STM tip is a very bright electron source, this is not a problem in terms of lithographic performance.) Therefore, the problem of shot noise will arise at far smaller feature sizes with low voltage e-beam lithography.

STM technology provides a very compact low voltage electron source. As a result, a low voltage systems can be made far more compact than a conventional e-beam writer. This simplifies system design and engineering, especially in terms of sensitivity to mechanical vibration and stray magnetic fields. This suggests that the STM approach would be attractive for the development of a low voltage e-beam tool for nanometer scale lithography. One could envision such a tool being used in much the same way that converted SEM's are currently used for lithography at many institutions pursuing research and development. It is worth noting that such a STM based lithography tool would involve significantly less expenditure in terms of capital and maintenance costs than the conventional high voltage alternative.

5. SUMMARY

Technologically useful low voltage e-beam lithography can be performed with the STM. The lithographic exposure of a commercially available e-beam resist and the transfer of the pattern into GaAs with a reactive ion etch has been demonstrated. The combination of the imaging and patterning capabilities of the STM indicate that STM based lithography will be particularly suited to a mix and match technology. Issues related to transferring such lithographic techniques out of the research lab have been discussed. Significant challenges, particularly related to the speed of the lithography, are apparent. However, the technology exists to significantly impact these problems. Whereas a lithography tool capable of commercial mask making will require a significant investment for development, smaller scale lithography suited to the research and development environment is more realistic.

ACKNOWLEDGEMENTS

The work described here has been performed over the past three years at the Naval Research Laboratory. Over that time we have had the benefit of numerous stimulating interactions and discussions with colleagues at the Lab and in the research community. In particular, we would like to thank F.K. Perkins, J.M. Calvert, C.S. Dulcey, T.S. Koloski, S.L. Brandow, R.J. Colton, D.P. DiLella, K. Lee, J.M. Murday, C. Sandroff, J.E. Griffith and M.C. Peckerar.

Certain commercial equipment, instruments and materials are identified in this paper in order to specify adequately the experimental procedure. Such identification does not imply recommendation or endorsement by the National Institute of Standards and Technology or does it imply that the materials or equipment identified are necessarily the best available for the purpose.

REFERENCES

[1] Processing specified in "Microposit SAL 600 e-Beam Process", Shipley Corporation.
[2] R.J. Colton, C.R.K. Marrian, A. Snow and D. DiLella, **J. Vac. Sci. Technol. B5**, 1353 (1987).
[3] For example, E.A. Dobisz and C.R.K. Marrian, **J. Vac. Sci. Tech. B9**, 3024 (1991).
[4] E.A. Dobisz, C.R.K. Marrian & R.J. Colton, **J. Appl. Phys. 70**, 1793 (1991).
[5] C.R.K. Marrian, E.A. Dobisz, and R.J. Colton in **Scanned Probe Microscopies**, K. Wickramasinghe ed., AIP Press, **241**, 408 (1992).
[6] M.A. McCord and R.F.W. Pease, **J. Vac. Sci. Tech. B4**, 86 (1986).
[7] A. Majumdar, P.I. Oden, J.P. Carrejo, L.A. Nagahara, J.J. Graham and J. Alexander, **Appl. Phys. Lett. 61**, 2293 (1992).
[8] E.A. Dobisz, C.R.K. Marrian, L. Shirey and M. Ancona, **J. Vac. Sci. Tech. B10**, 3067 (1992).
[9] C.R.K. Marrian, E.A. Dobisz & R.J. Colton, **J. Vac. Sci. Technol. B9**, 1367 (1991).
[10] D.P. DiLella, J.H. Wandass, R.J. Colton and C.R.K. Marrian, **Rev. Sci. Instrum. 60**, 997 (1989).
[11] H.J. Mamin and D. Rugar presented at "STM 91" Interlaken, Switzerland August 12-16, 1991.
[12] T.R. Albrecht, S. Akamine, M.J. Zdeblick and C.F. Quate, **J. Vac. Sci. Tech. A8**, 317 (1990).
[13] B. Michel, G. Travaglini, H. Rohrer, C. Joachim and A. Amerein, **Z. Phys. B 76**, 99 (1989) and S. Lindsay, Y. Li, J. Pan, T. Thundat, L.A. Nagahara, P. Oden, J.A. DeRose and U. Knipping, **J. Vac. Sci. Technol. B9**, 1096 (1991).
[14] F.K. Perkins, private communication
[15] R.F.W. Pease and C.D. Wilkinson, private communications

[16] J.M. Schnur, M.C. Peckerar, C.R.K. Marrian, P.E. Schoen, J.M. Calvert, and J.H. Georger, **US Patent** 5,077,085 (1991) and **US Patent** 5,079,600 (1992) and J.M. Calvert, P.E. Pehrsson, C.S. Dulcey, and M.C. Peckerar, **Materials Research Society Proceedings,** 260, in press.

[17] see previous reference and J.M. Calvert, C.S. Dulcey, J.H. Georger, M.C. Peckerar, J.M. Schnur, P.E. Schoen, G.S. Calabrese, and P. Sricharoenchaikit, **Solid State Technology** 34 (10), 77 (1991).

[18] H.J. Mamin, S. Chiang, H. Birk, P.H. Guenther and D. Rugar, **J. Vac. Sci. Tech.** B9, 1398 (1991).

Nanolithography and Atomic Manipulation on Silicon Surfaces by STM

F. Grey, A. Kobayashi, H. Uchida, D. H. Huang and M. Aono
Aono Atomcraft Project, ERATO, JRDC, Tsukuba, Ibaraki 300-26, Japan

Abstract:

By applying a voltage in the range 3–6V between the metal tip of a scanning tunneling microscope (STM) and a Si(111) surface, it is possible to form nanometer-scale grooves on the surface and even to extract individual Si adatoms. Analysis of the current and voltage dependence of this process shows that it is field-induced. The process is found to depend sensitively on tip material and tip preparation. Variations in the extraction probabilities for different Si adatoms are explained in terms of local chemical bonding differences.

1. Introduction

The STM is a promising tool for engineering on the nanometer scale, and even for fabricating devices atom by atom. Examples of potential applications include extremely high-density data-storage (see Fig. 1) and electronic circuits with novel quantum properties due to their minute size. Before such technology can become a reality, a better understanding is needed of the physical processes that occur when an STM modifies a surface, in order to develop reliable techniques for nanolithography.

Fig. 1. Kanji (Chinese characters) written on Si(111) using a W tip at a tip bias of 4.0V and 0.25nA tunnel current. The characters 日 (sun, day) and 本 (origin) together signify Japan. The imaging conditions are sample bias 2V and tunnel current 0.25nA. The height of the characters is 70nm.

In this chapter, we present recent results obtained in our group for field-induced modifications of clean Si(111) surfaces using an STM in ultra high vacuum. There are two reasons for focusing on this system. The first is that Si is the basis of the current microelectronics industry, and is likely to remain so as the scale of devices shrinks further. The second reason, not unrelated, is that Si is one of the most carefully characterized materials, and the structure and properties of the clean (111) crystallographic surface are particularly well understood.

Though the scope of this chapter may seem narrow, we hope to compensate for this by placing emphasis on the use of simple, general physical models to analyze the dependence of the modification process on such parameters as the tip material, the current-voltage characteristics of the tunnel junction, and the bonding energies of the surface atoms. The models should prove useful for the analysis of STM nanolithography experiments on other systems.

The next section provides a brief review of STM field-induced modification of semiconductors and other materials. In sections 3-7, theoretical expressions are derived for the electric field dependence of the STM's tunneling characteristics. These are used in sections 8-10 to analyze experimental results for the formation of nanometer-scale grooves on Si(111). Sections 11-13 deal with experimental results for single atom extraction from Si(111) and their theoretical interpretation. Section 14 concludes this chapter with a brief look to the future of STM field-induced nanolithography on silicon surfaces.

2. Field-induced Nanomodification

A number of physical mechanisms have been proposed to explain how an STM can modify a surface on the nanometer scale [1,2]. Gomer[3] was the first to point out the theoretical possibility of field ion emission, since even at a bias of only a few volts, the electric field between tip and sample is very high, of the order of 10^{10}V/m. The first report of field-induced modification on a semiconductor surface at the atomic scale was that of Becker et al.[4], who made atomic protuberances on Ge(111) by briefly increasing the bias between the W tip and the sample above 3V, at constant tunneling current. It was argued that the protuberances were Ge atoms which had been previously deposited on the tip by allowing it to make contact with the sample at a different point on the surface.

The field-induced emission of ions from a sharp tip has been well-documented using the field ion microscope (FIM)[5,6], and occurs when the electric field at the surface of a sharp, free-standing tip exceeds a critical value which lies in the range 20-70 V/nm for most materials. Evidence for field ion emission has also been found in a number of STM nanomodification studies. Mamin et al.[7] have used a Au tip to deposit or remove mounds of Au 10–20nm in diameter on a Au surface, by applying sub-microsecond 3.6V pulses to the sample. Using a similar approach, McBride et al.[8] made 20nm mounds on a Pt surface using a W tip and deduced a critical field of 2.3V/nm.

Rabe et al.[9] have made holes and hillocks on a Ag(111) film on mica by pulsing the sample at 3–7V in both polarities. Indentations in WSe_2 have been made by Schimmel et al.[10] by pulsing a Au tip at -5V. In these cases, too, critical fields of only a few V/nm can be estimated, assuming a tunneling gap of about 1nm. Critical fields in STM geometry are thus about an order of magnitude less in STM geometry compared to FIM.

Becker et al. were unable to modify Si(111) by the same process used for Ge(111), even for biases up to 20V and several nA current[4]. Lyo and Avouris[11] succeeded by moving the tip to within 0.3nm of the surface and applying a 3V pulse. Single Si atom manipulation was achieved for the first time in this way. A field dependence of this process was measured, with a critical field of about 10V/nm. However, due to the close proximity of tip and sample during this process, direct chemical interaction between tip and sample is expected to play an important role.

Chemical interaction complicates the physical analysis of the modification process considerably. Recently we have been able to modify Si simply by increasing the bias between tip and sample to a value in the range 3–6V[12]. Fig. 1 is an example of the result. The modification is made at constant current, so that a relatively large gap of about 1nm is maintained. The success of this approach depends sensitively on tip preparation, which may explain why similar results were not reported earlier[4]. Single atom manipulation has been achieved using the same approach[13]. In the rest of this chapter, the experimental and theoretical aspects of this modification technique will be described in detail.

3. Theory of the Medium Voltage Regime

In the theoretical analysis of STM images, it is usually assumed that the energy, eV, which an electron acquires due to the applied voltage V is much smaller than either the tip or sample workfunctions: $eV << \phi_t, \phi_s$. This low voltage regime is appropriate for typical imaging conditions, where the bias is less than 1V, while typical workfunctions are of the order of 4eV. On the other hand, literature concerned with electron field emission focuses on the high voltage regime, where $eV >> \phi_s, \phi_t$. However, in most STM nanolithography experiments, the bias is raised to a value of several volts in order to modify the surface, and so $eV \cong \phi_s, \phi_t$. One of the few works to describe this medium voltage regime, and to compare it in a unified way with the low and high voltage regimes, is due to Simmons[14]. Here, we briefly review the physical basis of this theory, and derive results relevant to the analysis of the STM modification experiments.

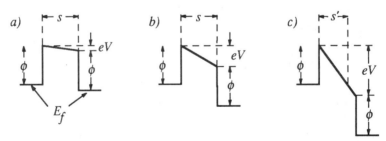

Fig. 2. Schematic diagram of the different tunneling regimes. a) low voltage b) medium voltage c) high voltage. E_f indicates the Fermi level.

Fig. 2 shows a one dimensional model of the three different voltage regimes. Two conducting media, assumed initially to have the same workfunction ϕ, are separated by an insulating gap s. With no voltage applied, equal and opposite tunneling current density components of magnitude j cancel out, so that the net current density, J, is zero. The magnitude of j is given by[14]:

$$j = \frac{e\phi}{2\pi h s^2} exp\left\{-\frac{4\pi s}{h}\sqrt{2m\phi}\right\} \qquad (1)$$

where e and m are the charge and mass of the electron and h is Planck's constant. The exponential in Eqn. 1 is characteristic of the transmission coefficient for tunneling through a potential barrier of height ϕ.

When a small bias is applied (Fig. 2b), the barrier for electrons traveling in either direction is no longer the same, and so there is a net current density. This can be estimated by substituting $\phi - eV/2$ for the average barrier height for electrons traveling from the high potential region and $\phi + eV/2$ for electrons traveling from the low potential region. Although this is a very crude approximation, it contains the essential physics. The expression for the net current density J is then:

$$J = \frac{e}{2\pi h s^2}\left[(\phi-\frac{eV}{2})exp\left\{-\frac{4\pi s}{h}\sqrt{2m(\phi-\frac{eV}{2})}\right\} - (\phi+\frac{eV}{2})exp\left\{-\frac{4\pi s}{h}\sqrt{2m(\phi+\frac{eV}{2})}\right\}\right] \qquad (2)$$

For $eV << \phi$, expanding to first order in $eV/2\phi$ gives:

$$J = \frac{e}{2\pi h s^2}\left[eV\frac{2\pi s}{h}\sqrt{2m\phi} - eV\right] exp\left\{-\frac{4\pi s}{h}\sqrt{2m\phi}\right\} \qquad (3)$$

For typical values of $s=1$nm and $\phi =5$eV, the second term in square brackets in Eqn. 3 is about 5% of the first, and so it can be ignored, yielding:

$$J = \frac{e^2 V}{h^2 s}\sqrt{2m\phi}\ exp\left\{-\frac{4\pi s}{h}\sqrt{2m\phi}\right\} \qquad (4)$$

Eqn. 4 has the same functional form as that derived by Tersoff and Hamann[15] for the analysis of STM images. The main difference is that Eqn. 4 does not explicitly include the dependence on the local density of states or on the tip radius, terms which appear in the exponential prefactor in the more detailed analysis[15].

For the case $eV>>\phi$, the probability of electron tunneling from the low potential region is negligible. Due to the steep potential gradient, the effective width of the potential barrier is reduced. From Fig. 2c, this width can be estimated geometrically as $s'=s\phi/eV$. The mean height of the barrier is $\phi/2$. Substituting in Eqn. 2 gives:

$$J = \frac{e^3 F^2}{4\pi h\phi}\ exp\left\{-\frac{4\pi \sqrt{m}}{heF}\phi^{3/2}\right\} \qquad (5)$$

Where $F=V/s$ is the electric field across the gap. This result is known as the Fowler-Nordheim relation for electron field emission[16]. It is interesting to contrast the behavior described by Eqns. 4 and 5. At constant gap s, Eqn.4 describes an ohmic junction, while in Eqn. 5, current increases exponentially with voltage. At constant current, s increases only logarithmically with V in Eqn. 4, so that a bias increase produces a considerable increase in the field, $F=V/s$. In Eqn. 5, on the other hand, the field is constant at constant current.

Clearly, the low and high voltage regimes are very different. Since it would be unphysical to have an abrupt transition between the regimes, Eqn. 2 must be used in its full form to describe the medium voltage regime. In terms of the trapezoidal potential of Fig. 2, the transition between medium and high voltage regimes occurs at $V=\phi/e$. Above this bias, s' begins to differ from s. For a more realistic potential barrier, the transition will not be abrupt, and so the medium voltage equation should be an adequate approximation even above $V=\phi/e$. However, beyond $V_{max}=2\phi/e$, Eqn. 2 becomes complex. Eqn. 2 also breaks down at small s, because J becomes negative at positive bias. The spacing at which this happens can be estimated from Eqn. 3 as $s_{min} = h/2\pi\sqrt{(2m\phi)}$.

4. Current-Voltage Characteristics

The three voltage regimes are best contrasted by comparing their different current-voltage (I-V) characteristics. This is done in Figs. 3a-c for a junction with ϕ=4.5eV and three different values of s. For illustration purposes, the low and high voltage regime equations are plotted outside of their range of validity in Figs. 3a and 3c.

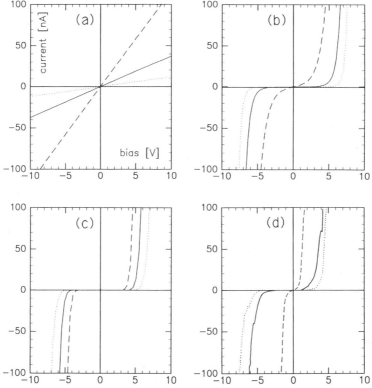

Fig. 3. Theoretical I-V curves for low (a) medium (b) and high (c) voltage regimes. The curves in (a) correspond to a gap of s=0.95nm (dashed), s=1.0nm (solid) and s=1.05nm (dotted). The curves in (b) and (c) correspond to a gap of s=0.8nm (dashed), s=1.0nm (solid) and s=1.2nm (dotted). Experimental results are shown in (d), the set voltage being 1.0V and the set current 5nA (dashed), 0.5nA(solid) and 0.05nA (dotted).

Fig. 3d shows typical experimental STM *I-V* curves for a W tip and a Si(111) sample. These are obtained by first setting a fixed gap *s*, turning off the feedback loop for the vertical piezo motion, and then scanning the voltage. The value of the gap is determined by the voltage and current settings prior to turning off the feedback loop, which are given in the figure caption.

Clearly, the low voltage ohmic regime cannot describe the experimental results at all. Of the other two, the medium voltage *I-V* curves more closely resemble the experimental results. It should be noted that, to compare the theoretical current density with the measured current, it is necessary to introduce an effective tunneling area *A*, such that $J=I/A$. This parameter sets the overall scale of the curves in Fig. 3.

For Fig. 3b and 3c, the value of *A* is $10^3 nm^2$. This seems large, since tunneling in an STM is normally from a single atom. It is likely, however, that this discrepancy occurs because the current density *J* is underestimated due to the simple model used. In practice, the potential barrier is not the abrupt trapezoidal function illustrated in Fig. 2. In a classical picture, the attractive image potential of the electron in the metal will lead to a rounding and overall lowering of the potential. Analytic treatment of such a potential barrier is considerably more complex[14], but the net effect is to lower the average potential and hence increase the current density.

5. Field Ion Emission

Field ion emission from a sharp tip in a field ion microscope (FIM) occurs when the tip is raised to a positive potential of several keV[5,6]. The process is thermally activated, with an evaporation rate, *R*, having the Arrhenius form:

$$R = v \, exp \, (-Q/kT) \tag{6}$$

where *v* is a frequency factor, *Q* is a field-dependent activation energy, *k* the Boltzmann constant and *T* the absolute temperature. For experimental purposes, a critical field F_c is defined as the field at which the evaporation rate is one atom per second per site affected by the field.

A model for predicting the field dependence of the activation energy was first developed by Müller[17]. This is called the image-hump model, and is illustrated in Fig. 4. In the absence of a field, the activation energy needed to ionize a neutral atom to the nth ionization state, remove it to infinity, and return n electrons to the Fermi level of the surface, is

$$Q(F=0, x) = \Lambda + I_n - n\phi - n^2 e^2 / 16\pi\varepsilon_0 x \tag{7}$$

where Λ is the sublimation energy for a neutral atom, I_n is the energy needed to ionize it to the n+ ionization state, and $-n\phi$ the energy gained by returning n electrons to the Fermi level. The term $-n^2 e^2/16\pi\varepsilon_0 x$ is the attractive electrostatic image potential of the ion due to the metallic surface, where the coordinate *x* is measured from the charge plane of the surface. For negative ion emission, I_n is replaced by χ_n, the electron affinity, and the sign of $-n\phi$ is reversed[18].

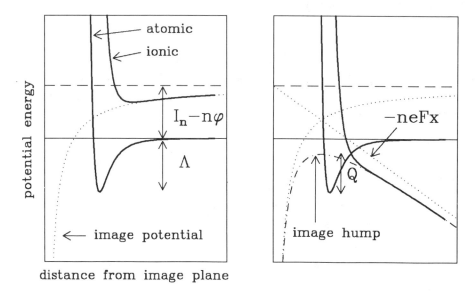

Fig. 4. Sketch of potential diagram for image hump model. In (a) there is no field, and the ionized state potential lies above the neutral atom potential at all distances from the surface. Steep repulsive potentials are assumed near the surface. In (b), the potential of the ion due to the applied field is shown (dotted). The neutral atom can escape into the ionized state over an activation barrier Q. The image hump is indicated (dash-dot).

In the presence of a uniform field F, the activation energy of the ion is reduced by $-neFx$. The combination of this term with the image potential produces an activation barrier known as the image hump (see Fig. 4) which has a maximum value of $-(n^3e^3F/4\pi\varepsilon_0)^{1/2}$ at $x_m = (ne/16\pi\varepsilon_0 F)^{1/2}$. The activation energy $Q(F,x)$ is a function of field and position. In the image hump model, the activation energy is defined, somewhat arbitrarily, as the value at x_m, where it is largest. At the critical field, the activation energy is:

$$Q(F=F_c, x=x_m) = \Lambda + I_n - n\phi - (n^3e^3F_c/4\pi\varepsilon_0)^{1/2} \qquad (8)$$

From Eqn. 6, the activation energy can also be expressed as:

$$Q(F=F_c) = -kT \log(R/\nu) \qquad (9)$$

where $R=1 s^{-1}$ at F_c. The attempt frequency, ν, of an atom to escape over the activation barrier can be estimated from the thermal vibration frequency, which is of the order of $10^{13} s^{-1}$, so that $Q(F_c) \cong 0.8 eV$ at 300K. Combining Eqns. 8 and 9 the critical field can be expressed as:

$$F_c = 4\pi\varepsilon_0 \, n^{-3} e^{-3} \{ \Lambda + I_n - n\phi + kT \log (R/v) \} \tag{10}$$

All terms in Eqn. 10 are known or can be estimated, so F_c can be calculated. The results for positive ion emission are within 15% of the values observed in FIM studies for most materials[6]; the experimental values are in the range 20-70 V/nm for all materials. We note that the calculated F_c values are not very sensitive to the exact values of R and v.

The image-hump model is controversial, and other models have been proposed[19,20]. Nevertheless, the image-hump model is retained because of its success in predicting the critical fields. Recently, a modified version of this model has been proposed to account for the much lower critical fields observed in STM[7,18]. This involves including the image potential of the ion due to the counter electrode. This additional potential lowers the activation barrier. However, calculation shows that the effect is less than 10% of F_c for an electrode gap of s=1nm.

The image potential is a classical concept, and is not strictly applicable at small separations. Simulations using a jellium model for the electrode surfaces[21] indicate that the critical field at separations of about 0.3nm may be substantially reduced. One reason is that an atom may be transferred between the two electrodes while at all times in a state that is only partially ionized. This sort of mechanism may account for the low critical fields seen when the tip is moved very close to the surface[11]. However, there is as yet no satisfactory explanation for the low critical fields estimated in STM experiments at separations of about $s \cong 1$nm.

Concerning the experimental estimates of critical fields, a parallel-plate model of the STM junction is often assumed, with $F=V/s$. The geometrical justification for this is that the radius of curvature of sharp tips is usually 100nm or more, so that the gap resembles a parallel-plate arrangement on the nanometer scale, as illustrated in Fig. 5. The field at the tip surface may be slightly larger than estimated by the parallel-plate model, due to the curvature of the tip. However, since the condition $\int F ds = V$ must hold across the gap, the field at the surface will be correspondingly reduced. In either case, deviations from the parallel-plate value should be small.

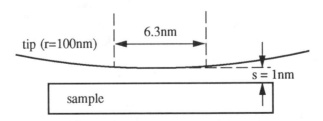

Fig. 5. Sketch of appearance of tip and sample on the nanometer scale. Within the region 6.3nm wide, the gap s varies by less than 10% of its minimum value of 1nm.

6. Tunneling at Constant Field

In STM geometry, the field is determined both by the applied bias and indirectly by the tunneling current, since this also controls the separation of tip and sample. It is therefore interesting to derive how the tunneling current varies with bias for the condition of constant field. To do this, we substitute $s=V/F$ in Eqns. 2, 4 and 5, and assume a constant value of F. The resulting current-voltage relationships are plotted in Fig. 6 for the three voltage regimes with $F=6$V/nm, $\phi=4.5$eV and $A=1$nm^2.

The behavior in the low voltage and medium voltage regimes is similar, while the high-voltage regime shows that at constant field, the current is practically independent of voltage. This is expected since the high voltage regime is that the current is constant at constant field (see Eqn. 5).

The medium voltage equation is intermediate to the other two, being nearly identical to the low voltage equation at small bias, but with decreasing slope at larger bias, suggestive of the high-voltage regime. As noted in section 3, the medium voltage equation is not valid for $V>2\phi/e$ or for $s<h/2\pi\sqrt{(2m\phi)}$, in other words $V < Fh/2\pi\sqrt{(2m\phi)}$. For the conditions of Fig. 6, the range of validity is 0.5–9eV.

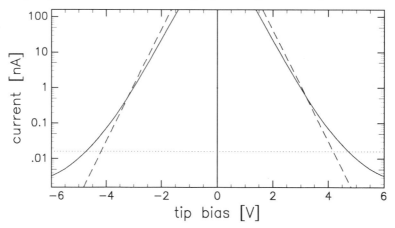

Fig. 6. Variation of tunneling current with applied bias for a constant field across the gap. Results for low (dashed), medium (solid) and high (dotted) voltage regimes.

7. Consequences of Workfunction Difference

Up to now, we have considered the case where the workfunctions of both electrodes are the same. However, a workfunction difference will up an electric field across the junction[3], so the magnitude of the net electric field across the junction is:

$$F = ((\phi_s - \phi_t)/e - V) / s \qquad (11)$$

This field will either assist or hinder field evaporation, depending on the polarity of the applied bias, as indicated in Fig. 7. Since workfunction differences between different materials can be of the order of 1eV, this effect is significant at applied biases of a few volts.

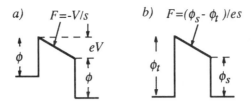

Fig. 7. a) A junction between two materials of the same workfunction $\phi=2eV$ with a bias of $-1V$ applied to the left side b) A junction between two materials of workfunction $\phi_t=3eV$ and $\phi_s=2eV$ in equilibrium. The field, which is equal to gradient of the potential in the gap, is the same in both cases.

The result of including the induced field due to the workfunction difference can be investigated directly by substituting for s from Eqn. 11 into Eqn. 2 for the medium voltage regime. The workfunction ϕ that appears in Eqn. 2 can be replaced by the mean value $\phi' = (\phi_t+\phi_s)/2$. The result is shown in Fig. 8 for three different values of the tip workfunction, $\phi_t =4.0eV$, $4.5eV$ and $5.0eV$, the sample workfunction being $\phi_s =4.5eV$ in each case. At negative tip bias, the effect of increasing the tip workfunction is to decrease the external applied field required to reach a given critical field, as illustrated in Fig. 7. The opposite is true at positive tip bias. Note, though, that the curves in Fig. 8 are not symmetrical about the origin, since the average barrier height ϕ is different in all three cases.

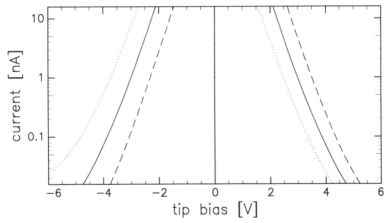

Fig. 8. Dependence of constant field condition on the workfunction difference of tip and sample in the medium voltage regime. $\phi_s=4.5eV$ and $\phi_t=4.0eV$(dotted), $4.5eV$(solid) and $5.0eV$(dashed). The constant field is $F=6V/nm$ in each case and $A=1nm^2$.

It is well known that the potential difference due to the different workfunctions also sets up a Schottky barrier which has rectifying properties[22]. Simmons[23] has incorporated this effect in his analysis, but its effect on the I-V characteristics of the junction is significant only in the high voltage regime.

8. Groove Formation

In this section we discuss the experimental evidence for STM field-induced groove formation on Si(111). In the sections 9 and 10, the experimental results are analyzed using the theory developed in the previous sections.

Samples of n-type Si(111) wafers, resistivity 0.01Ω-cm, were cleaned in ultra-high-vacuum (UHV) by repeated flash heating to 1150 °C, using direct heating through the sample. Low energy electron diffraction images revealed a diffraction pattern with sharp spots characteristic of the 7x7 reconstruction of Si(111).

Sharp W tips were prepared by electrochemical etching in a 1M solution of KOH. The tips were rinsed and then introduced into the ultra-high vacuum chamber of a VG-2000 STM. The base pressure was 5×10^{-11} Torr. The typical radius of curvature of the as-etched tips is estimated to be 100nm or more by scanning electron microscopy. The tips were electron bombarded in the UHV chamber, using an electron current of 2mA from a Th-coated W filament applied for a period of 1 min.

Electron bombardment is a crucial preparation step to obtain reproducible modification results with W tips. Tips that were not electron bombarded did not behave reproducibly, and often could not modify silicon surfaces even for tip biases of up to 10V.

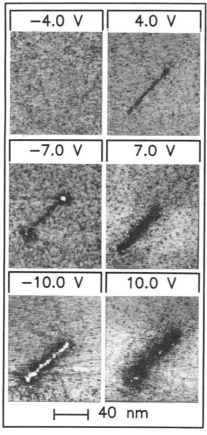

Fig. 9. Images of grooves formed on Si(111) at three different biases in both polarities. The constant current during the groove formation is 0.25nA. The imaging conditions are -1.95 V tip bias and 0.25nA. The height range is set automatically by the STM and is about 5.0Å from black to white. The grooves are probably deeper than this. The grainy appearance of the Si(111)7x7 surface is due to the limited spatial resolution of the scan.

Atomic-scale resolution of the Si(111) 7x7 reconstruction was usually obtained after raising the bias of the tip several times to a voltage of about 7V in either polarity. The exact effect of this process is not known, but it probably removes any remaining contamination from the tip, and perhaps helps to sharpen the tip locally, so that a single atom becomes more prominent than its neighbors.

To modify the surface, the magnitude of the tip bias was increased and the tip was scanned at 50nm/s over a length of 60nm to 100nm. The same area was then imaged to detect modification. As the tip bias during the modification scan increases, there is a critical voltage, V_c, above which a dark line is observed along the scanned region. Typical examples are shown in Fig. 9 for a W tip at both polarities.

The dark features in Fig. 9 appear to be grooves in the surface, where Si has been removed. An electronic effect due to some deposited material cannot be ruled out from a single image. But it is observed that the grooves appear dark at all scanning biases in the range +3V to -3V, whereas common adsorbates, such as O [24] and H [25] should appear bright at some biases in this range. This is strong evidence that the features are topological in nature.

The periodic structure of the Si(111)7x7 reconstruction is not distinguishable on the scale of Fig. 9. However, in Fig. 10, a small region is shown where a modification has been made by applying +3V for 2s at a tunneling current of 0.25nA. The result is imaged at several tip biases in both polarities, and the central modified area always appears dark.

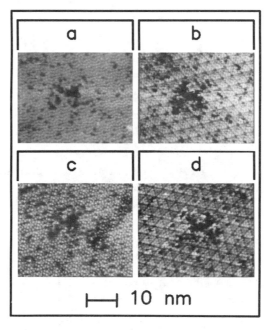

Fig 10. Images of a modified region about 10nm in diameter for a tunneling current of 0.25nA and biases of a) -1.95V b) -0.98V c) 1.15V d) 1.95V. The depth of the feature is about 4Å, corresponding to removal of the adatom layer and first Si bilayer.

Attempts to observe structure inside the groove have so far proved unsuccessful. The nominal depth of the grooves, based on the motion of the z-piezo controller, is 0.7nm-1.2nm, or about two to three Si layers. It is possible that the grooves are deeper, but that this cannot be seen due to the large opening angle of the end of the tip.

Several volts above the critical voltage, protrusions are observed in the grooves. The nature of these protrusions is not known at present. Similar protrusions have been observed before [11]. It has been argued that the protrusions are Si that had been drawn up by the high field directly under the tip [11]. We note that in Fig. 9 there is a dependence of protrusion formation on polarity, the protrusions being much clearer at negative tip bias. A possible explanation is that the protrusion is in fact material deposited from the tip, which is more easily evaporated as negative ions than as positive ions. The possibility that the protrusions are W is intriguing, since it suggests a way of depositing nanometer-scale metallic wires on silicon. In the following, though, only the grooves occurring near V_C will be discussed.

Groove formation using the tip preparation method described in the previous section was very reproducible. Over 100 different tips have been tested, and the onset of groove formation occurred at a voltage that varied by less than 10%. This reproducibility indicates that the modification process does not depend sensitively on the microscopic tip geometry, which can vary from tip to tip. This further supports the parallel-plate model of the tip-sample geometry which was argued for in section 5.

No deterioration of the ability of a tip to modify the surface has ever been observed, even though some tips have been used to make more than 100 grooves. However, we observe occasionally that a newly bombarded tip behaves irreproducibly, sometimes making large deposits on the surface. This behavior is similar to unbombarded tips, and suggests that electron bombardment is not 100% effective.

9. Dependence of Groove Width on Bias and Current

The grooves in Fig. 9 become wider as the absolute value of the tip voltage is increased. In Fig. 11 the average groove width is shown as a function of tip bias at constant tunneling current. The depth of the grooves cannot be estimated reliably from the STM data, but taking the average width as a measure of the material removed, the process is clearly strongly voltage-dependent, consistent with field ion emission.

In contrast, the inset of Fig. 11 shows that the current dependence of the average groove width is weak: less than 30% change when increasing the current two orders of magnitude. This rules out mechanisms such as electromigration [26] local melting [18], or electron-beam-induced chemical reaction [27], which vary strongly with current. The slow increase of groove width with tunneling current I is consistent with field emission, since in the tunneling regime, $s \propto -\log(I)$ [8], and so $F \propto -1 / \log(I)$.

The definition of the critical voltage used in practice here is that at V_C, the groove becomes continuous. Below V_C, the groove is not continuous but patchy. About 1V below V_C, only the endpoints of the groove are seen. This is because the tip spends a significantly longer time at the end points, and possibly due to transient effects when the voltage is ramped up and down. At lower biases, no modification can be detected. Given the scanning speed and the size of the groove at V_c, the evaporation rate can be estimated as $R=10^2$-10^3 s^{-1}. This is different from the value of $R=1$s^{-1} used to define F_c, but as noted in section 5, F_c is not very sensitive to the exact value of R.

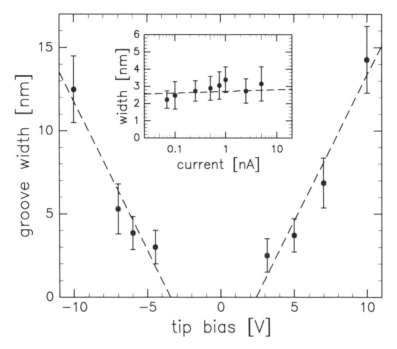

Fig. 11 The average groove width as a function of sample bias, at a fixed tunnel current of 0.25nA. Inset: current dependence of groove width.

The above results are consistent with field emission from the sample in both polarities, in other words, emission of Si cations when the tip is biased negatively, and Si anions at positive bias. In FIM studies, generally only cation emission is observed, since at large negative voltages the electron field emission current melts the tip[27]. In FIM, unlike STM, current and bias cannot be controlled independently. However, emission of weakly adsorbed species as anions at lower voltages has been observed by FIM[28], showing that negative ion emission is certainly a possible mechanism.

Our results show that sample material is more easily evaporated than tip material in both polarities. In FIM studies of cation emission, the critical field of W is 57V/nm, compared to 30V/nm for Si[5]. For the case of anions, calculations using the formalism of section 5 give a critical field of 84V/nm for W and 32V/nm for Si[11]. In other words, the observation that Si is more easily field evaporated than W in STM geometry is in qualitative agreement with FIM results.

The critical voltage, V_c, depends on current and tip polarity. We have measured V_c for W tips at both polarities in the current range 0.05 to 5 nA. The results are shown in Fig. 12.

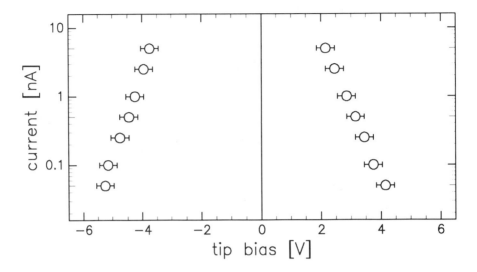

Fig. 12 Variation of Vc with tunneling current for both tip polarities. For a given current the data points define the bias above which a continuous groove is formed. Error bars are based on variation for observations with over 10 different tips and samples.

The main feature of Fig. 12 is that $V_c \propto \log(I)$. Since $s \propto \log(I)$, this means that V_c/s is constant, in other words, modification occurs at a current-independent critical field, F_c. Fig. 12 is qualitatively very similar to the theoretical results of Fig. 6 for the medium voltage regime. The variation of V_c with current contrasts with electron beam induced desorption, which occurs at a fixed electron kinetic energy, in other words, a constant voltage threshold. Such a process has been proposed for the STM-induced removal of H from H-terminated Si(111) [30].

10. Dependence of Critical Voltage on Tip Material

To investigate the dependence of groove formation on tip material, sharp Au, Ag and Pt tips were prepared by chemical etching. The etchant was a 1M solution of HCl for Au, a 1M solution of nitric acid for Ag and a molten mixture of sodium nitrate and sodium chloride in the ratio 4:1 for Pt. After rinsing, the tips were introduced into the UHV chamber. Ag and Pt tips were electron-bombarded in the UHV chamber as described above. Au tips, however, could be used without prior bombardment and behaved reproducibly.

For Au and Pt tips, atomic-scale resolution of the Si(111)7x7 was obtained routinely. For Ag tips, the atomic detail was rarely distinct, although the periodicity of the 7x7 structure was always observed. More than 10 tips were tested for all materials, and the results for the critical voltage were reproducible to within ±1V in almost all cases, and usually to within less than ±0.5V.

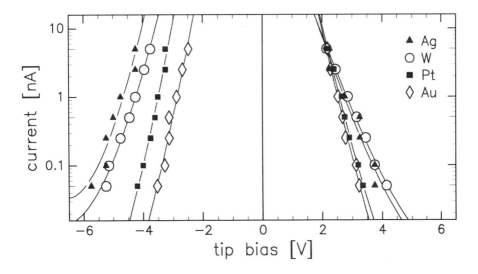

Fig. 13. *Dependence of critical voltage on tip material, for Ag(ϕ_t=4.3), W(ϕ_t=4.5), Pt(ϕ_t=4.8) and Au(ϕ_t=5.2). Curves are fits to the data as described in the text.*

The dependence of the critical voltage on tip material is shown in Fig. 13. The error bars on the data are not indicated, but are ±0.3V or less in all cases, based on a statistical average of many observations. The curves represent best fits to the data, based on Eqn. 2 for the medium voltage regime. Standard values of the workfunctions of the materials are given in the figure caption. Qualitatively, the theoretical dependence of V_c on workfunction discussed in section 7 is reproduced in the results, especially clearly at negative tip bias.

In fitting the medium voltage equation to the data, there are two unknown parameters: the effective tunneling area A and the critical field F_c. Since the critical field for emitting cations will in general be different from that for anions, the data at positive and negative polarity are fitted separately. A non-linear least squares fitting procedure was used. The fit parameters are tabulated in Table 1, along with the standard values of the workfunctions[31] and the nominal variation of tip-sample separation over the appropriate voltage range, as determined from Eqn. 11.

The values of F_c in Table 1 are all considerably lower than the value of 30 V/nm found in field ion emission experiments for positive Si ions[11]. The F_c values at negative bias are within errors the same for all tips, which agrees with the simple picture that the critical field is an intrinsic property of the Si surface, independent of the tip material. There is more scatter at positive bias, but the nominal tip-sample separation given in Table 1 is also considerably smaller in this polarity, so chemical interaction cannot be ruled out at positive bias.

The values of A are of the order of 1nm^2 or less at positive bias, reasonable for tunneling from a single atom. The values at negative bias are much larger. An increase at negative bias is expected, because the nominal tip-sample separations are larger, so the tunneling current will fan out over a larger area. It would be overly simplistic, though, to interpret these differences solely in terms of the tunneling area. As discussed in section 4, the value of A may be underestimated in this analysis due to the simplicity of the model potential barrier that is used.

tip bias	tip material	ϕ_t [eV]	F_c [V/Å]	$\log_{10}(A\ [\text{Å}^2])$	s [Å]
−	Ag	4.3	0.31 ± 0.10	6.9 ± 2.5	12.0 − 16.7
−	W	4.5	0.39 ± 0.04	4.8 ± 0.8	8.9 − 12.6
−	Pt	4.8	0.31 ± 0.05	6.5 ± 1.4	10.4 − 13.5
−	Au	5.2	0.38 ± 0.08	4.8 ± 1.6	7.6 − 10.3
+	Ag	4.3	0.67 ± 0.16	1.0 ± 1.2	4.1 − 6.3
+	W	4.5	0.74 ± 0.04	0.4 ± 0.3	3.3 − 6.0
+	Pt	4.8	0.48 ± 0.03	1.7 ± 0.6	4.4 − 7.0
+	Au	5.2	0.48 ± 0.04	0.8 ± 0.4	3.7 − 5.9

Table 1. Fit parameters for curves in Fig. 13.

The large difference between A values at positive and negative tip bias may be due to the different density of states at the Fermi level of the emitter in both polarities. This density of states appears in the exponential prefactor of the tunneling current in more detailed calculations for the low voltage regime [15]. Such a term would be subsumed in the value of A. Because it is much larger when the metal tip is the emitter than when silicon is, this could explain the larger A values at negative bias. To further investigate such effects, a more detailed theory of the medium voltage regime is needed.

In the analysis A, but not F_c, couples strongly to the values of ϕ_t and ϕ_s. Quoted values for workfunctions vary considerably[31]. The values used here for the metal tips are based on contact potential measurements, related to a standard value for W[31]. For Si, values range widely, but a frequently-quoted value of 4.8eV was chosen[6]. However, the workfunctions are clearly a source of uncertainty in the results. The standard errors in Table 1 include a ±0.2V variation in the workfunction values.

11. Single Atom Manipulation

By reducing the time a tip spends above a certain position at a bias equal or greater than V_c, it is possible to reduce the amount of material removed. Ultimately, single atoms can be extracted from the surface in this way. In the following three sections, we present results for field-induced atom extraction by this procedure.

The experiments discussed here were performed in a JEOL JSTM-4000 XV commercial STM. Si(111) n-type wafers were cleaned by repeated flash heatings at 1200°C. The base pressure in the chamber was 1×10^{-8}Pa. The STM tip was a 0.1mm W wire etched in a 0.5M solution of KOH. Electron bombardment heating of the tip, raising the tip temperature to above 1200°C, was performed in the UHV chamber.

Silicon adatoms were modified as follows. A clean, flat area of the sample was imaged at a sample bias of +2V and -2V and a tunneling current of 0.6nA. The tip was then moved to a point in the imaged region, and a voltage pulse of either +6V or -6V was applied to the sample at a nominal current of 0.6nA for 10ms. Because of the short duration of the pulse, the piezo feedback cannot satisfy the condition of constant current, so the supplied current is somewhat greater than the nominal value demanded. After the pulse, images of the same area are again taken at +2V and -2V. Examples of the results are shown in Fig. 14.

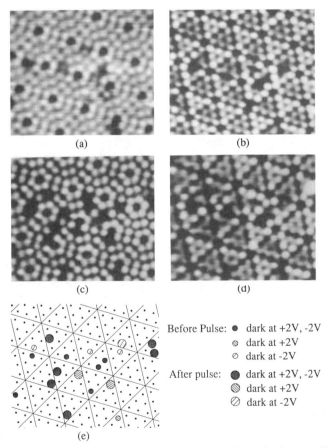

Fig. 14. Appearance of Si(111)7x7 surface before (a, b) and after (c, d) a -6V 10ms pulse is applied. Imaging sample bias is +2V (a and c) and -2V (b and d). In (e) a schematic diagram of the resulting modification is shown.

Certain adatom positions are modified, the dominant form of modification being a dark spot in both polarities. As mentioned in the last section, it is unlikely that these features are adsorbate induced, since common adsorbates are known to appear light in one or both polarities. The interpretation that the dark spots are vacancies is also consistent with the observation of groove formation when the bias is applied for longer periods.

Other sorts of defects, which do not appear dark in both polarities, are occasionally generated by the voltage pulse (see Fig. 14). The exact origin of these defects is not known yet, though it is possible that they are due in part to material field evaporated from the tip onto the surface[13].

The extent of modification on the Si surface varies with the magnitude of the pulse. Of the order of 10 vacancies are produced by a -6V pulse to the sample, as shown in Fig. 14, and the result is similar for a +6V pulse. For a -4V pulse, fewer vacancies are formed and frequently a single adatom can be extracted from a predetermined position. Such site specific manipulations are possible due to the very low thermal drift of the STM (< 1Å per minute). An example where a pulse resulted in the extraction of a single atom is illustrated in Fig. 15. For a +4V pulse to the sample no vacancies are formed. This agrees with the observation that higher voltages are required to form grooves when the tip is biased negatively[5].

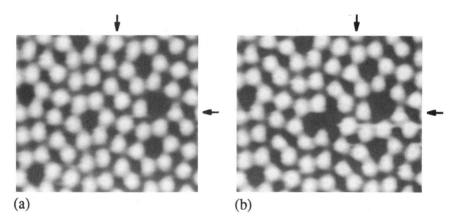

Fig. 15. Single atom modification made by a -6V pulse, imaged at +2V before(a) and after(b) pulsing.

Single adatom extraction as in Fig. 15 is successful in about 30% of attempts: in other cases a pulse will either not produce an adatom vacancy, or produce several at once. This is consistent with field ion emission being a thermally activated process, where emission of an adatom is a probabilistic event. In Fig. 14, the extracted adatoms are spread over a fairly large area. This is consistent with the separation of tip and sample, and hence the electric field, varying only little over a region several nm wide (see Fig. 5). Nevertheless, the fact that single atoms can frequently be removed from predetermined positions on the surface suggests that the atomic-scale protrusion on the tip which forms the tunnel junction may also increase the electric field locally, facilitating field emission from the position directly below it.

12. Statistics of Adatom Extraction

The adatoms on the Si(111)7x7 surface can be divided into four crystallographically distinct types, according to whether they are corner adatoms or center adatoms, and whether they reside on the faulted or unfaulted half of the 7x7 unit cell. The distinction between corner and center adatom is made in Fig. 16. The stacking faulted half of the 7x7 unit cell is apparent in scans at negative sample bias (Fig. 14b and 14d), where it appears darker than the unfaulted half.

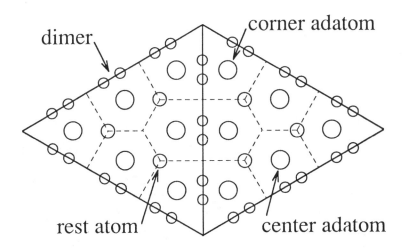

Fig. 16. Schematic diagram of Si(111)7x7 unit cell.

To obtain relative extraction probabilities for these four different groups, pulse experiments similar to that shown in Fig. 14 were repeated many times. The position of the tip during the pulse was chosen at random. In total, 246 adatoms were extracted for -6V pulses to the sample and 230 for +6V pulses. The breakdown according to adatom type is given in Table 2. There are two important aspects to the results. Firstly, the probability of extraction from the faulted and unfaulted halves of the unit cell was the same, within error. Secondly, center adatoms were more frequently extracted than corner adatoms by a ratio of 1.6±0.2, for both pulse polarities.

The difference between center and corner adatoms cannot be accounted for by geometrical effects. There are equal numbers of center and corner adatoms in each unit cell. If the unit cell is partitioned by the bisectors of lines joining nearest neighbor adatoms (dashed lines in Fig. 16), the area associated with corner adatoms is 4% larger than that of center adatoms. Assuming the probability of adatom extraction to be proportional to this area, corner adatom extraction should be slightly preferred. No other reasonable geometrical definition of an adatom cross-section can produce the large difference in favor of center adatom extraction that is observed experimentally.

(a) Negative Voltage Pulse : -6V 10msec

Corner Si Adatom		Center Si Adatom	
Faulted Half	Unfaulted Half	Faulted Half	Unfaulted Half
51(21%)	48(19%)	73(30%)	74(30%)
99(40%)		147(60%)	

(b) Positive Voltage Pulse : +6V 10msec

Corner Si Adatom		Center Si Adatom	
Faulted Half	Unfaulted Half	Faulted Half	Unfaulted Half
41(18%)	46(20%)	79(34%)	64(28%)
87(38%)		143(62%)	

Table 2. Adatom extraction statistics for (a) negative and (b) positive voltage pulses. The number of extracted atoms is given; in parenthesis is the relative percentage.

Local variation of the electric field at different adatom sites is also an unsatisfactory explanation of the experimental result. On the basis of simple electrostatic considerations, the field should, if anything, be higher at the corner adatom sites, in virtue of their position at the relatively sharp corners of the 7x7 cells, and also because they are calculated to protrude 0.05Å further out of the surface than center adatoms[32].

The configuration of nearest neighbors around a corner adatom and a center adatom is different. Correlation effects, which might make it easier to remove an adatom neighboring a vacancy, and thus possibly skew the statistics, can be eliminated by counting only those vacancies formed in areas isolated from other vacancies or defects. This subtotal is 91 for center adatoms and 60 for corner adatoms, giving a ratio of 1.5±0.25, the same within errors as for all vacancies taken together.

13. Activation Energy Model

We suggest that the difference in extraction probabilities reflects a difference in the activation energy for field ion emission of the center and corner adatoms. Given that corner and center adatoms have identical bonding geometries, their vibrational properties should be very similar, so it is unlikely that the exponential prefactor, v, in Eqn. 6 can account for the relatively large difference in evaporation rates. On the other hand, a small difference in activation energy will produce a large effect. Defining Q_{corner} and Q_{center} as the activation energies of corner and center adatoms, and equating the observed ratio of extracted corner and center adatoms to the ratio of evaporation rates, then from Eqn. 6, $Q_{corner} - Q_{center} = kT \log(1.6) \cong 0.01$ eV, assuming that the sample remains at 300K during the process. The magnitude of the activation energy can be estimated very crudely from an evaporation rate of 10 atoms per 10ms = 10^3 s^{-1} and a typical vibration frequency of 10^{13}s^{-1}. From Eqn. 6, $Q_{corner} \cong Q_{center} \approx 0.8$ eV.

The observation that the extraction probability ratio is identical within errors for both polarities suggests that the difference between the corner and center adatoms is related to the polarity-independent binding energy term Λ in Eqn. 10. Thus a small 0.01 eV difference in binding energy between corner adatom and center adatom can account for the observed difference in extraction probabilities. This difference should be compared with the binding energy of Si adatoms, which has been estimated to be about 0.6-0.8eV/adatom [33,34]. Recently, first principles calculations of the full Si(111)7x7 structure have been made, though the difference in the binding energies between center and corner adatoms is not reported[32]. However, Meade and Vanderbildt[34] have calculated the energies of several Si adatom structures, such as 1x1, $\sqrt{3}$x$\sqrt{3}$R30° and 2x2, which have different local adatom geometries. Energy differences as small as 0.03 eV per 1x1 unit cell were found in certain cases, similar in magnitude to the difference between center and corner adatoms estimated here.

Recent results for the chemical reactivity of adatom sites on Si(111)7x7 display an intriguing trend, which suggests a relation to the adatom extraction discussed here. For NH_3 reacting with a Si(111)7x7 surface, the center adatom sites are more reactive than the corner adatom sites by a ratio of more than 4:1 [35]. Similar behavior is seen for H_2O [35]. Enhanced chemical reactivity of center adatoms has also been seen for several metals adsorbed in small amounts on Si(111), for example Cu [36], Ag [36], and Pd [37]. In the case of metals, in contrast to NH_3 there is a pronounced preference for reaction with the unfaulted half of the unit cell. A notable exception to the trend of enhanced reactivity at center adatom sites is for O, which shows a preference for corner adatoms[24]. However, the mechanism is believed to be quite different in this case, with the O atom absorbed in a corner adatom backbond, rather than reacting with a dangling bond[24].

It is reasonable to expect that the chemically more reactive sites will also be the structurally less stable ones. Thus the larger activation energies of center adatoms for chemical reaction and for field ion emission are probably not unrelated. The reason for the greater reactivity of center adatoms is not yet understood fully. It may be related to a greater degree of charge transfer from the center adatoms to the surface rest atoms, which are indicated in Fig. 16 [35]. Strain relaxation may also play a role, since corner adatoms are adjacent to two Si dimers, while center adatoms are adjacent to only one. Clearly, such differences will also affect the adatom binding energy.

14. Conclusions and Future Perspectives

In the previous sections, we have described a low voltage field-induced modification process for nanomodification and atomic manipulation of Si(111). A coherent picture has been presented relating the experimental results to simple theoretical models of the tunneling junction and of the field ion emission process. These results raise a number of intriguing questions. In this section we discuss ongoing experimental research in our group to further test the interpretation of the results proposed here. We also look more generally at what lessons can be drawn from our experience for the future development of STM nanolithography.

For the groove formation studies, as further confirmation of the modification process discussed here, we have made preliminary measurements on p-type Si(111) and on Si(100) surfaces and find identical groove formation with a similar variation of the critical voltage. Slight differences in the critical voltage variation, compared to the case of n-type Si(111), can be related to variations in the rectifying behavior of the junction, due to Schottky barrier formation, as discussed in section 7. Work is also in progress to image the inside of the grooves and to identify the material that forms the protrusions seen at high bias.

For the atomic manipulation results, an interesting issue is where the atoms from the surface go after evaporation. There is evidence that some extracted Si adatoms are scattered on the surface, rather than leaving it completely[38]. On Si(100), it is found that the probability of extraction of atoms from monatomic steps on the surface depends on the type of step. There are two inequivalent types of step on the 2x1 reconstructed surface of Si(100), and preliminary results suggest that the type that is known to be structurally less stable is also more easily modified. This result agrees with the model proposed in section 13.

As described in section 8, the ability to modify the Si(111) surface at low voltages is very strongly dependent on tip preparation. This may explain why, in their pioneering study, Becker et al. were not able to modify Si(111)[4]. Although this effect is not yet properly understood, it is worth speculating briefly on its origin.

Good tunneling junction characteristics are usually obtained with electron-bombarded tips, even when they are approached to the surface for the first time. In contrast, unbombarded tips frequently crash into the surface, or form an unstable junction. Transmission electron microscopy reveals that an oxide layer several nanometers thick is present on as-etched tips[39]. The assumption is that a break must first be made in the insulating oxide layer by physical contact, before tunneling can occur. This would explain the unstable behavior.

It is reasonable to assume that the oxide layer is removed over a much wider area due to electron bombardment. As mentioned in section 10, Au tips were able to modify Si(111) reproducibly without electron bombardment. This result is reasonable since Au is very inert, and so is unlikely to have a significant oxide layer. The fact that unbombarded Au tips modify the surface in a similar manner to the bombarded W tips can also be taken as evidence that the electron bombardment process itself is not having some extraneous effect, such as evaporation of filament material onto the tip.

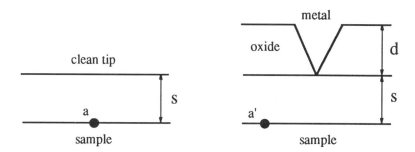

Fig. 17. Sketch of the difference between a clean tip and a tip with an oxide layer several nanometers thick.

In Fig. 17 we sketch the possible appearance of a clean tip and one with an oxide layer in the vicinity of the conducting part of the tip. If the oxide layer has permittivity κ and thickness d, then it is straightforward to show that the ratio of the electric field F' at point a' and F at point a is $F'/F = 1/(1+d/\kappa s)$, which is always less than unity. In the close vicinity of the conducting region on the tip, the field will depend sensitively on the exact geometry of the break in the oxide layer, which may account for the irreproducible behavior of such tips. In Fig. 17, the field on the metallic protrusion should be higher than V/s, due to the local curvature of the conducting surface. However, as argued in section 5, the requirement that $\int F ds = V$ over the gap implies that the field at the surface directly below the metallic protrusion will be correspondingly less than V/s. Thus the effect of the dielectric is to reduce the field at all points on the sample surface.

At higher applied bias, another effect may prevent field ion emission. As discussed in section 3, once the junction is in the high voltage regime, further increase of tip bias at constant current will not increase the field. Instead, the tip will simply move proportionally further away from the surface to preserve constant current. Thus, if the critical field is not reached in the medium voltage regime, it may not be reached at all, even for much higher voltages.

Experiments where the tip is modified in a controlled way would be very useful in order to test these ideas. Indeed, perhaps the greatest uncertainty in STM nanolithography experiments at present stems from the ill-defined state of the STM tip.

When the STM is used uniquely for imaging, the state of the tip is of secondary importance, as long as it "works". Common tip preparation techniques, such as cutting a tungsten wire with pliers, are exceedingly crude compared to the effort that is lavished on sample preparation. These contrasting attitudes to the two sides of the tunnel junction are clearly inappropriate in the case where material is being exchanged between them.

The current situation may be likened to the early days of surface science, before ultra-high vacuum was available. Much of what was learned then had to be revised radically when cleaner surfaces could be prepared. In a similar way, these are the early days of tip science. Though our studies clearly show the importance of tip preparation, we cannot yet pretend to know, let alone control, the state of the tip on the nanometer or atomic scale. Once tip science comes of age, it may be necessary to revise many of the results on STM-induced surface modification reported to date, including those presented here! Therefore, the most enduring message of this chapter is that careful tip preparation is crucial to obtaining reproducible results, and thus to developing STM nanolithography into a reliable technology.

References
[1] J. A. Stroscio and D. M. Eigler, Science **254**, 1319 (1991)
[2] C. F. Quate in *Highlights of Condensed Matter Physics and Future Prospects*, L. Esaki editor, Plenum, New York, 573 (1991).
[3] R. Gomer, IBM Journ. Res. Develop. 30, 428 (1986)
[4] R. S. Becker, J. A. Golovchenko and B. S. Swartzentruber, Nature **325**, 419 (1987)
[5] T. Sakurai, A. Sakai and H. Pickering in *Atom Probe Field Ion Microscopy and its Applications*, Academic Press (1989)
[6] *Atom-Probe Field Ion Microscopy*, T. T. Tsong, Cambridge University Press (1990)
[7] H. J. Mamin, P. H. Guethner and D. Rugar, Phys Rev. Lett **65**, 2418 (1990)
[8] S. E. McBride and G. C. Wetsel, Appl. Phys. Lett. **59**, 3056 (1991)
[9] J. P. Rabe and S. Buchholz, Appl. Phys. Lett. **58**, 702 (1991)
[10] T. Schimmel, H. Fuchs, S. Akari and K. Dransfeld, Appl. Phys. Lett. **58**, 1039 (1991)
[11] I.-W. Lyo and P. Avouris, Science **253**, 173 (1991)
[12] A. Kobayashi, F.Grey, S. Williams and M. Aono, Science **259**, 1724 (1993)
[13] H. Uchida, D. H. Huang, F. Grey and M. Aono, Phys. Rev. Lett. **70**, 2040 (1993)
[14] J. G. Simmons, Journ. Appl. Phys. **34**, 1793 (1963)
[15] J. Tersoff and D. Hamann, Phys. Rev. B31, 805 (1985)
[16] R. N. Fowler and L. Nordheim, Proc. Roy. Soc. Lond. **A119**, 173 (1928).
[17] E. W. Müller, Phys. Rev. **102**, 618 (1956)
[18] T. T. Tsong, Phys. Rev. **B44** 13703 (1991)

[19] R. Gomer and W. Swanson, J. Chem. Phys **38**, 1613 (1963)

[20] H. J. Kreuzer and K. Nath, Surf. Sci. **183**, 591 (1987)

[21] N. D. Lang, Phys. Rev. **B45** 13599 (1992)

[22] E. H. Rhoderick and R. H. Williams in *Metal-Semiconductor Contacts,* 2nd ed. Clarendon Press, (1988)

[23] J. G. Simmons, J. Appl. Phys. **34**, 2581 (1963)

[24] I.-W. Lyo, P. Avouris, B. Schubert and R. Hoffmann, J. Phys. Chem. **94**, 4400 (1990)

[25] J. Boland, J. Vac. Sci. Tech. **B9**, (1991)

[26] D. M. Eigler, C. P. Lutz and W. E. Rudge, Nature **352**, 600 (1991)

[27] E. A. Dobisz and C. R. K. Marrian, J. Vac. Sci. Tech. **B9** 3024 (1991)

[28] T. T. Tsong, private communication.

[29] E. Bramer-Weger, S. Thiebes and F. W. Röllgen J. Phys. Coll. (Paris) **50**, C8-159 (1989)

[30] R. S. Becker, G. S. Higashi, Y. J. Chabal and A. J. Becker, Phys. Rev. Lett. **65**, 1917 (1990)

[31] *American Institute of Physics Handbook*, D.E. Gray editor, McGraw Hill, New York (1972)

[32] K. Brommer, M. Needels, B. Larson and J. Joannopoulos, Phys. Rev. Lett. **68** 1355 (1992)

[33] J. E. Northrup, Phys. Rev. Lett. **57** 154 (1986)

[34] R. D. Meade and D. Vanderbildt, Phys. Rev. **B40** 3905 (1989)

[35] R. Wolklow and P. Avouris, Phys. Rev. Lett. **60** 1049 (1988); P. Avouris and R. Wolklow, Phys. Rev. **B39** 5091 (1989)

[36] S. Tosch and H. Neddermeyer, J. Microscopy **152**, 415 (1988)

[37] U. Koehler, J. Demuth and R. Hamers, Phys. Rev. Lett. **60**, 2499 (1988)

[38] D. H. Huang, H. Uchida and M. Aono, Jpn. Journ. Appl. Phys. **12B**, 297 (1992)

[39] J. Garnaes, F. Kragh, K. A. Morch and A. R. Tholen, J. Vac. Sci. Tech. **A8**, 441 (1990)

SCANNING TUNNELING MICROSCOPE-BASED FABRICATION AND CHARACTERIZATION ON PASSIVATED SEMICONDUCTOR SURFACES

J. A. Dagata and J. Schneir

*Precision Engineering Division
National Institute of Standards and Technology
Gaithersburg MD 20899*

ABSTRACT

Recent results employing scanning tunneling microscope (STM)-based fabrication and characterization techniques on passivated semiconductor surfaces are reviewed. The preparation and characterization of silicon and gallium arsenide surfaces for STM patterning of ultrashallow, oxide masks and the use of these masks in selective-area heteroepitaxy and reactive ion etching are discussed. Methods for performing meaningful spectroscopic characterization of GaAs surfaces and structures are considered in detail. An ultrahigh-vacuum-compatible STM system constructed in our laboratory specifically for lithographic process development is described.

INTRODUCTION

The scanning tunneling microscope (STM) and other scanned probe techniques will play a significant role in forthcoming technological developments. Many of these applications will utilize the unique surface sensitivity of the STM to probe or modify surfaces during processing. This extreme surface sensitivity, while relatively easy to achieve under standard surface science conditions, is incompatible with existing applications which subject the surfaces to chemically and mechanically harsh conditions, thereby disrupting the electrical, chemical, and structural perfection of the surface. This situation applies for most materials simply upon the removal of the surface from ultrahigh vacuum. Maintaining surface quality on the nanometer scale in realistic processing environments will be a daunting challenge for proximal probe research for some time to come.

In this paper, we review efforts in our laboratory over the past four years to develop processes, methodology, and instrumentation which are allowing us to fabricate and characterize nanostructures on semiconductor substrates.[1-16] The fabrication of nanometer-scale structures for nanoelectronic device applications involves the control of electrical characteristics and chemical composition on the atomic, or near-atomic, scale. A key concern for us is the persistence of these characteristics in air as well as other processing ambients. It is our belief that this control is a prerequisite for the integration of STM-based modification processes into the microelectronics processing environment and an essential step towards recognizing the fundamental *processing* issues facing the implementation of promising STM modification schemes. The aim of this paper is to emphasize some of the technological requirements which STM-based modification processes must satisfy in order to have a significant impact in the field of nanostructure fabrication.

THE ROLE OF THE STM IN NANOMETER-SCALE TECHNOLOGY

The STM and related proximal probe instruments offer a new and unique concept for fabrication and characterization on the nanometer scale.[17] It is important at this early stage in the evolution of the STM toward application-oriented research to identify the fundamental issues within the scope of one's research. We have chosen to emphasize several important issues in the integration of STM-based techniques into electronics device processing.[9] These are: (1) substrate preparation, (2) nanostructure pattern generation, (3) pattern development, (4) substrate and nanostructure characterization, and (5) process integration. As such, this work is not focused directly on the equally crucial parallel hardware issues of multi-tip arrays[6,7], required for adequate throughput, or problems related to scanned probe metrology.[18] It is evident that breakthroughs in all these areas will be required before truly significant application of this concept will be possible.

SUBSTRATE PREPARATION AND CHARACTERIZATION

The preparation of chemically and topographically uniform, passivated silicon surfaces has been discussed extensively in the literature.[19] Because of the unique characteristics of the oxide and the fortuitous hydrogen-terminating chemistry of dilute fluorine etches, the apparent surface roughness of H:Si(111) surfaces can now be controlled to atomic dimensions. Our STM lithography (see next section) was performed on silicon surfaces which were not prepared by the most advanced methods[16], so that the topographic roughness, due to 5-nm dia. hillocks remains, as indicated by the R_{app} value given in Table I. (The *apparent* roughness, R_{app}, measured by the STM can be an extremely useful diagnostic of the surface uniformity since changes in the local conductivity of various surface phases, such as oxides, will produce topographical contrast in the STM image.) This residual roughness limits the ultimate resolution to perhaps 10 nm.

Whereas hydrogen-terminated silicon surfaces are remarkably stable in air making them ideal substrates for STM lithography studies, there was no corresponding treatment in existence for passivating GaAs surfaces when we began our attempts at patterning compound semiconductors.[12] The chemical inhomogeneity of GaAs substrates may be reduced by growing an ordered arsenic cap following GaAs molecular beam epitaxy (MBE) or by treatment of the substrate by a dilute, aqueous P_2S_5 solution. Methods for preparing each of these surfaces have been given previously.[10-12] In both cases the surface is passivated by a stable, ordered ultrathin oxide which is highly resistant to further oxidation and segregation. The dramatic reduction in R_{app} with the P_2S_5 solution has lead to several collaborations aimed at better understanding the electrical and chemical properties of these surfaces, since the surface preparation issue is one which impacts on existing technologies such as etching and epitaxial regrowth, as well as on

TABLE I

Comparison of *apparent* roughness, R_{app}, for variously prepared semiconductor surfaces. P-V (peak-to-valley) and rms (root-mean-square) roughness values were calculated from representative STM image data.

Sample	Stable?	Roughness	
		P-V (nm)	rms (nm)
H:Si (111)	yes	3-5	< 0.5
As:GaAs(100) (MBE-grown)	yes	4-6	1.0
etched GaAs (100)	no	48-55	7.4-8.8
$(NH_4)_2$S-treated GaAs(100)	no	12-16	2.1-2.8
P_2S_5-treated GaAs(100)	yes	1.0-1.4	0.3-0.4

future nanoelectronics.

The characterization of the substrate using STM spectroscopy (STS) is a fundamental issue since it is imperative to understand the details of the tip-sample interaction during the lithography process. Furthermore, the local probing of nanostructures will require that this information be well in hand. STS has played an important role in our development of the P_2S_5 passivation method.[1,5] For example, the interplay between STM imaging and spectroscopy is shown in Figure 1. The uniform termination of the GaAs surface by a stable, ultrathin gallium-rich oxide permits us to obtain current-voltage (I-V) characteristics representative of the bulk band structure of sample, Figure 1 (a). The surface defects on the etched surface, Figure 1 (b), deplete the surface, making spectroscopy impossible, leading to large R_{app} values.

The ability to carry out the initial stages of combined STM/MBE/reactive ion etching (RIE) investigations by incorporating *ex situ* sample transfer[9] will depend on reliable methods for preparing and passivating semiconductor surfaces and interfaces with bulklike electrical character. Considerable work needs to be done in this area. The STM can play an important role as a sensitive local diagnostic, in conjunction with other surface techniques. Serving the needs of the electronics and opto-electronics community in this way will only hasten its wide acceptance in industry.

PATTERN GENERATION ON SEMICONDUCTOR SURFACES

Local chemical modification of hydrogen-passivated silicon and As-capped GaAs surfaces by an STM operating in air has been used to write permanent, nanometer-scale oxide patterns on these substrates[14-16]. We have demonstrated previously that the modification of these freshly-passivated surfaces can be modulated by adjusting the bias voltage between the tip and sample and that the modification requires the presence of an oxygen ambient. The lateral resolution and thickness of these patterns depend on the electric field between the tip and surface. The field strength is ultimately determined by the tip geometry, tip-sample distance, setpoint current I, and the bias voltage V_b. An example of a pattern written and subesquently imaged by an STM on H-passivated silicon substrates is illustrated in Figure 2.

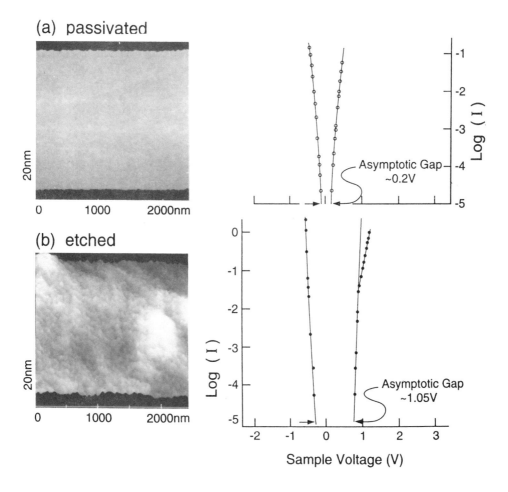

Figure 1. STM image and I-V characteristics of (a) passivated and (b) etched n-type GaAs (100) surfaces. The image of the etched surface reveals that the thin oxide covering the surface possesses a highly nonuniform conductivity compared to the passivated surface. The spectroscopy indicates that current transport through the etched surface oxide is completely dominated by surface states, whereas the passivated surface exhibits I-V characteristics which are representative of the bulk semiconductor properties.

The role of oxygen in the modification process was established by O_2/N_2 partial pressure experiments, which indicated that the process is oxygen coverage-dependent. Imaging time of flight secondary ion mass spectrometry (TOF SIMS) ion maps have shown that the STM-patterned regions are sources of enhanced oxygen ion yield, relative to the unpatterned substrate and the thickness of the patterned regions on the order of one to two monolayers[16].

Arsenic-capped, MBE-grown GaAs samples were also patterned[12] using the STM, as shown in Figure 3. The resolution limit of the lithography is clearly determined in this case by the electrical, rather than the topographical, uniformity of the surface. Indeed, we found that the thin As_xO_y-rich layer formed on this surface only partially inhibits the segregation reactions which usually occur on GaAs surfaces. However, by growing a 12 monolayer-thick, epitaxial InGaAs layer between the GaAs and the arsenic cap, we were able to better control the oxidation of the underlying GaAs (see next section).

We also observed that the As_xO_y is only loosely bound to the substrate and is easily swept out of the STM scan region at the sample bias voltages used for imaging, $V_b \sim 3 - 4$ V.[9] This action is very different from the lithographic process which requires $V_b \sim 4.5 - 6.5$ V. The patterns formed were again identified by TOF SIMS as a stable oxide phase, probably Ga_2O_3.

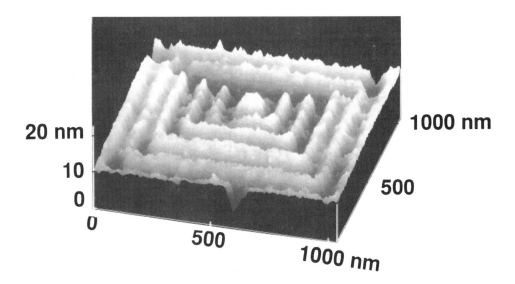

Figure 2. Feature patterned on H-passivated, n-Si (111), $\rho = 10$ Ω-cm, using a STM operating in air. The feature shown in this STM surface plot was produced using a bias voltage $V_b = 3.0$ V (tip positive) and a setpoint current I = 3.0 nA. The feature was then imaged at $V_b = 1.7$ V, I = 3.0 nA. It consists of 35-nm wide lines (FWHM) converging to a 65-nm wide square. The average apparent peak-to-valley height of the lines as measured by the STM is roughly 2 nm.

PATTERN TRANSFER

The chemical modification of a surface by an STM has technological significance only if it can be *accessed*, for example, in a subsequent processing step. In this section we discuss two examples which illustrate pattern transfer using ultrathin oxide masks on silicon and GaAs.

For the first example of pattern transfer, we have employed the STM-patterned oxide as a mask for selective-area GaAs heteroepitaxy on silicon, an essential step in mating

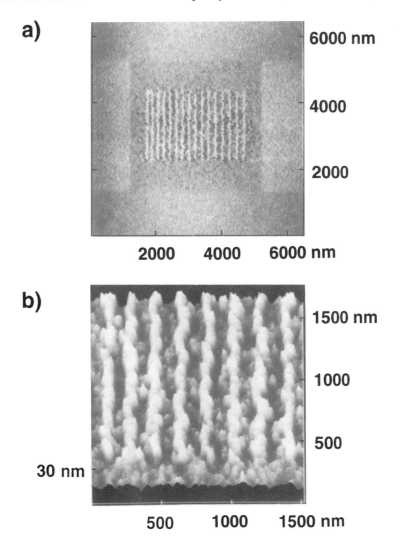

Figure 3. (a) STM image of an oxide mask patterned onto an As-capped, MBE-grown GaAs (100) surface. Writing was performed with V_b = 5.5 V and I = 0.4 nA, and imaging was carried out at V_b = 3.8 V and I = 0.4 nA. (b) Detail of the lower lefthand corner.

GaAs and silicon device technologies[15]. After an intial GaAs MBE growth, verified by reflection high-energy electron diffraction, the sample was removed from the growth chamber and examined by SEM. The gallium atoms have a much greater mobility on the SiO_x relative to the bare silicon surface and therefore GaAs is not formed on the oxide. SEM and TOF SIMS clearly revealed that no GaAs grew where 1x1-μm^2 square

a) STM image

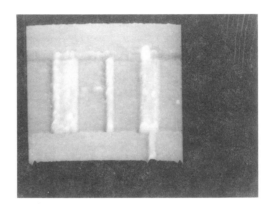

b) SEM (after etching)

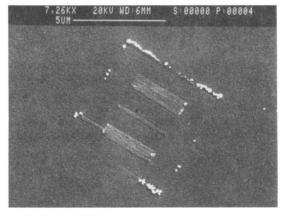

c) SEM (detail)

Figure 4. (a) STM image of an STM-patterned oxide mask on an As:InGaAs-capped GaAs (100) surface. (b) and (c) SEM images of the RIE-etched substrate.

masks had been produced by exposing the silicon substrate to a dose of at least $V_b = 3.2$ V.

Oxide masks are already being employed in use for *in situ* RIE etching of e-beam and focused ion beam direct write patterns. We have shown that the STM-patterned oxide on GaAs can survive the RIE processing environment as well.[7,9] Figure 4 (a) shows an STM image of an oxide mask STM-patterned on an As:InGaAs-capped GaAs sample grown by MBE. The lithography was carried out with a bias voltage $V_b = 5.5$ V and imaged with $V_b = 3.8$ V. The sample was then placed in an RIE chamber and etched using BCl_3. The sample was subsequently coated and imaged by SEM, Figure 4 (b) and (c).

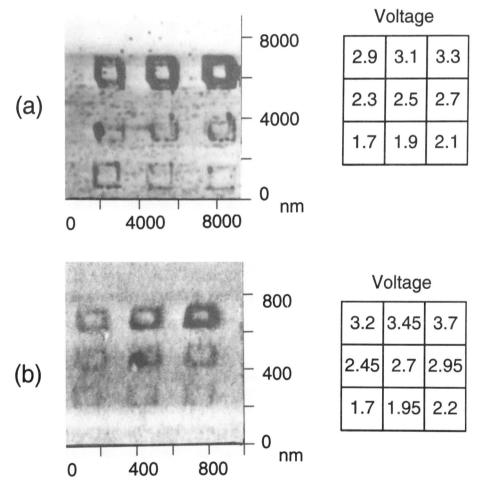

Figure 5. Open squares patterned on H-passivated Si using the lithography system described in Figure 5. (a) Lithography was carried out using a mechanically cut PtIr tip, whereas in (b) lithography was performed using an electrochemically polished tip. At the high bias voltages employed in lithography, it is essential that the electric field between the tip and sample be as symmetric as possible.

Although this work is only in the most preliminary stage of development, we feel that a key point has been made by these demonstrations: not only is it possible to refine the STM lithographic process to produce smaller features, but that it is possible to routinely examine and further modify these structures by traditional means.

PROCESS INTEGRATION

In principal, STM lithography offers a number of advantages in environmental flexibility and instrumental simplicity over e-beam and focused ion beam methods. Coupled with hardware evolution, such as the multi-tip array concept[6,7], it is possible that proximal probe lithography may become the technology of choice for certain operations. Ultimately, however, all processing must be carried out under highly controlled conditions. With this in mind we have designed a novel STM and control electronics system to investigate STM-based nanostructure fabrication.[2] The design goals for this project include: (1) operation throughout a pressure range extending from ultrahigh vacuum to atmospheric; (2) ability to exchange tips and samples; (3) lateral stability of 0.1 nm; (4) a 10-μm scan range or better with high scan speeds; (5) accurate positioning of the tip over a 1-cm^2 area of the sample; (6) pattern generation software and electronics; and, (7) compability with commercially available STM imaging systems. The lithography system generates the x,y displacement vectors and outputs the appropriate voltages to the piezotube scanner electrodes. The sample bias voltage, writing speed, and number of repetitions are also controlled by the lithography software, as shown in Figure 5. Note the lateral scales of each figure. The silicon oxidation process, described above, was peformed at the bias voltages indicated on the righthand side of the figure, after which the entire area was imaged at $V_b = 1.7$ V. The line writing speeds for Figures 6 (a) and (b) are 1.7 μm/s and 64 nm/s, respectively.

These unique operational and instrumental characteristics of our STM are now being exploited for nanostructure fabrication.

SUMMARY

Nanometer-scale STM pattern generation on H-passivated silicon and As-capped GaAs substrates has been presented. TOF SIMS and SEM analysis have shown these features to consist of an oxide layer, approximately one to two monolayers deep. Two examples of pattern transfer were discussed: First, the use of the patterned oxide as a mask for selective-area GaAs MBE, and second, for reactive ion etching of GaAs. A novel system for performing STM-based lithography developed in our laboratory is described and preliminary tests using the oxide patterning of silicon were described. Although the field of STM applications is a new one, we have shown that the STM is a flexible tool for the fabrication and characterization of semiconductor surfaces and nanostructures.

ACKNOWLEDGMENTS

The authors acknowledge the significant and ongoing contributions to this work made by our colleagues at NIST and NRL: W. Tseng, J. Bennett, M. Chester, C. R. K. Marrian, O. Glembocki, and E. Dobisz.

REFERENCES

1. "Scanning Tunneling Microscopy of Passivated GaAs under Ambient Conditions", J. A. Dagata, W. Tseng, and R. M. Silver, *J. Vac. Sci. Technol. A* 11, xx-xx (1993).

2. "STM-based Nanostructure Fabrication System", J. Schneir, J. A. Dagata, and H. H. Harary, *J. Vac. Sci. Technol. A* 11, xx-xx (1993).

3. "X-ray Photoelectron and Auger electron Spectroscopy Study of UV/ozone oxidized, $P_2S_5/(NH_4)_2S$ treated GaAs (100) Surfaces", M. Chester, T. Jach, and J. A. Dagata, *J. Vac. Sci. Technol. A* 11, xx-xx (1993).

4. "Optical Characterization of the Electrical Properties of Processed GaAs", O. J. Glembocki, J. A. Dagata, E. S. Snow, and D. S. Katzer, *Appl. Surf. Sci.* 63, 143-152 (1993).

5. "Ambient Scanning Tunneling Spectroscopy of n- and p-type Gallium Arsenide", J. A. Dagata and W. Tseng, *Appl. Phys. Lett.* 62, xx-xx (1993).

6. "E-beam Lithography with the Scanning Tunneling Microscope", C. R. K. Marrian, E. A. Dobisz, and J. A. Dagata, *J. Vac. Sci. Technol. B* 10, 2877-2881 (1992).

7. "STM Lithography: A Viable Lithographic Technology?", C. R. K. Marrian, E. A. Dobisz, and J. A. Dagata, *Proc. SPIE* 1671, 166-177 (1992).

8. "The Effects of P_2S_5 Surface Passivation on Dry-etched GaAs", O. J. Glembocki, J. A. Dagata, E. A. Dobisz, and D. S. Katzer, *MRS Symp. Proc.* 236, 217-222 (1992).

9. "Integration of STM-based Nanolithography and Electronics Device Processing", J. A. Dagata, W. Tseng, J. Bennett, E. A. Dobisz, J. Schneir, and H. H. Harary, *J. Vac. Sci. Technol. A* 10, 2105-2113 (1992).

10. "STM Imaging of Passivated III-V Semiconductor Surfaces", J. A. Dagata, W. Tseng, J. Bennett, J. Schneir, and H. H. Harary, *Ultramicroscopy* 42-44, 1288-1294 (1992).

11. "P_2S_5 Passivation of GaAs Surfaces for Scanning Tunneling Microscopy in Air", J. A. Dagata, W. Tseng, J. Bennett, J. Schneir, and H. H. Harary, *Appl. Phys. Lett.* 59, 3288-3290 (1991).

12. "Nanolithography on III-V Semiconductor Surfaces Using a Scanning Tunneling Microscope Operating in Air", J. A. Dagata, W. Tseng, J. Bennett, J. Schneir, and H. H. Harary, *J. Appl. Phys.* 70, 3661-3665 (1991).

13. "STM Pattern Generation on Silicon and GaAs Surfaces", J. A. Dagata, J. Schneir, H. H. Harary, C. J. Evans, M. T. Postek, J. Bennett, and W. Tseng, *Scanning* 13, I71 (1991).

14. "Pattern Generation on Semiconductor Surfaces by a Scanning Tunneling Microscope Operating in Air", J. A. Dagata, J. Schneir, H. H. Harary, J. Bennett, and W. Tseng, *J. Vac. Sci. Technol. B* 9, 1384-1388 (1991).

15. "Selective-area Growth of Epitaxial Gallium Arsenide on Silicon Substrates Patterned using a Scanning Tunneling Microscope Operating in Air", J. A. Dagata, W. Tseng, J. Bennett, C. J. Evans, J. Schneir, and H. H. Harary, *Appl. Phys. Lett.* 57, 2437-2439 (1990).

16. "Modification of Hydrogen-passivated Silicon by a Scanning Tunneling Microscope Operating in Air", J. A. Dagata, J. Schneir, H. H. Harary, C. J. Evans, M. T. Postek, and J. Bennett, *Appl. Phys. Lett.* 56, 2001-2003 (1990).

17. C. F. Quate, in **Highlights of the Eighties and Future Prospects in Condensed Matter Physics**, edited by L. Esaki, NATO ASI (Plenum NY 1991).

18. E. C. Teague, in **Proc. of Scanned Probe Microscopy: STM and Beyond**, (AIP NY 1992).

19. An excellent review (in Japanese) concerning the use of the STM as a diagnostic for the preparation of silicon surfaces appeared in *Nikkei Microdevices* (April 1992), pp. 64-80, by K. Kajimura. The article contains an extensive set of up-to-date references in English.

Scanning Tunneling Microscope-Based Nanolithography for Electronic Device Fabrication

Joseph W. Lyding, Roger T. Brockenbrough, Patrick Fay, John R. Tucker, and Karl Hess

University of Illinois, Department of Electrical and Computer Engineering and Beckman Institute, Urbana, Illinois, 61801

Ted K. Higman

University of Minnesota, Department of Electrical Engineering, St. Paul, Minnesota

Abstract:

The scanning tunneling microscope (STM) has evolved from a new surface science tool into one capable of performing surface modifications down to nanometer and even atomic dimensions. This suggests the possibility of developing STM-based lithography strategies for the fabrication of electronic device structures with nanometer scale features. On this size scale, it should be possible for quantum interference and coulomb charging effects to influence device function at 77 K or even higher temperatures. This chapter reviews our progress towards merging STM nanolithography with conventional processing technologies to create devices whose electronic function depends on their nanofabricated features. We have successfully patterned the gate regions of silicon MOSFET devices which were fabricated with provisions for STM modification.

1. Introduction

The scanning tunneling microscope (STM) is evolving from a basic surface science tool[1] into one capable of performing lithography down to the atomic size scale.[2] Simultaneously, the endless tyranny of shrinking integrated circuit feature sizes is putting great stress on conventional lithography schemes. Currently it remains unclear as to whether schemes based on optical projection, electron beam, or x-ray lithography will perform satisfactorily as the sub-0.1μm regime is entered over the next decade. Furthermore, device function will change in this new size regime as quantum size and discrete charging effects begin to play a significant role. These factors have influenced our efforts to develop STM nanolithography into a tool compatible with technologically important device substrates, i.e. silicon and III-V compound semiconductors. In shaping this effort, we seek to merge the STM nanolithography capabilities with established semiconductor device fabrication technology and theory, as depicted schematically in figure 1.

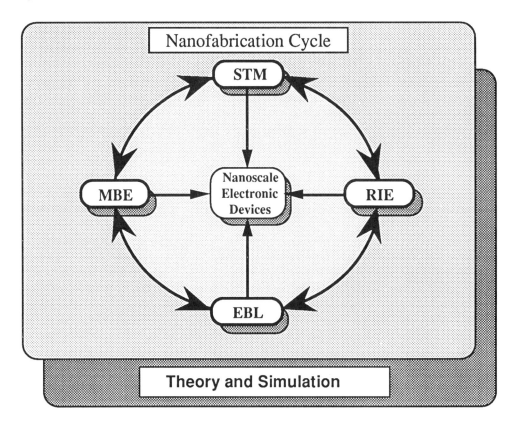

Fig. 1: Nanofabrication cycle depicting the merger of STM nanolithography with established semiconductor growth and processing techniques including molecular beam epitaxy (MBE), reactive ion etching (RIE) and electron beam lithography (EBL). This merger relies on a strong foundation of theory and device simulation to create new nanoscale electronic devices.

Several important technological considerations must be addressed in order to achieve the interplay depicted here. First, and foremost, the STM must be equipped with a coarse translation capability in order to locate devices that have been pre-patterned by other techniques, such as electron beam lithography (EBL). In general, the atomic resolution scan area of the STM is quite small (typically 10 μm x 10 μm) on the scale of a typical substrate. In the absence of coarse translation, this would necessitate precise mounting of device substrates on the STM sample holder in order to register individual devices within the STM scan area. Although the STM scan range can be extended to ~100 μm x 100 μm, atomic resolution work becomes nearly impossible, and device registration would still be difficult for experiments performed under remote conditions, as in an ultrahigh vacuum (UHV) chamber. Recently, we have developed a new inertial coarse translation system[3] that permits two-dimensional lateral motion of the 10 μm x 10 μm scan area of our STM over a 3 mm diameter area. Although lateral translation is a straightforward consideration for air-operated STMs, it can be quite a complex issue for UHV or other remotely operated instruments. Our translation mechanism works equally well for both UHV and air operated STMs, and we routinely use it to locate device structures for STM-patterning experiments, as shown in section 3 of this chapter.

Another key issue is maintaining surface integrity while transferring samples between different processing and analysis systems. Only a few materials such as graphite and MoS_2 can withstand ambient exposure and be subsequently imaged in the STM. Unfortunately, most of these materials are generally not used as electronic device substrates. On the other hand, silicon and GaAs-based substrates degrade immediately upon ambient exposure, and must be protected prior to transfer between separate systems. One means of surface protection is to use a portable ion pump or "vacuum suitcase" to carry samples between systems. This has been demonstrated for the case of GaAs, in which samples were transferred from a molecular beam epitaxy (MBE) growth system to a UHV-STM for characterization.[4] Other sample protection schemes are based on chemical passivation of the surface, or rely upon physically "capping" the surface with a relatively inert barrier layer. Several of these methods are compatible with STM imaging, and, in some cases, the surface treatment plays an integral role in the lithography. As an example, Dagata, et al.[5] have developed wet chemical methods for passivating GaAs and InGaAs substrates. The have shown that the passivation layer acts as an inorganic resist that can be "exposed" by STM, resulting in pattern definition down to ~45 nm. Arsenic capping of MBE grown III-V compound semiconductor substrates has also been demonstrated as a suitable surface protection scheme. After transferring into a UHV-STM system, the capping layer can be removed by heating to a relatively low temperature, exposing the substrate for further processing. We have used this technique to prepare GaAs(100) surfaces for atomic resolution STM imaging as shown in figure 2. The technique works for capped samples that are stored under dry nitrogen conditions for periods ranging from weeks to several months.

Silicon substrates can also be protected from ambient exposure. Standard wet chemical cleaning procedures use an HF etch to remove surface oxide while hydrogen terminating the dangling bonds at the surface. Becker, et al.[6] have demonstrated atomic resolution STM imaging of a cleaved Si(111) sample that was wet chemically passivated and subsequently transferred into a vacuum STM. As with the III-V substrates, Dagata[7] has shown that chemically passivated silicon is amenable to STM-patterning on the 100 nm size scale. In these experiments, the electric field of the STM locally breaks down the surface passivation layer, stimulating the growth of a thin oxide.

Fig. 2: An atomic resolution STM image of a GaAs(100) surface that was As-capped in the MBE growth system. These samples are typically stored for several weeks in a dry N_2 box. Sample mounting and transfer operations are carried out under ambient conditions. (Sample courtesy of Prof. K. Y. Cheng)

To date a wide variety of STM lithographic schemes have been demonstrated. Many of these are described in the chapters of this book. Our objective has been to build on this base nanolithography schemes that are suitable for the processing of silicon substrates. In the next section we describe the STM nanolithography scheme that we have adapted for this purpose, and in section 3 we demonstrate its application to silicon MOSFET device structures. In Section 4 we discusses future directions for the development of STM nanolithography.

2. STM Nanolithography on Silicon

Over the past few years a number of groups have explored STM lithography as a method to modify a wide variety of surfaces. The techniques used have ranged from direct mechanical modification to the use of electric fields and electron flow to directly alter the surface or stimulate local chemistry with an adsorbed or deposited layer. Our objective is to develop a robust, STM-based patterning scheme suitable for processing chemically passivated silicon surfaces under normal ambient conditions. In addition to resolution and repeatability, this scheme must enable pattern transfer into the silicon substrate. We have developed such a scheme in which the STM is used to stimulate the local growth of oxide. Our technique for patterning passivated silicon surfaces is based on the work of Dagata's group at NIST.

Dagata's et al. at NIST [5,7,8,9] have used STM nanolithography to chemically modify passivated Si and III-V compound semiconductor surfaces. Their work opens new opportunities for nanolithography on semiconductor electronic devices. On silicon substrates they were able to write patterns with ~100 nm resolution by raising the probe voltage from a lower imaging level to a patterning bias of ~3-4 V (tip positive). This causes the passivation layer to break down, promoting the formation of a 1-2 ML thick SiO_x layer as verified by TOF-SIMS measurements. They have also achieved pattern transfer by demonstrating selective nucleation of GaAs during MBE growth outside of the SiO_x regions on STM-patterned silicon substrates. In our work, we use a negative tip bias of ~2 V for patterning and 1 V for imaging. Figure 3 shows an STM image of a H-passivated substrate patterned in this manner. The line pitch is ~100 nm, however it is clear that the linewidth is much narrower, on the order of 20 nm. The patterned lines appear dark, indicating that either material has been removed from the substrate or that an electronic change consistent with the formation of a more insulating region has occurred. Using a variety of methods we have determined that the STM- patterned areas are actually topographic protrusions composed of oxidized silicon. Professor C. F. Quate, at Stanford, has performed AFM measurements on our STM-patterned samples to determine that the STM-patterned lines extend 2-3 nm above the surrounding substrate. To determine their chemical nature, large areas were patterned and the substrates were then subjected to a repeated sequence of ion milling and Auger analysis. These experiments indicated the presence of SiO_x of ~7-10 nm thickness. The apparatus was calibrated using standard oxidized silicon wafers of known SiO_2 thickness. This result is in reasonable agreement with the AFM measurement since the ion milling rates for SiO_x and SiO_2 might differ, and the Si/SiO_2 interface moves into the substrate during oxide growth, thereby making the AFM measured corrugation smaller than the true oxide thickness. Experiments are currently in progress to determine the mechanism of this STM-induced oxidation. So far, our experiments indicate that the electric field of the tunnel junction plays a more dominant role than electron energy in stimulating local oxidation. We are as yet unsure whether the reaction mechanism involves just oxygen or, in addition, adsorbed water as was demonstrated for the case of graphite.[10,11]

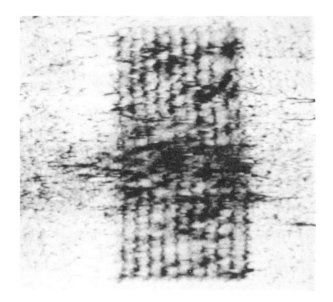

Fig. 3: STM image of STM-patterned H-passivated silicon surface showing dark patterned lines on a 100 nm pitch. Patterning was performed at 2 V (tip negative) and the image was acquired at 1 V.

We have also demonstrated pattern transfer for STM-patterned silicon substrates. Figure 4 is an SEM image showing clear evidence of the STM-patterned lines after the growth of a 15 nm thick thermal oxide and subsequent metallization for SEM purposes. Recently, we have also shown that the STM patterning serves as an effective mask for chlorine reactive ion etching.[12]

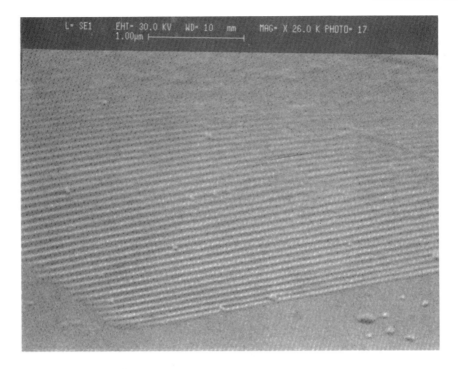

Fig. 4: SEM image of STM-patterned H-passivated silicon surface after a 15 nm thick dry oxide is grown and the surface is metallized for SEM imaging.

In this section we have demonstrated an STM-based scheme that is suitable for patterning and pattern transfer on silicon surfaces. In the next section, we apply this technique to the modification of a silicon MOSFET.

3. Silicon MOSFET

The silicon MOSFET is the primary building blocks for most of today's integrated circuits. It is particularly amenable to high resolution lithography schemes since current flow is controlled at the oxide/silicon interface. In addition to scaling down device dimensions, these devices also present the opportunity to explore new realms of device function. For example, a periodic modulation of the gate oxide thickness can be used to establish a corresponding potential modulation in the channel region. This modulation can produce coherent quantum interference, leading to negative resistance when the wavelength of the carriers coincides with that of the imposed grating. These 'surface superlattice' effects have already been observed at ~1 K in Si MOSFETs by

Warren et al.[13] in which EBL was used to create a 200 nm period buried gate. It should be possible to observe these effects at higher temperatures by reducing the period. To pursue this goal using STM lithography, we have fabricated MOSFET structures that incorporate STM-patterning into the fabrication sequence. Figure 5 shows the device crossections leading up to the STM-patterning step. The source/drain n+ ion implantation is masked by patterned silicon nitride. Thermal annealing activates the implant while simultaneously forming oxide over the doped regions. After etching down to bare silicon, the doped regions appear as ~200 Å deep depressions in the surface. The substrate is then ready for STM-patterning of the channel region, followed by thermal oxidation and contact metallization. Figure 6 shows a diagram of the top view of a finished device, an SEM micrograph of a device ready for STM-patterning, and an optical micrograph of a finished, metallized device.

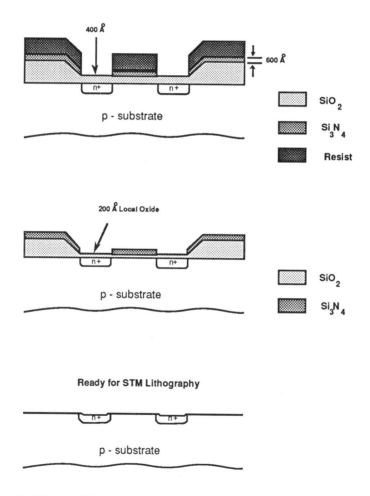

Fig. 5: Silicon MOSFET fabrication sequence in preparation for STM-patterning after removal of the nitride and oxide layers.

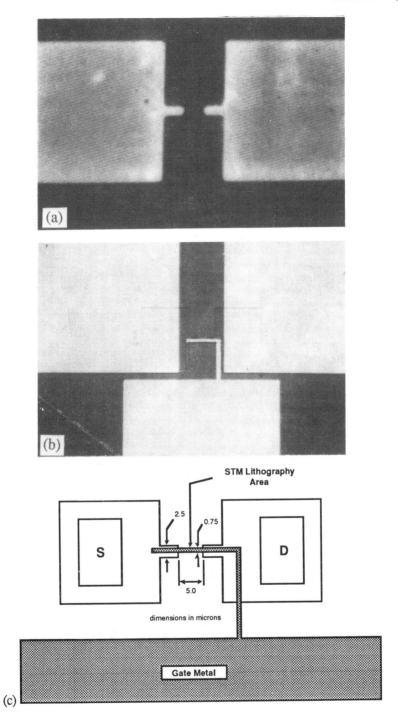

Fig. 6: a) SEM micrograph of a MOSFET device ready for STM-patterning, b) optical micrograph of a finished device after metallization, and c) diagram of the topview of the MOSFET.

Using the coarse translation capability of our STM, we can locate the highly doped source/drain depressions and then center the scan area over the channel region as shown in figure 7a. Figure 7b shows an STM image of a 100 nm period transverse grating written by STM in the channel region using a tip voltage of -2 V.

a)

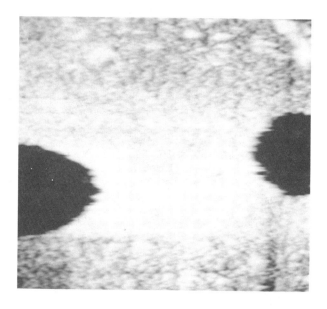

b)

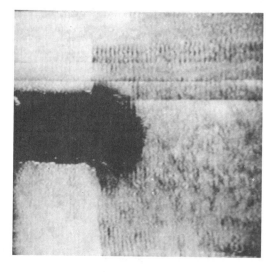

Fig. 7: a) A 10 μm x 10 μm STM image of the channel region of a MOSFET located using the STM coarse translation capability. b) STM image of a STM-patterned transverse grating written in the channel region.

After STM-patterning, device processing continues with the growth of a thermal gate oxide followed by contact metallization. Figure 8 shows SEM images of the gate region of a device after these steps. The STM-patterning is clearly visible, even through the 45 nm thick gate metal.

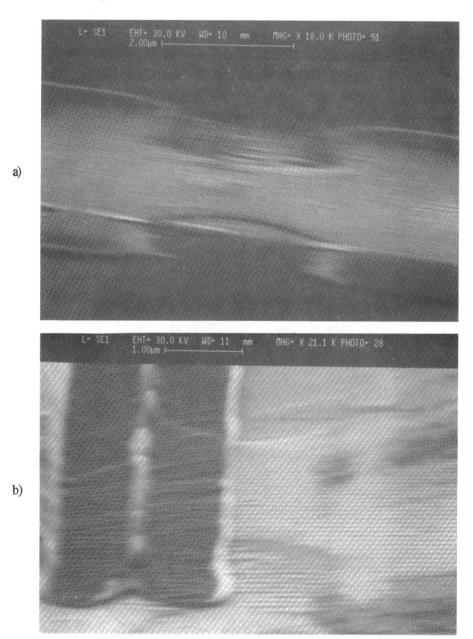

Fig. 8: SEM images of a) the MOSFET gate region after STM-patterning and subsequent growth of 15 nm of thermal oxide, and b) the gate region after 45 nm thick metallization.

An eventual goal of this effort is to pattern transverse gratings on a 10 to 20 nm period in order to observe electron interference effects at temperatures above liquid 4.2 K. To date, we have achieved a linewidth of ~20 nm on a pitch of 80 nm. In the near term, however, we are experimenting with longitudinal patterned MOSFETs where the STM-patterned grating is written parallel to the channel region. The simplest model of the finished device would be two electrically parallel MOSFETs, each having a different threshold voltage that is determined by the thickness modulation of the patterned oxide. Figure 9 shows a comparison of the room temperature electrical characteristics of a longitudinal STM-patterned MOSFET and a control device fabricated on the same test chip. There is very little difference between these characteristics, and both devices have about the same

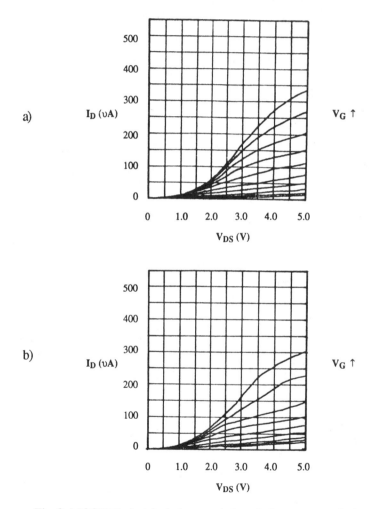

Fig. 9: MOSFET electrical characteristics; drain current vs drain-to-source voltage for a) a control device, and b) an STM-modified device with a longitudinal grating in the channel region, on the same chip.

threshold voltage, $V_T \sim 1.1$ V. For an oxide thickness modulation of ~5nm, and substrate doping level (10^{15} cm^{-3}), we expect $\Delta V_T \ll kT$ for these room temperature measurements. However, this result does show that the STM-patterning does not adversely effect the electrical characteristics of the MOSFET. Recently, we have fabricated similar devices using highly doped substrates (10^{17} cm^{-3}), for which we expect $\Delta V_T > kT$. Electrical measurements are now in progress for these devices.

4. Future Directions

There are many avenues for STM-based and other scanned probe lithography schemes as evidenced by the chapters in this book. Our effort is being shaped by the desire to merge STM nanolithography with conventional device substrates and processing techniques. Coarse lateral translation of the STM scan area allows us to locate and pattern devices, such as the silicon MOSFETs described in Section 3. An important direction of this research is toward smaller feature sizes. We feel that UHV conditions are best suited for the highest resolution lithography on device substrates. Under UHV conditions, surface and probe preparation and chemistry can be precisely controlled. UHV-based device fabrication schemes are also under consideration by industry to combat the spiraling costs associated with providing ultra-clean environments for sub-μm processing.

We are currently developing a multiple chamber UHV-STM system that allows for UHV transfer of samples and probes between preparation and experimental facilities. A diagram of this facility is shown in figure 10. At present, the UHV cryogenic STM is still under construction as is chamber B, which will be devoted to sample and tip preparation. We have initiated our UHV-STM lithography work on silicon substrates in this system. Figure 11 shows an example of the letters 'UI' patterned into a Si(111) 7x7 surface using one of the UHV-STMs in chamber A. This pattern was written using a sharp W(111) probe with a tip radius of curvature of ~25nm, after chemical etching and ion milling. The pattern is composed of overlapping dots, each of which was written at a tip bias of -8 V, 50 pA, for a total dose of 5×10^{-10} C. The linewidth is ~3 - 4 nm and the pattern appears to form as a result of the regrouping of Si adatoms in the high field region near the tip apex. No variations in tip quality for patterning ability are observed during the course of several days of experimenting with the same tip and sample.

124 / TECHNOLOGY OF PROXIMAL PROBE LITHOGRAPHY

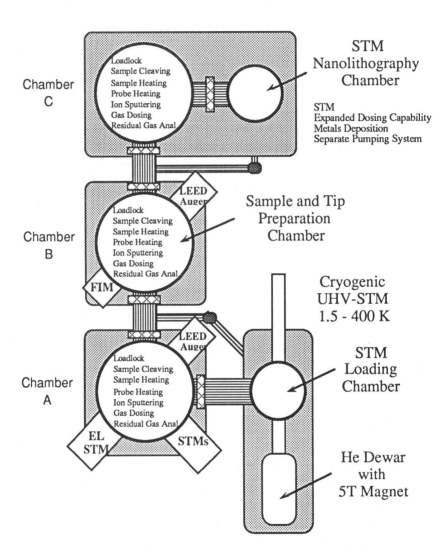

Fig. 10: Diagram of the multi-chamber UHV-STM system under construction at the University of Illinois for the development of high resolution STM nanolithography schemes.

Fig. 11: Si(111) 7x7 surface with the STM-patterned letters 'UI' written at a tip bias of -8 V, 50 pA, and a dose of 5×10^{-10} C for each of the overlapping dots that compose the pattern. This experiment was performed in chamber A of the system diagrammed in figure 9.

A longer term objective of our nanolithography effort is to fabricate electronic devices that use the coulomb blockade effect to control the switching of logic levels. In effect, an STM tunnel junction is perhaps the world's smallest two-terminal device. By separating tunnel junctions with a capacitively biased small metal island, it should be possible to create a single-electron transistor as proposed by Likharev[14]. Recently, Tucker[15] has proposed a similar scheme which uses sizeable numbers of electrons and therefore may prove easier to implement in the near term. Furthermore, he proposes a planar fabrication scheme that is well suited for STM lithography.

Acknowledgements

We have benefited from discussions and interactions with John Dagata, Christie Marrian, and Cal Quate. We thank Phil Scott for the preparation of single crystal tungsten tips and assistance with analyzing the STM-patterned oxide on silicon. We thank K.-Y. Cheng for the As-capped GaAs(100) samples. Funding for this work has been provided by the Office of Naval Research under contract N00014-92-J-1519 and the National Science Foundation under contract NSF ECD 89-43166.

References

1. G. Binnig and H. Rohrer, Helv. Phys. Acta **55**, 726 (1982).
2. For a review of STM nanolithography techniques see Shedd, G. M. and Russell, P. E., Nanotechnology **1**, 63 (1990).
3. R.T. Brockenbrough and J.W. Lyding, in preparation for publication.
4. Pashley, M. D., Haberern, K. W., Friday, W., Woodall, J. M. and Kirchner, P. D., Phys. Rev. Lett. **60**, 2176 (1988).
5. Dagata, J. A., Tseng, W., Bennett, J., Schneir, J., and Harary, H. H., Appl. Phys. Lett., **59**, 3288 (1991).
6. R. S. Becker, G. S. Higashi, Y. J. Chabal, and A. J. Becker, Phys. Rev. Lett. **65**, 1917 (1990).
7. Dagata, J. A., Schneir, J., Harary, H. H., Evans, C. J., Postek, M. T., and Bennett, J., Appl. Phys. Lett. **56**, 2001 (1990).
8. Dagata, J. A., Tseng, W., Bennett, J., Evans, C. J., Schneir, J., and Harary, H. H., Appl. Phys. Lett. **57**, 2437 (1990).
9. Dagata, J. A., Tseng, W., Bennett, J., Schneir, J., and Harary, H. H., J. Appl. Phys. **70**, 3661 (1991).
10. Albrecht, T. R., Dovek, M. M., Kirk, M. K., Lang, C. A. , Quate, C. F., and Smith, D. P. E., Appl. Phys. Lett. **55**, 1727 (1989).
11. C. F. Quate, private communication.
12. P. Fay, R. T. Brockenbrough, P. Scott, G. Abeln, and J. W. Lyding, in preparation for publication.
13. Warren, A. C., Antoniadis, D. A., Smith, H. I., and Melngailis, J., IEEE Electron Device Lett. **EDL-6**, 294 (1985).
14. K. K. Likharev, IEEE Trans. Magn. **MAG-23**, 1142 (1987).
15. J. R. Tucker, J. Appl. Phys. **72**, 4399 (1992).

Arrayed Lithography Using STM Based Microcolumns

T.H.P. Chang, L.P. Muray[†], U. Staufer[††], M.A. McCord and D.P. Kern

IBM T.J. Watson Research Center, Yorktown Heights, New York 10598, U.S.A.

[†] National Nanofabrication Facility at Cornell University,
Knight Laboratory, Ithaca, New York 14853, U.S.A.

[††] Institute of Physics, University of Basel,
Kingelbergstrasse 82, 4056 Basel, Switzerland

ABSTRACT

This paper outlines a novel method based on arrays of electron microcolumns for lithography in the 100nm and below linewidth regime. Throughput on the order of 10 to >50 wafers per hour for 100nm lithography on 200 mm wafers is believed to be achievable depending on the number of columns employed. It requires no mask and can be extended to the sub-100Å linewidth regime. It offers the potential of a breakthrough for a low cost high throughput manufacturing process for the new generation of ultra-high density devices.

The proposed microcolumns are to be based on a new concept which combines scanning tunneling microscope (STM), micro-fabricated lenses and field emission technologies to achieve a performance that is expected to surpass the conventional column. The proposed approach will embody an array of these microcolumns, each with its own pattern generation capability, operating in parallel at low voltage to achieve high throughput. The low voltage operation is attractive because proximity effect corrections may not be needed. In addition, an arrayed microcolumn system also has the potential of reducing the cost of the overall system through the compaction of the mechanical system.

INTRODUCTION

If current trends of integrated circuit development are to continue, sub-100nm minimum feature size lithography will be required in the not-too-distant future. Lithographic options to achieve linewidths around 100nm and below are very limited. Optical lithography is not expected to be extendable to this regime due to limitations of source, lens material and reflective optical issues. X-ray lithography [1,2] with a mask in close proximity to the sample - about ≤10 μm for sub-100nm lithography - is one of the techniques being explored. Although with a vanishingly small gap (contact), features as small as 20nm have been printed, accurate gap control needed for manufacturing use presents a challenge. Moreover, achieving accurate 1 x masks at sub-100nm dimensions and accuracy with aspect ratio larger than ≃4:1 on membranes transparent to x-ray is still a difficult task. Projection x-ray lithography[3] which utilizes reflective optics coated with dielectric multi-layers to enhance x-ray reflection may overcome some of the membrane mask problems since a reflective mask (solid) may be used and the projection optics involves demagnification. However, designing, building and assembling of these multilayer mirrors needed for the reflective optics remains a formidable, unproven task. Projection electron and ion beam systems have received renewed attention in recent years with the development of a prototype commercial ion projection system [4] and the development of a novel mask[5] for electron projection, but difficulties related to control of distortion in projection optics and level to level overlay remain.

Scanning electron beam lithography remains today the only demonstrated method for complex circuit fabrication in the sub-100nm regime. It suffers, however from serious throughput limitations and high system cost which limit its use mainly to mask making and for prototyping of experimental devices.

To overcome the throughput limitation, several schemes utilizing arrays of electron beamlets for lithography have been explored in the past, but they have so far not been successful. Some of these schemes used a single cathode to illuminate an array of lenslets to form a beamlet array[6,7] Performance of such an approach can be limited by the illumination system which, in some cases, has to cover a large number of lenslets distributed over a wide area. It is difficult in these cases to achieve beamlets with adequate resolution, current density, uniformity and stability. Multiple beam systems based on multiple cathodes and columns have also been explored. One scheme uses an array of field emission tips (evaporated metal [8] or etched silicon[9,10] which are fabricated together with self aligned extraction electrodes. By incorporating additional micro-electrodes to such a structure, arrays of probe forming columns at very high packing density can be obtained. A disadvantage of this scheme is that, since the tip and the extractor are fabricated together, it is difficult to independently control the radius, height and material of the tip and to apply the necessary tip annealing process to ensure properly defined and stable field emission. Another scheme suggests the use of an array of STMs[11] operating in the field emission mode in close proximity to the target to form an array of beamlets at the sample. Nanometer scale beam diameter with extremely high current density has been shown to be achievable[12,13]. However, the use of the sample as a part of the probe forming optics leads to severe tip/target interaction problems and, in addition, dependence of this scheme on precision mechanical scanning can result in severe throughput limitations. Efforts to eliminate the problems with the aforementioned schemes have led to the development of a new approach[14-24] which combines STM, field emission and microlenses (electrostatic or magnetic) to form microcolumns that are amenable to arrayed operation.

THE ARRAYED MICROCOLUMN LITHOGRAPHY

In a significant departure from conventional technologies, the proposed microcolumns are to be based on a novel STM aligned field emission (SAFE) concept[14-17] and microfabricated lenses to achieve a very significant reduction in physical dimensions (column length and diameter) to the order of millimeters. This miniaturization also results in reducing electron optical aberrations so that the performance of such a microcolumn is expected to surpass conventional columns. The proposed approach will embody an array of these microcolumns, each having a field emission tip as the source, with individual STM sensors and controls for emission current, height and x-y positioning, in combination with microminiaturized electron optical elements to form a focused beam which can be deflected and blanked to perform the task of an individual scanning probe. An array of these micro-beams can be used to generate patterns in parallel, one or more columns per chip, to achieve high throughput for nanolithography. It is envisaged

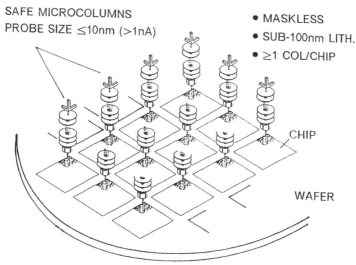

Figure 1: The basic concept of arrayed microcolumn lithography.

that one or more such microcolumns will be used per chip with pattern writing, overlay and column to column alignment (in the multi-column per chip case) performed on a chip by chip basis. Patterns can be written either in the vector or the raster mode with the beam scanned over only a narrow stripe (say ≤100μm width) and with a continuously moving laser controlled stage in the orthogonal direction to build-up the complete pattern using "stitching". Figure 1 shows the basic concept of this arrayed microcolumn approach[25]. The array can also be used for metrology, testing, inspection and other applications where the use of multiple beams for parallel processing is desired. While the micro-columns are not restricted to low voltage operation, low voltage, in the order of 1 keV has obvious advantages in terms of ease of miniaturization and in improving reliability and cost. Low voltage can also be attractive for lithography because proximity effect corrections may not need to be applied. In addition, major mechanical components of the system such as table, work-chamber, vacuum system and anti-vibration platform etc. can be significantly reduced in size due to the compactness of the column assembly and a much smaller table motion requirement. This can result in a compact overall mechanical system with important cost benefit.

The technical issues relating to this proposed approach will now be discussed in three parts: (1) the SAFE-microcolumn technology, (2) low voltage lithographic processes and (3) throughput considerations.

(1) The SAFE-MICROCOLUMN TECHNOLOGY

The basic configuration of a SAFE-microcolumn is shown in Figure 2. In this approach, the STM feedback principle is used for precision x, y, and z alignment of a field emission tip to a microlens to form a microsource which can be used by itself or in conjunction with other microlenses to form a focused probe of electrons. The microlenses can be made using many of the standard integrated circuit fabrication techniques on silicon or other suitable substrates with critical dimensions reduced to micrometer scale. As the lens aberrations generally scale with lens dimensions, such microlenses can be designed to have negligible aberrations resulting in exceptionally high brightness and resolution. The STM control provides several important additional advantages. Emission stability is improved by the STM feedback principle which automatically adjusts the z position through the piezo-element. This is particularly effective for stabilizing against low frequency drift and

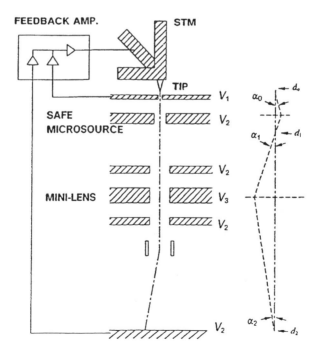

Figure 2: The basic configuration of a SAFE-microcolumn.

changes. The low extraction voltage ($\simeq 100$ V) sufficient for the microsource can significantly reduce the sputtering effects of ions bombarding the tip thus reducing noise. The STM also allows to choose tips from a wide range of materials, to move them away from the lens for tip annealing processing, to exchange tips and to accurately position them in close proximity to the microlens. In addition, there is also the possibility of using STM adjustments in x and y to compensate for lateral drift. As will be shown, SAFE columns are typically only a few millimeters in length and diameter. They are therefore highly amenable to arrayed operation and more easily isolated and shielded against the external mechanical, acoustic, and electromagnetic interferences.

Electron Optical Analysis

The overall performance of a microprobe system can be analyzed by first characterizing the two main components: the electron source and the probe forming lens (Figure 2). This involves potential calculations by the finite element method for the various source and lens geometries, and the use of paraxial trajectories and third order aberration theory to determine the optical properties[26-27]. Following the procedure generally used for the study of conventional field emission sources, the properties of the microsource are analyzed by replacing the emitter by a virtual source, d_o, of nominal properties[27] at the specified tip locations and the dual electrode accelerator structure performs as an imaging lens to form an intermediate image of the source - the errors introduced by this simplified assumption have been shown to be minor[27]. The intermediate image of diameter d_1 so formed represents the effective source size of the microsource and can be expressed as

$$d_1 = M_1(d_o^2 + d_d^2 + d_{so}^2 + d_{co}^2)^{1/2} \qquad (1)$$

where M_1 is the magnification of the microsource. d_o is the size of the virtual source generated by the tip field. $d_d (= 1.5/\alpha_0\sqrt{V_1})$ is the diameter (in nanometers) of the diffraction disk with V_1 being the voltage of the first electrode and α_0 being the beam semiconvergent angle at the tip. $d_{so}(= 0.5C_{so}\alpha_0^3)$ is the spherical aberration disk with a coefficient C_{so} referred to the object space. $d_{co}(= C_{co}\alpha_0\Delta V/V_1)$ is the chromatic aberration disk also referred to the object space with ΔV being the energy spread of the electron beam. The final probe size, d_2 can be expressed as

$$d_2 = (M_2 d_1^2 + d_{s2}^2 + d_{c2}^2)^{1/2} \qquad (2)$$

where M_2 is the magnification of the probe forming lens. $d_{s2}(= 0.5C_{s2}\alpha_2^3)$ and $d_{c2}(= C_{c2}\alpha_2\Delta V/V_2)$ are the spherical and chromatic aberration disks, respectively, of the probe forming lens with C_{s2} and C_{c2} being the spherical and chromatic aberration coefficients of the lens referred to the image space and V_2 being the final beam potential. Substituting d_1 in eq. (2) by eq. (1) and by replacing α_2 by

$$\alpha_2 = \frac{\alpha_1}{M_2} = \frac{\alpha_0}{M_1 M_2}\sqrt{\frac{V_1}{V_2}} \qquad (3)$$

an expression relating d_2 to the source can be established. In this expression, the only parameters that can be varied are α_0 and M_2. The parameters relating to the probe forming lens, C_{s2} and C_{c2}, remain fairly constant over a limited range of M_2 values, while the parameters relating to the source, M_1, C_{so}, and C_{co}, are fixed values for a given design of the source. Since the beam current I_2 is related to α_0 by

$$I_2 = \pi \alpha_0^2 dI/d\Omega_o \tag{4}$$

where $dI/d\Omega_o$ is the angular emission density of the tip, one can therefore relate d_2 to I_2 with M_2 being the only variable. An optimum M_2 value can be found for each I_2 value to give a minimum for the probe diameter d_2. The design procedures therefore involve, first of all, the optimization of the lens parameters for the source and the probe forming lens to achieve the lowest possible values for the aberrations followed by an optimization of the overall system. It should be noted that the current analysis does not include the space charge and electron-electron interaction effects. Due to the much reduced path length in the microprobe column, the interaction effects are expected to be less than those in the conventional system at comparable currents. However, as a much higher current appears to be feasible in the microprobe, these effects need to be studied in more detail.

The SAFE Microsource

As aberrations scale with lens geometry, reductions in aberrations can be achieved by miniaturizing the lens dimensions. For a field emission source consisting of a tip and a two-electrode immersion lens, traditional scaling requires that both the tip - lens spacing and the lens dimensions (electrode bore diameters and spacing) be uniformly reduced. In the SAFE approach, the reduction of tip - lens spacing can be achieved using the STM. In miniaturizing the immersion lens, two practical constraints need to be considered. (1) The minimum bore diameter of the electrode must be compatible with the tolerance of the microfabrication techniques employed. Taking 5 to 10 nm as the tolerance in edge roughness achievable using the best resist, electron beam lithography and RIE etching, a minimum bore diameter of approximately $1\mu m$ is considered as the practical limit. Since conventional lenses typically use a minimum bore diameter in the order of 1 mm, this constraint imposes a limit of scaling to approximately 1000. (2) Following the guideline practiced in conventional electron optical designs, the electric field between the electrodes needs to be restricted to a value not exceeding 10^4 V/mm to avoid breakdown. As many of the conventional lenses are designed fairly close to this field strength limit, very little room is therefore left for scaling. To overcome this problem, a selective scaling method has been developed. In this method the scaling is applied non-uniformly to two regions of the source:

(1) A strong reduction factor of approximately 1000x is applied to the region consisting of the first electrode and the tip, reducing the electrode thickness, bore diameter, and the tip spacing to the micrometer range.

(2) A relatively weaker reduction factor of a value ranging from a few times to 100x is applied to the second region, namely the second electrode (primarily the bore diameter) and the electrode spacing. The actual scaling factor used for the electrode spacing can be tailored to meet the guideline on field strength.

The end result of this selective scaling is an improvement in aberration coefficients (spherical and chromatic) by approximately three orders of magnitude from a conventional source while allowing the same potential to be applied. The reduced aberrations lead to an improvement in effective brightness by 2 to 3 orders of magnitude. Figure 3 compares a conventional 1 kV field emission source and the corresponding selectively scaled microsource for the same potential. The differences in geometry of these two sources are apparent. Table 1A and 1B give the spherical and chromatic aberration coefficients (C_{so} and C_{co} referenced to the object space) for the selectively scaled source and the conventional source, respectively, at three potentials, 200 V, 1 kV and 10 kV. The three orders of magnitude improvement in aberration coefficients for the selectively scaled sources can be readily seen. The reasons for the working of selective scaling can be attributed to many factors. Qualitatively, one can point out that the major lens action of the source appears to be taking place at the first electrode region. Reduced extraction voltage ($\simeq 100$ V) generates low energy electrons which, coupled with high field and field gradients in this region, causes the bulk of the lens aberrations. The reduced geometry confines this region to the micrometer range thus leading to low aberrations. The second electrode, typically being at a higher potential, acts mainly as an accelerator for the electron and contributes little to the aberrations. Its position and diameter are therefore less critical.

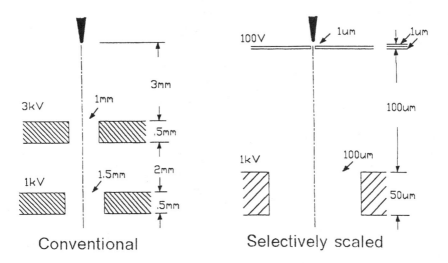

Figure 3: A comparison of the geometries of a conventional source and a selectively scaled SAFE microsource at 1 kV.

(A) MICRO-SOURCE:

V_1	V_2	D_1, t_1, Z_0	S	D_2	C_{so}	C_{co}
100V	200V	$\simeq 1\mu m$	$10\mu m$	$10\mu m$	$0.3\mu m$	$0.3\mu m$
100V	1kV	$\simeq 1\mu m$	$100\mu m$	$100\mu m$	$3.5\mu m$	$3.7\mu m$
100V	10kV	$\simeq 1\mu m$	$1000\mu m$	$1000\mu m$	$9.4\mu m$	$9.0\mu m$

(B) CONVENTIONAL F.E. SOURCE:

V_1	V_2	D_1, t_1, Z_0	S	D_2	C_{so}	C_{co}
3kV	200V	1-3mm	3mm	1.5mm	1400mm	200mm
3kV	1kV	1-3mm	3mm	1.5mm	70mm	20mm
3kV	10kV	1-3mm	9mm	1.5mm	10mm	2mm

Table 1: Spherical and chromatic aberration coefficients of selectively scaled microsources (1A) and conventional sources (1B). D_1, t_1, D_2 and t_2 are the diam. and thick. of the 1st and 2nd electrodes, s the electrode spacing and z_0 the tip distance.

Using the aberration coefficient values in Table 1 for the conventional source and the selectively scaled microsource at the three potentials, the effective brightness can be computed. The effective brightness is the brightness at the exit of the source with the aberrations of the source taken into consideration. As these aberrations depend on the semiconvergent angle α_0 at the tip, the results are plotted against this angle and shown in Figure 4. At low values of α_0, the diffraction effect dominates the performance yielding a relatively low effective brightness for both the conventional and the microsource. As α_0 increases, so does the effective brightness initially but it eventually begins to decline when the effects of the other aberrations (chromatic and spherical) start to exert themselves. The peak effective brightness is reached when the effective source size is at a minimum. The conventional source with much higher aberrations reaches the peak much earlier than the microsource. Thus, at the three potentials studied, the microsource offers a 2 to 3 orders of magnitude improvement in peak effective brightness over the conventional source.

The above analysis is based on a conventional tip with a tip radius of $\simeq 100$ nm, an angular emission intensity of the order of 1×10^{-4} A/ster. and an energy spread of 0.2 eV. Further improvements in the effective brightness can be achieved with a better tip technology. A recent breakthrough in such a technology was reported [19, 28]. This involves the formation of ultrasharp monatomic tips which terminated with a single atom on single crystal W<111> wires. With these ultrasharp tips, an extraction voltage of only a few tens of volts is needed for field emission. The unique electric field distribution at the tip also channels the electrons into a narrow cone of emission ($\simeq 2°$ half-angle). The superior performance of the single atom tip can lead to an additional improvement in brightness over the same source based on a conventional tip.

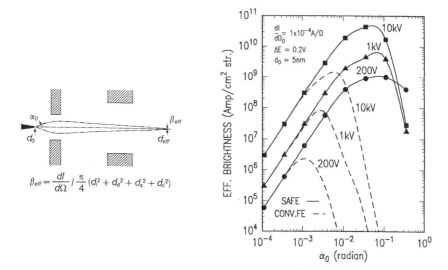

Figure 4: The effective brightness of SAFE and conventional source.

The Miniaturized Probe Forming Systems

In principle, the probe forming lens can be either a magnetic or an electrostatic lens. Magnetic lenses are the ones most commonly used in the conventional systems because they have lower aberrations, especially the spherical aberration at short working distances than their electrostatic counterparts. In addition, they are generally more reliable and easier to maintain and the magnetic body provides a useful added shielding against external electromagnetic interferences. Electrostatic lenses, on the other hand, have the advantages of simplicity, compactness, dissipating no power and, above all, they are more compatible with the microfabrication processes used for integrated circuits and are therefore amenable to a much higher degree of miniaturization. As miniaturization is an important theme of this study, electrostatic systems only will be addressed.

An electrostatic probe forming lens has been designed using symmetrical, einzel lens for the microcolumn, although other lens configurations such as asymmetrical einzel lenses or various forms of multiple-element lenses can also be used. A reduction in aberrations can be achieved in all these options by scaling the lens geometries. However, the extent by which the geometries can be beneficially reduced is set by factors other than aberration considerations. In the case of a symmetrical einzel lens, which has been extensively analyzed in the literature[29], it can be shown that the working distance in combination with minimum electrode spacing needed for a given potential, sets a limit to the miniaturization.

Figure 5 shows the characteristics of a symmetrical einzel lens with geometries of $R_1 = R_2 = P_1 = G/2$ and $P_2 = G$. Such a lens can be operated in either an accelerating or decelerating mode. As working distance W is a prime design consideration, it was found helpful to plot the relevant data, as in Figure 5, using W as the reference. It should be pointed out that a range of other lens geometries with R_1, R_2, P_1, and P_2 independently varied from G/4 to G have also been analyzed in the same manner and all yield fairly similar results to those shown in Figure 5. Therefore, Figure 5 can be taken as a reasonable representation for this group of designs. In a symmetrical einzel lens, the spherical aberration coefficient is always larger than the chromatic aberration coefficient. This is not necessarily a major limitation for a field emission system, since the chromatic aberration becomes increasingly important in defining the ultimate resolution especially at low voltages.

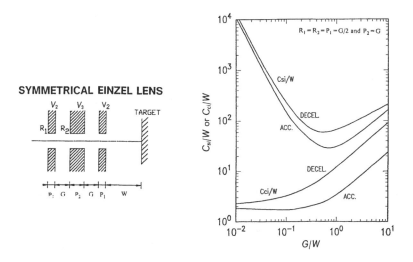

Figure 5: The spherical and chromatic aberration coefficients of a symmetrical einzel lens plotted against the electrode spacing G with the working distance W as the reference.

For a given operating potential, a minimum G value can be determined based on the guideline discussed earlier. Thus for a certain working distance W, a minimum G/W value can be established and the corresponding values of C_{si} and C_{ci} can be determined from Figure 5. Table 2 summarizes the geometries and performance of the einzel lenses for the miniaturized column at the working distances and potentials specified using this procedure. It can be seen that at low voltages, 200 V, and short working distances, a highly miniaturized lens with electrode radius in the 5 μm range and total lens thickness ($\simeq$4G) measuring 40 μm would be required. The size of the lens increases with increasing potential and working distance reaching an electrode radius of 0.5 mm and lens thickness of 4 mm at the other end. For comparison purpose, the performance of conventional magnetic lenses for the same working distances is also listed in Table 2. For the comparison to be meaningful, very aggressive parameters, including the use of 1 mm design rule for bore diameter and gap, have been applied and this set of data should represent an optimistic estimate of the best possible for the magnetic lenses at these working

MINIATURIZED EINZEL LENS ($R_1 = R_2 = t_1 = S/2$, $t_2 = S$)

	W(mm)	S(mm)	R(mm)	S/W	Csi(mm)	Cci(mm)
200V	0.1	0.01	.005	0.1	15	0.18
	1	0.1	.05	0.1	150	1.8
	10	1	.5	0.1	1500	18
1kV	0.1	0.1	.05	1	6.5	1.2
	1	0.1	.05	0.1	150	1.8
	10	1	.5	0.1	1500	18
10kV	0.1	1	.5	10	22	9
	1	1	.5	1	65	12
	10	1	.5	0.1	1500	18

CONVENTIONAL MAGNETIC LENS

W(mm)	Csi(mm)	Cci(mm)
0.1	5	2
1	10	3
10	20	10

Table 2: Performance of the probe forming lenses.

distances. As can be seen, the performance of the miniaturized einzel lenses is superior to that of the conventional lenses in terms of the chromatic aberration coefficient in the low potentials (200 V and 1 kV) and short working distances (0.1 mm and 1 mm) regimes and is comparable in the other regimes. Though the spherical aberration coefficients are higher, their effect on the ultimate resolution is not as serious as the numbers may have implied as pointed out earlier.

Based on the data for the microsource and the einzel lens for the three potentials and working distances, complete column configurations and performance can be defined using a program that takes the lens as well as the source aberrations into consideration and computes the optimum beam diameter for each beam current value by allowing the system magnification to be varied. The results are shown in Figure 6 together with a summary of the minimum beam diameters and the column lengths of the miniaturized columns for the three potentials and working distances. In all cases, the beam diameter reaches a minimum value which is dominated by diffraction in combination with chromatic aberration, in some cases also spherical aberration. This minimum value remains unchanged over a wide range of current at the low end as expected since the beam forming semi-convergent angle can be kept at the same optimum value by varying the system magnification. At the high current end, spherical aberration dominates and the beam diameter increases with increasing current. A knee shaped region representing the transition of these two effects occurs at a beam current range of approximately 10 to 100 nA. It can be seen that the optical performance of the miniaturized columns exceeds or approximately equals that of conventional systems at currents that are 2 to 3 orders of magnitude higher for the range of potentials and working distances studied.

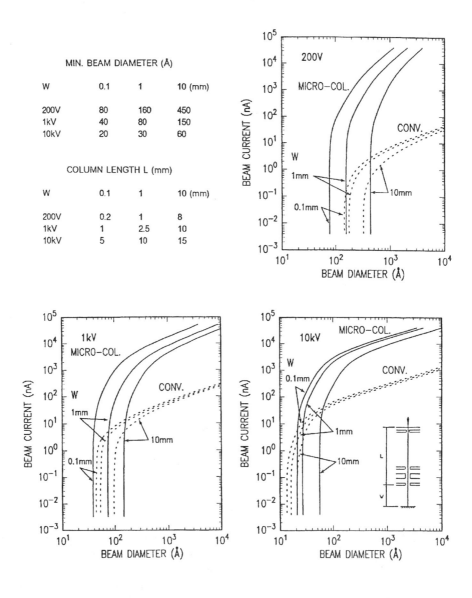

Figure 6: The performance of microcolumn and conventional field emission systems at three potentials (200V, 1kV and 10kV) and three working distances (0.1, 1 and 10 mm).

Experimental Evaluations

To demonstrate the feasibility of the SAFE concept and to evaluate the performance of a miniaturized electron beam column, a series of experiments have been conducted[22-24]. A block diagram of the complete test system is shown in Figure 7. The system consists of a UHV chamber which houses the STM, the microcolumn and the detectors. The electronic system can be broken into four

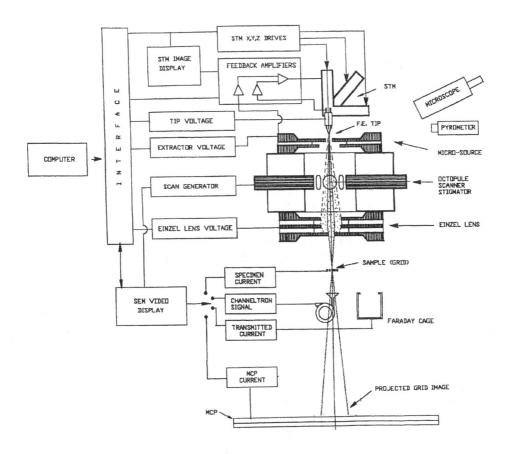

Figure 7: Block diagram of microcolumn test system.

functional sub-systems, all controlled by the host microcomputer. These are i) power supplies; ii) feedback and positioning hardware, coarse and fine; iii) scan generation electronics; and iv) image formation electronics, including detection and display components. All images were acquired digitally. An optical microscope was positioned outside the UHV chamber to assist tip alignment and a pyrometer was used to measure the tip temperature. A microchannel plate (MCP) was also incorporated into the system to allow projection studies of the column operation. Scanning images can be formed in transmission mode using a channeltron positioned behind the sample. Alternately, a specimen current amplifier can also be used for imaging in the reflective mode. All tests were performed in UHV at pressure below 5×10^{-9} torr. In the course of this experimental evaluation, processes for the fabrication of microlenses with bore diameter down to 1 μm based on silicon membrane technology, electron beam lithography, and RIE have been developed and a special field emission tip preparation and processing method tailored for microsource applications has been successfully formulated. Dual electrode microsources operating under the STM alignment principle have been demonstrated and prototype 1 kV micro-columns measuring only 2.5 mm in length have been successfully fabricated and demonstrated to be fully operational. While much remain to be learned, the work is rapidly entering a new phase where novel applications such as multi-beam lithography can be explored.

(a) Microsource and Lens Fabrication

As has been shown in the electron optical analysis, the critical dimensions, bore diameter and thickness, of the electrodes of the microlens need to be reduced to the 1 μm range. These electrodes, therefore, have to be fabricated from thin conducting membranes. A membrane material that has been studied for this application with encouraging result is boron-doped silicon membranes. These membranes provided a strong mechanical base; high conductivity; and compatibility with ultra-high resolution techniques. Furthermore, because the membranes were single-crystal material, in principle, atomically smooth round holes could be fabricated. The membranes were also not prone to surface diffusion and "whisker" growth as observed in some metal films. Metal-coated nitride films were not considered desirable primarily because of adhesion problems under high-field stress.

A schematic of the lens fabrication process is shown in Figure 8. The (100)-oriented Si membrane windows, 2 to 3 μm thick, were formed by anisotropically etching high concentration boron-diffused wafers in a mixture of pyrocatechol, ethylene diamine and water. For the patterning step, a 200 nm masking layer of silicon dioxide was deposited in a low-pressure chemical vapor deposition system and coated with 1 μm of poly-methyl methacrylate (PMMA). Using an electron beam system equipped with polar coordinate hardware, round patterns of the electrode holes were delineated. Calibration of the field size to one part in 10,000, produced holes with a roundness of better than 0.1% ($\Delta r/r$). Etching the patterns into the membranes was accomplished in three steps of reactive ion etching (RIE). First, the oxide was etched using the PMMA as a mask in CF_4/CHF_3(37%). This etch produced good selectivity, acceptable (20 nm/min) etch rates and good anisotropy. Next, the Si membrane was etched to a depth of 1-1.5 μm in $CBrF_6/SF_6$(16%). The final step was a backside etch to remove residual oxide and thin down the membrane to the desired thickness. A requirement for the process was that Si and SiO_2 have the same etch rates; otherwise, any residual oxide would lead to "grass" formation in the window. This requirement was

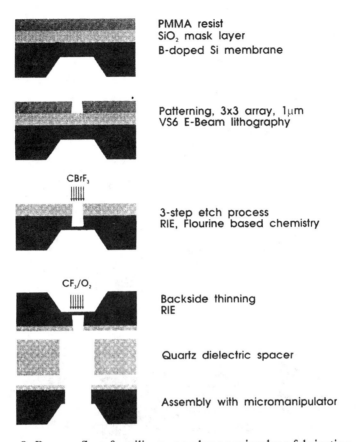

Figure 8: Process flow for silicon membrane microlens fabrication

satisfied with CF_4/O_2(10%), at 30 mTorr and 0.4 Watts/cm^2, with the additional benefit that the backside roughness was reduced. Figure 9 shows a typical 1μm diameter electrode with edge roughness less than 10 nm in silicon membrane etched using this procedure. The 1 kV dual electrode microsource was assembled by placing a 100 μm quartz spacer with a 2 mm hole between facing 1 μm and 100 μm electrodes. The electrodes were concentrically aligned to within 1 μm under a high-power optical microscope and fastened at the edges with UHV compatible epoxy. A similar assembly procedure was followed for the einzel lens. In the experimental version of the prototype column, an asymmetrical design of the einzel lens was adopted for reason of simplicity instead of the symmetrical design used in the analysis - implications of this design in terms of performance will be discussed later. This lens consisted of a 100 μm hole in the front and center electrode, and a 5 μm hole for the exit electrode. The exit electrode also served as the limiting aperture of the system. The center electrode was fabricated from ultra-thin silicon (100) wafers, 2-3μm thick. The wafers were mounted on a 1″ silicon carrier and coated with 1μm of a novolak-based resist. The 100μm diameter hole was patterned using electron beam lithography followed by RIE using $CBrF_3/SF_6$(16%) gas. The electrodes were then transferred from the carrier in between two quartz spacers, 100μm thick with a 2mm central hole,, which had been previously coated with a composite layer of Cr(20nm)/Al(200nm)/In(900nm). Heating the stack above the melting point of In layer, ($\simeq 160°C$) bonded the electrode to the spacers. After a backside etch in CF_4/O_2 (10%) for thinning, the electrode stack was assembled between two silicon membrane electrodes.

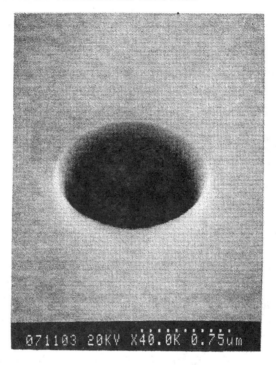

Figure 9: SEM micrograph of 1μm hole in first electrode

(b) Tip Materials and Processing Techniques

A field emission electron source[30-32] offers the highest brightness of all electron sources for a given potential. This feature has been successfully exploited in the high performance SEMs, STEMs, SAMs and other electron beam systems for microscopy and spectroscopy for material analysis as well as for a wide range of other applications. A conventional field emission source uses a single crystal tungsten tip, <310> and <111> being the two most frequently used orientations, operating in ultra-high vacuum at room temperature. Such a source is commonly referred to as a cold field emission (CFE) source. To obtain field emission, a field strength in the order $\geq 10^7$ V/cm is required. This is achieved by the use of a very sharp tip, typical tip radius ≤ 100 nm, and the application of an extraction voltage between the tip and the first electrode. In conventional sources, an extraction voltage in the range of 1 to 5 kV is usually required with the electrode to tip spacing of the order of a few millimeters. High brightness is achieved by a combination of high angular emission density and a very small source size. CFE source, however, suffers from a well known emission stability problem. Fluctuation of emission current, which directly affects the beam current downstream, occurs, in part, due to emitter surface work function changes introduced by migration of adsorbed gas molecules and, in part, due to local change in radius of curvature induced by ion bombardment. The effect of ion bombardment is particularly worrisome as it usually manifests itself in the form of current spikes, the amplitude of which can be quite large, and, in severe cases, can cause emitter failure by the initiation of a vacuum arc. In addition, a beam positional stability problem also arises in conventional columns employing such a source. It has been shown that the circle of least confusion of the electron bundle emitted from the tip has a diameter of approximately 5 nm[32] which is 2 to 3 orders of magnitude smaller than a thermionic source. The probe forming optics following the source is therefore operating either in a magnifying or in a very low demagnifying mode. In this mode, any motion of the tip due to thermal drift, column bending, acoustic, vibration, and external electromagnetic interferences will be amplified or directly transmitted to the final probe. Conventional systems with a relatively long column length are particularly susceptible to all these perturbations resulting in the positional stability problem. Thermionic field emission source[33] based on a zirconiated tungsten tip can improve the emission stability with, however, a significant increase in energy spread.

Typically, field emission tips were prepared from single crystal tungsten wire using a standard electrochemical etch. To remove surface contaminants and oxides and to heal defects induced by the etching process, conventional processes use an annealing temperature of $\simeq$ 2000 K over a duration of minutes. These processes, however, also dull the tip by self diffusion of tungsten atoms from the apex to the shank. The tip radius changes typically from $\lesssim$ 200 Å immediately after the etching to about 1000 Å which is too large for the microsource application. In order to avoid dulling but still get a thermally cleaned tip, a new low temperature field assisted annealing process[24] has been developed. This process involves a low temperature (900 K) preclean followed by flash anneal at a temperature of 1600 K in the presence of a strong electric field to achieve a clean, sharp built-up tip with properties attractive for SAFE operation. The process is highly reproducible and does not need the continuous monitoring of an FIM. The key for the success of this procedure is in the preclean step in which tungsten oxide is not removed. Hence, in the flash anneal phase the oxide helps to facilitate the build-up, an effect that had been reported by others[34-35]. This process works particularly well with W<111>, but is also applicable for W<310>. The electron emission pattern of W<111> tips processed with this new method exhibited a single lobe of emission with an easily recognizable triangular shaped cross-section reflecting the crystalline symmetry as shown in Figure 10. The electron emission of these tips was found to be confined to a smaller angle (10°) than for a conventional thermally annealed tip (20°) and this can lead to improvements in the angular emission intensity by a factor of three to four.

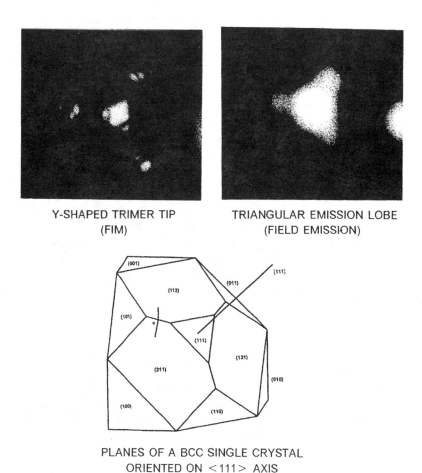

Figure 10: FIM image of an annealed W <111> tip showing the characteristic "Y" shaped apex formed by the three (211) planes (top-left). Schematic of a built-up tip showing the pyramid formed by the {211} planes truncated by the {111} plane (bottom). Field emission electron microscope image of same tip with triangular-shaped emission pattern (top-right).

(c) Microsource Performance

Dual electrode microsources have been successfully fabricated and operated with an STM controlled tip in the SAFE mode of operation. The source performance has been evaluated in terms of the emission characteristics of the tip and the focusing properties of the dual electrode lens.

The transmission efficiency, defined as the ratio of the transmitted current I_t to the total emission current I_e is shown in Figure 11a. The efficiency rises sharply when z_0 is below 5 μm and approaches 95% at $z_0 \simeq 0$ μm. Calculations of the radial

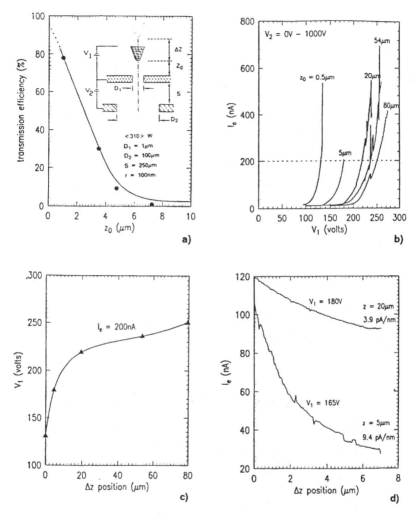

Figure 11: Emission characteristics of microsource. a) transmission efficiency of 100nm tip. b) Emission current as a function of tip position and extraction voltage. c) Z-dependence of extraction voltage. d) Feedback sensitivity at different z_0

distribution of the field further predicted a maximum efficiency at a value of $z_0 > 0~\mu m$. Below this value, current would be lost to the first electrode through side emission. Based on Figure 11a, z_0 in the range of 1-5 μm was considered acceptable for operation.

Figure 11b shows the emission characteristics of the microsource as a function of extraction voltage. As the tip approached the electrode, the I_e curves shifted toward lower voltages. In terms of electron emission, this shift corresponds to a lower extraction voltage for a fixed emission current. An example of this effect is shown in Figure 11c, for I_e = 200 nA. Extraction voltages of in the order of 100 V were achievable with sub-50 nm tips. Figure 11d shows the converse relationship, at fixed extraction voltage, I_e falls off rapidly as the tip is withdrawn. Defining sensitivity as $\Delta I_e/\Delta z$, it is evident that the sensitivity of the system is reduced or, equivalently, the gain of a feedback system must be increased, as z_0 is increased. The strong dependence of I_e on both extraction voltage and tip position suggests that there are two independent means of feedback available for current stabilization. The extraction voltage can be used to compensate for high frequency fluctuation and the z-position for low frequency.

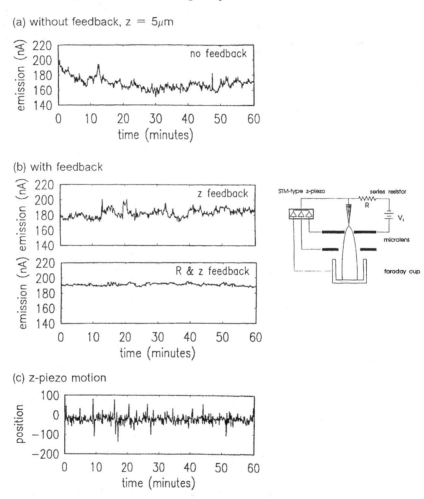

Figure 12: Noise stability of microsource with and without compensation

An important feature of the SAFE microsource was the use of STM-type feedback of the tip position to control drift in the emission current. With this method, in conjunction with resistive feedback, stable emission was observed for several hours. Figure 12a shows a typical stability curve for one hour without compensation. This trace shows two undesirable effects: i) a 20% long term drift, and ii) a $\sigma = 5\%$ fluctuation. Utilizing just z feedback, the drift is eliminated (Figure 12b). Including both z and R feedback, the stability is greatly improved,

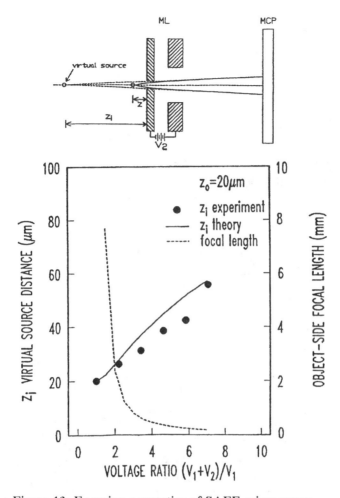

Figure 13: Focusing properties of SAFE microsource

with $\sigma \leq 1\%$. This resistor was chosen to be 1×10^8 Ω, as determined from the $z = 5$ μm curve in Figure 10b. The correction signal applied to the piezo, Figure 12c, clearly tracks the small fluctuations in the stabilized emission current. The present system is not yet optimized and further improvements are expected.

Focusing properties of the microsource were measured using the projected images of the first electrode on to a distant micro-channel plate (MCP) screen. As the focusing properties of the microsource change with the voltage, V_2, applied to the second electrode, the projected image will therefore also change with V_2. These changes can be traced back through the microsource as a shift of the virtual source location. This situation is plotted in Figure 13 as a function of the ratio of the total acceleration potential $(V_1 + V_2)$ to the extraction voltage (V_1). In this case, the voltage ratio of 7.5 corresponded to 1 kV operation with a 150 V extraction voltage. Good agreement existed between the experimental points (solid

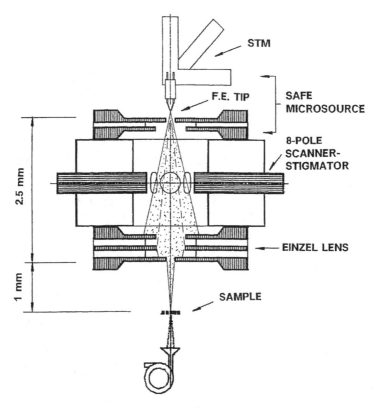

Figure 14: Cross section of the experimental 1kV SAFE microcolumn.

dots) and the calculated data (solid line). These calculations were made using the measured dimensions of the microsource, rather than the design values. Plotted along with z_i is the calculated object-side focal length of the same source. The focal length drops from 8 mm to less than 100 μm as the lens is activated.

d) Operation of 1kV microcolumns

An experimental 1 keV micro-column measuring 2.5 mm in length consisting of all electrostatic lenses fabricated from (100) boron doped silicon membranes has been fabricated. A cross section of the column is shown in Figure 14. The column consists of a filament-attached <111>-oriented tungsten tip mounted on an STM, a dual electrode microsource and an einzel lens as described earlier, a precision machined octupole, positioned between the source and the lens, for beam deflection and astigmatism corrections. Dimensions of the column were 1μm bore diameter in the first electrode in the source, 5μm diameter in the exit electrode of the einzel lens and 100μm diameter in all remaining electrodes. Spacing between electrodes were 100μm and separation between source and lens was 2mm. The 5μm diameter electrode served also as the limiting aperture and provide a beam forming semiconvergent angle of 2.5mrad. for a sample positioned 1mm away from the lens. Alignment of the column was achieved using a specially designed precision stage equipped with a three axes micromanipulator. Alignment was performed under a 320x optical microscope to an accuracy better than 1μm. The sample used in the initial evaluation was an 120nm thick Al grid of 1μm squares on 3μm centers, supported by a 400 mesh copper TEM grid (63.5μm period). These grids were fabricated by floating off evaporated Al from an electron beam patterned PMMA sample. The sample, consisting of two superimposed grids, could be imaged in either transmission or front surface modes.

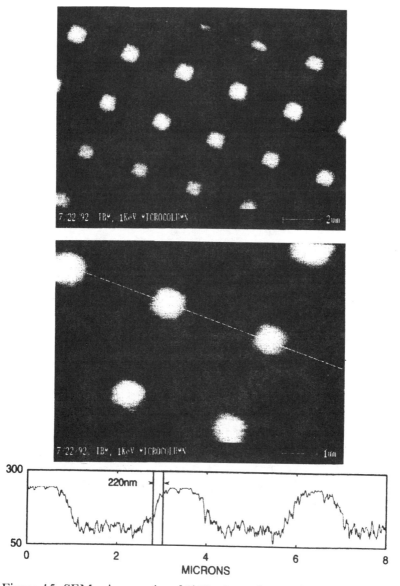

Figure 15: SEM micrographs of 1kV microcolumn obtained in transmission mode at two magnifications. The graph at the bottom is a line scan obtained from the white line in the lower image.

Successful operation of the complete column at 1kV in forming a focused probe of electrons at a working distance of 1 mm has been successfully demonstrated [22]. Preliminary performance of this microcolumn together with scanning images in the transmission mode have been evaluated [22]. Figure 15 shows one such scanning image of a 1μm grid sample. Images were acquired with scan fields ranging from 10 to 100μm and preliminary resolution measurements indicated a Gaussian beam diameter of approx. 200nm. Since the present design of the microcolumn uses a asymmetrical einzel lens with the exit electrode serving also as the limiting aperture, the aberrations of such a lens is expected to be considerably higher than the symmetrical design used in the analysis. This is because the electron trajectories are no longer confined to the paraxial regime and are much more sensitive to imperfections in the lens geometry - a very small imperfection in the bore geometry of the 5μm exit electrode can cause a significant change in trajectories resulting in a significant increase in aberrations. More detailed evaluation is in progress and significant improvements in performance are expected with a column based on a symmetrical einzel lens with a separate limiting aperture, better column tuning and experience. The results clearly demonstrate the basic concept and are achieved by a combination of a number of new technologies both in the design and in the fabrication of the microcolumn.

(2) LOW VOLTAGE LITHOGRAPHIC PROCESSES

Low voltage electron beam lithography has been discussed[36] before. At low beam energies of 1 to 2keV, the penetration range of electrons in typical resist materials is significantly reduced, ~60nm at 1keV in PMMA[37]. This means that one cannot expect to directly pattern thick resist layers with low voltage electrons. Even in thin layers with thickness comparable to the electron range, resolution will be significantly deteriorated by electron scattering in the resist. To achieve high resolution, new top surface processes which restrict the imaging process of the low energy electrons to an extremely thin, ~10 -20 nm, top layer of the resist, as shown in Figure 16a, need to be developed. The many ways to achieve this are still in the research phase in particular from a materials point of view. They may include silylation or metallation of the top surface of a resist; a two-layer system composed of a very thin top layer of silicon or metal containing resists over a thicker resist or other protective material layer; a three-layer system composed of an ultra-thin top layer of conventional electron resist, such as Langmuir-Blodgett films, for pattern transfer into an intermediate thin hard mask over a thicker resist or other protective material layer; or electron beam induced surface modification such as electron assisted chemical deposition, etching or seeding for subsequent electrochemical processes.

Top surface imaging at low voltage offers several advantages: electron scattering in the imaging layer becomes negligible; the lateral range of backscattered electrons is drastically reduced from the high voltage situation, at least to the total

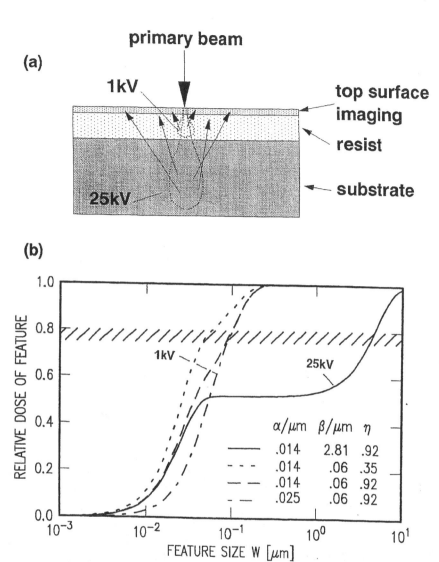

Figure 16: Lithography and proximity effect at low voltage.

range of the primary electrons; and, in addition, since the electrons never reach the high Z substrate material, the total amount of backscattering may be reduced as well. This can lead to an important benefit i.e. a significant reduction of proximity effect. While modeling and experiments regarding low energy electron interaction with resists and substrates are still incomplete, the following discussion offers a qualitative assessment of what can be expected. Taking the effective exposure distribution due to the primary electrons, small angle scattered and backscattered electrons to be described by a double Gaussian function[38−39],

$$F(r) = \frac{1}{\pi(1+\eta)} \{ \frac{1}{\alpha^2} \exp[-r^2/\alpha^2] + \eta \frac{1}{\beta^2} \exp[-r^2/\beta^2] \}$$

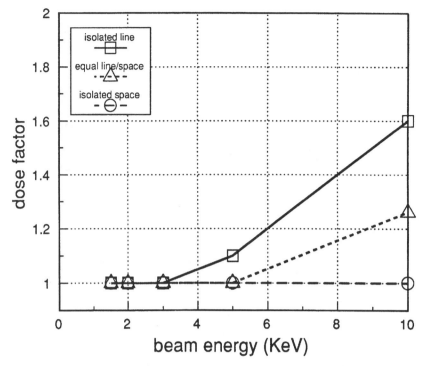

Figure 17: A plot of proximity dose correction factors required for 0.25μm features to print at their nominal size as a function of electron beam exposure energy.

the dose received in the center of a square feature with side length W is calculated relative to the dose in an infinitely large shape. Here α describes the lateral extent of the exposure due to primary and small angle scattered electrons, β the extent of the backscattered electron exposure and η the relative contribution of the backscattered electron exposure to the total exposure. In Figure 16b, this relative dose is plotted versus feature size for different assumptions. First, as a reference, the typical 25kV lithography case is shown with values for α, β, and η determined for thin resist on silicon[39]. It shows that as the feature size, W, approaches 2β, the dose of the isolated feature falls with decreasing feature size, reaching a plateau of $1/(1+\eta)$ until W becomes comparable with 2α. Since in a real pattern, the actual dose of the feature may vary between 1 (if close to large shapes) and the value plotted here (if isolated), proximity correction has to be applied for features smaller than about 3 to 4μm, or when the relative dose falls below .75 to .8. For the case of 1kV exposure using top surface imaging, it is reasonable to assume that α could maintain approximately the same value as in the 25kV case, i.e. 14nm, β to be approximately equal to the range of the 1keV electrons, namely 60nm, and η to take on a value approximately equal to that obtained at 25keV on very thick resist, where the electrons never reach the substrate[39], namely $\eta = .35$. The dotted curve in Figure 3b indicates that proximity effect correction should not be required down to feature sizes of ~60nm. This is due to the combined effect of reduced β and η.

If the value of η is increased to 0.92 (as in the case of 25keV), the dashed curve, and even if α is also increased to 25nm, the dash-dot curve, 100nm features should still be printed without proximity correction.

Experimental studies of low voltage resist exposure have been performed in order to verify the reduction in proximity effect[40-42]. In one case, 70 nm of PMMA resist was exposed using an electron beam with energies ranging from 1 keV to 10 keV. While the exposures were done using a conventional electron beam column, the results should be applicable to the microlens system. The minimum energy required to fully exposed the 70 nm thick resist was 1.5 keV, consistent with the penetration range at this energy. Exposures were made over a range of doses for each beam energy, and the resulting linewidths were measured and analyzed. In Figure 17 the experimental dose correction factors required for different feature types to print at their nominal linewidth of 0.25 μm are plotted as a function of beam energy. It is evident that for this case, proximity correction is not required at and below 3 keV. Features as small as 0.15 μm were patterned at 2 KeV and showed no sign of the proximity effect, and it would be expected that 0.1 μm features also would not need proximity correction when patterned at energies of 1.5 keV or less.

There were several other advantages noted at low energies in addition to the reduced proximity effect. The resist sensitivity at 2 keV is almost 4 times that at 10 keV due to the more efficient energy deposition of electrons at low energies. This results in an increase in throughput, or it allows the use of less sensitive resists and lower beam currents for improved resolution. In addition, the improved efficiency minimizes heating of the resist and substrate, which can affect resist profiles and placement accuracy. Finally, the greatly reduced range of the electrons can result in a large improvement in process latitude because of sharper definition between exposed and unexposed regions and in reduced radiation damage to the underlying substrate.

(3) THROUGHPUT CONSIDERATIONS

As mentioned earlier, throughput remains the main concern for the scanning electron beam lithography. To overcome this limitation, a great deal of effort has been applied in the past several decades. For the round beam systems, ultra-high exposure rates (in pixels/sec.) in excess of 100MHz have been developed[43-44] for systems equipped with thermionic field emission sources. Some degree of parallelism has been introduced in the variable shaped spot (VSS) system[45] which typically projects several tens of pixels at a time, and in the character or cell projection system[45-46] which projects several hundred pixels at a time to improve throughput. A pixel refers to a resolution element which equals 1/4 of the minimum linewidth. However, because of electron-electron interaction, the current density achievable in the shaped and cell projection systems is considerably lower than in the round beam system and the throughput improvement does not scale directly with the number of pixels projected. In addition, in the vector scan mode of pattern writing, settling time required between pattern jumps further reduces the throughput. Table 3 summarizes the throughput for 100nm lithography on 200 mm wafers based on pattern writing time only. In this analysis, 80% of the wafer area is assumed to be utilized as chips with a 25% pattern coverage on each chip. Four types of conventional scanning electron beam systems are considered: (1) round beam (raster), (2) round beam (vector), (3) VSS (vector) and (4) cell projection (vector). For each system, all key parameters contributing to the writing time are assigned a low and a high value to allow corresponding low and high throughput values to be derived. The lower values represent what is approximately achievable today and the higher values represent performance that maybe achievable in the future with considerable stretching of the current technologies. It can be seen that even with stretching, the throughput in the best case, the cell projection which has yet to demonstrate feasibility for actual device fabrication, barely exceeds 2 wafers per hour.

Micro-column arrays offer a solution to the throughput issue of scanning systems. By utilizing one or more columns per chip and by operating these columns in parallel as shown in Figure 1, a significant throughput improvement can be achieved. As the SAFE microcolumn has the potential of equaling or exceeding the performance of a conventional electron beam column, the throughput of an array of these microcolumns can, in the first approximation, be estimated by multiplying the throughput of a conventional column by the number of microcolumns employed in the array. Taking an average chip size of 20 mm x 20 mm, the number of chips per wafer (200 mm wafer) is approximately 60. In this proposal, the micro-columns are to be of the simple round beam design operating principally in the vector scan mode, although the raster scan mode will also be evaluated, for pattern generation. Column diameter is presently limited by the physical size of the miniaturized STM technology demonstrated todate[47-48] to roughly 1 to 2 mm, thus limiting the number of columns per chip to approximately 10. Table 3 shows the throughput of such a multi-beam system based on a one column per chip and 10-column per chip configurations. It can be seen that with multi-beams, attractive throughput approaching 10 wafers per hour can already be achieved using a simple one column per chip vector-scan configuration. This scales with the number of columns per chip reaching >50 wafers per hour in the 10 columns per chip configuration. Shaped beam optics for beam formation are also possible with miniaturized columns and can further improve throughput as indicated also in Table 3.

As throughput roughly scales inversely to the square of the linewidth, corresponding throughputs for lithography with other minimum linewidths can be derived. Using this procedure, attractive throughput for lithography with minimum linewidth ranging from 250nm to 25nm can be projected with arrayed microcolumns.

THROUGHPUT OF E/B SYSTEMS

(100nm lithography on 200 mm wafers)

	Round				Shaped			
	Raster		Vector		V.S.S.		Cell	
	Low	High	Low	High	Low	High	Low	High
CONVENTIONAL SYSTEMS:								
Coverage (%)	100	100	25	25	25	25	25	25
Resist ($\mu C/cm^2$)	10	10	10	10	1	1	1	1
j (A/cm^2)	1000	5000	1000	5000	25	100	10	25
Exposure rate (MHz)	100	500	100	500	25	100	10	25
Av. shape size (pixels)			36	64	36	64	500	1000
- cell coverage (%)							75	90
t_{settle} (ns)			100	25	100	25	100	50
Exp. time (10^3s/w)	400	80	100	20	11.1	1.56	8.4	1.0
Settle time (10^3s/w)	0	0	27.8	3.9	27.8	3.9	8.4	0.6
Σ Time (10^3s/w)	400	80	127	23.9	38.9	5.46	16.8	1.6
Throughput (w/hr.)*	.01	.05	.03	0.15	0.1	0.7	0.2	2.2
ARRAYED MICROCOLUMNS:								
(1) 1 column / chip (60 chips/wafer)								
Throughput (w/hr.)*	0.5	3	2	9	6	40		
(2) 10 columns / chip								
Throughput (w/hr.)*	5	30	20	90	56	>100		

(* Based on writing time only)

Table 3: Throughput of scanning electron beam systems for 100nm lithography on 200mm wafers based on writing time only assuming 80% of wafer area utilized as chips and pattern covering 25% of each chip. Four types of conventional systems: (1) round beam (raster), (2) round beam (vector), (3) VSS (vector) and (4) cell projection (vector) as well as arrayed microcolumns: (1) 1 column per chip and (2) 10 columns per chip are analyzed based on low and high values of key parameters. The low values represent current technology at its cutting edge and the high values represent performance with stretching of current technology.

This analysis, while somewhat speculative, indicates that the arrayed microcolumn approach has the potential of offering a sufficiently high throughput for manufacturing applications for lithography well below 100nm. This, coupled with the possible reduction in system cost due to compactness of the microcolumn assembly and the overall mechanical system (table and vacuum chamber now need only support a wafer motion of one chip width), suggest that this approach can be an attractive solution for lithography in the 100nm and below regime.

SUMMARY

An arrayed microcolumn approach for direct write (no mask) electron beam lithography has been evaluated. It has the potential of being a high throughput manufacturing system for lithography in the 100nm and below regime. It is a compact table top system with possible cost benefit. It operates at low voltage, 1 to 2 kV, and may require no proximity corrections. It has the potential of a breakthrough for the ultra-high density devices of the next generation.

There are many formidable technical challenges that need to be overcome before this approach can be realized. Maintaining reliable simultaneous operations of a large number of columns, accurate placement of columns with respect to each other, registration and resist problems at low voltages are some of these challenges. In addition, in the multiple column per chip configuration, data transfer rates, subfield stitching, field rotation and distortion correction are issues which need to be examined.

The arrayed concept is based on a novel SAFE- microcolumn technology which utilizes STM, micro-optics and field emission. Considerable progress in this technology including fully operational 1kV SAFE-microcolumn has been achieved in recent years.

ACKNOWLEDGEMENTS

The authors wish to acknowledge E. Bassous, K. Chen, J. Speidell, E. Anderson and S. Rishton for their collaboration and advice; J.M. Shaw, M. Yu and T.N. Theis for their support and encouragement. The collaboration of H.G. Craighead, M.S. Isaacson and E.D. Wolf of the National Nanofabrication Facility at Cornell University is also gratefully acknowledged.

References

1. D.L. Spears and H.I. Smith, Electron Letters 8, 102 (1972).
2. J. Warlaumont, J. Vac. Sci. Technol. B7, 1634 (1989).
3. N.M. Ceglio et. al., J. Vac. Sci. Technol. B8, 1325 (1990).
4. G. Stengl et al., J. Vac. Sci. Technol. B10, (Nov/Dec 1992).
5. S.D. Berger et al. J. Vac. Sci. Technol. B9, 2996 (1991).
6. I. Brodie, E.R. Westerberg, D.P. Cone, J.J. Muray, N. Williams, and C.I. Gasiorek, IEEE Trans. Electron Devices, ED28, 1422 (1981).

7. B. Lischke et al. Microelectroic Engineering 9, 199 (1989).
8. C.A. Spindt, I. Brodie, L. Humphrey, and E.R. Westerberg, J. Appl. Phys. 47,5248 (1976).
9. J.P. Spallas and N.C. MacDonald, IEDM, 209 (1991).
10. G.W. Jones, C.T. Sune, S.K. Jones, and H.F. Gray, SPIE (1991).
11. B. Binnig, H. Rohrer, Ch. Gerber, and E. Weibel, Phys. Rev. Lett. 49, 47 (1982).
12. M.A. McCord and R.F.W. Pease, J. Vac. Sci. Technol. B6, 293 (1988).
13. E.A. Dobisz and C.R.K. Marrian, J. Vac. Sci. Technol. B8, 1754 (1990).
14. T.H.P. Chang, D.P. Kern and L.P. Muray, J. Vac. Sci. Technol. B8, 1698 (1990).
15. T.H.P. Chang, D.P. Kern, M. McCord, and L.P. Muray, J.Vac. Sci. Technol. B9, 438 (1991).
16. T.H.P. Chang, D.P. Kern, and M.A. McCord, J. Vac. Sci. Technol B6, 1855 (1989).
17. M.A. McCord, T.H.P. Chang, D.P. Kern, and J. Speidell, J. Vac. Sci. Technol. B6, 1851 (1989).
18. L.S. Hordon and R.F.W. Pease, J. Vac. Sci. Technol. B8, 1686, (1990).
19. H-W. Fink, Physica Scripta, 38, 260 (1988).
20. B.D. Terris and D. Rugar, Proceedings of 50th EMSA Meeting, edited by G.W. Bailey, J. Bentley and J.A. Small (San Francisco Press, 1992) p.952.
21. D.A. Crewe, A.D. Feinerman, S.E. Shoaf, and D.C. Perng, J. Vac. Sci. Technol. B10 (Nov/Dec 1992).
22. L.P. Muray, U. Staufer, D.P. Kern, and T.H.P. Chang, J. Vac. Sci. Technol. B10 (Nov/Dec 1992).
23. L.P. Muray, U. Staufer, E. Bassous, D.P. Kern and T.H.P. Chang, J. Vac. Sci. Technol. B9, 2955 (1991).
24. U. Staufer, L.P. Muray, D.P. Kern and T.H.P. Chang, J. Vac. Sci. Technol. B9, 2962 (1991).
25. T.H.P. Chang, L.P. Muray and D.P. Kern, J. Vac. Sci. Technol. B10 (Nov/Dec 1992).
26. E. Munro, in Image Processing and Computer Aided Design in Electron Optics, edited by P.W. Hawkes (Academic, New York, 1993), p.284
27. D. Kern, Thesis, University of Tübingen, 1978
28. P.W. Hawkes and E. Kasper, Principles of Electron Optics Academic Press, New York, (1989).
29. Vu-Thien Binh, S.T. Purcell, G. Gardet and N. Garcia, in Atomic and Nanoscale Modification of Materilas, edited by P. Avouris, Kluwer, Dordrecht (1993).
30. T.E. Everhart, J. Appl. Phys. 38, 4944 (1967).
31. A.V. Crewe, M. Isaacson, and D. Johnson, Rev. Sci. Instr. 40, 241 (1969).
32. J.C. Wiesner and T.E. Everhart, J. Appl. Phys. 44, 2140 (1973).
33. L.W. Swanson, J. Vac. Sci. Technol. 12, 1228 (1975).
34. L.H. Veneklassen and B.M. Siegel, J. Appl. Phys. 43, 1600 (1972)
35. Vu-Thien Binh, J. Microsc. 152, 355 (1988).

36. Y.W. Yau, R.F.W. Pease, A.A. Iranmanesh, and K.J. Polasko, J. Vac. Sci. Technol. 19, 1048 (1981).
37. S.A Rishton, S.P. Beaumont, and C.D.W. Wilkinson, Microcircuit Engineering, 341 (1982).
38. T.H.P. Chang, J. Vac. Sci. Technol. 12, 1271 (1975).
39. S.A. Rishton and D.P. Kern, J Vac. Sci. Technol. B5, 135 (1987).
40. M.A. McCord and T.H. Newman, J. Vac. Sci. Technol. (Nov/Dec 1992).
41. Y-H. Yee, R. Browning and R.F.W. Pease, J.Vac. Sci. Technol. (Nov/Dec 1992).
42. P.A. Peterson, Z.J. Radzimski, S.A. Schwalm, and P.E. Russell, J. Vac. Sci. Technol. (Nov/Dec 1992).
43. D.S. Alles et. al., J. Vac. Sci. Technol. B5, 47 (1987).
44. F. Abboud et. al., J. Vac. Sci. Technol. B10, (Nov/Dec 1992).
45. H.C. Pfeiffer, IEEE Trans. Electron Devices, ED-26, 663 (1979)
46. Y. Nakayama, S. Okazaki, and N. Saitou, J. Vac. Sci. Technol. B8, 1836 (1990).
47. S. Akamine, T.R. Albrecht, M.J. Zdeblick, and C.F. Quate, IEEE Elec. Dev. Lett., 10, Nov. (1989).
48. J.J. Yao, N.C. MacDonald, and S.C. Arney, Nanostructure and Mesoscopic Systems, Wiley P. Kirk and Mark Reed, Editors, Academic Press, Dec. (1991).

Part II
Fabrication

FABRICATION

The papers in this section emphasize the fabrication and characterization of structures directly with a proximal probe. A wide range of modification mechanisms and material systems are described. The analytic capabilities of the probe allow the electrical and mechanical properties of the structure to be characterized in situ. Several authors describe novel lithographic techniques based on their development of proximal probe analytic methods. This serves to illustrate the tremendous versatility of proximal probes for basic research which is leading to the development of significant lithographic applications.

The first paper presents a review of the activities at the University of Basel, one of the institutions which has pioneered the use of proximal probes for material modification on the nanometer scale. Roland Wiesendanger describes material manipulation with both the STM and AFM and identifies several distinct mechanisms ranging from surface heating to direct probe surface interaction. The AFM experiments are described in greater detail by Thomas Jung (currently at IBM) and his collaborators later in the section.

An important example of STM induced material deposition is described by Alex de Lozanne and his colleagues from the University of Texas at Austin. They have shown that organometallic precursors can be broken down under the action of the STM to fabricate small metallic features. They have constructed a novel STM for their studies and can align STM induced deposition with existing features such as contact pads. Metal wires with close to 100% purity have been fabricated and characterized with electron transport measurements. An intriguing variant on this technique is described by Munir Nayfeh from the University of Illinois at Urbana-Champaign. He reports the results of laser assisted STM deposition of metallic clusters with sizes ranging down to a few atoms.

Charles Lieber (Harvard University) discusses the machining and manipulation of nanometer scale structures on the surfaces of layered compounds with the AFM. Small structures fabricated with the tip can be moved over distances of several hundred nanometers with a 'dragging' technique. Thomas Jung and his colleagues point out how AFM based techniques can provide valuable analytic information about a surface and lead to novel lithographic techniques. Determination of surface elastic and plastic properties has been developed into controlled nanoindentation and tribology studies have led to the demonstration of molecular level manipulation of absorbates.

Grover Wetsel and co-workers from the University of Texas at Dallas review the surface mound formation which can be achieved by pulsing the STM tip voltage. A key point here is the ability to control and predict the shape and size of the features. This paper also describes using STM for electronic measurements of quantum dot structures. The tip is carefully approached to the metallized contact of the dot and then used as an electrode for spectroscopy of the underlying dot. Hans Hallen (AT&T) has investigated the modification of buried interfaces under the action of voltage pulses applied with an STM. Atomic scale terraces are formed at the interface between a gold overlayer on silicon and can be detected by measuring the current reaching the substrate. This ability to create a structure with no free surfaces has some intriguing possibilities for data storage applications.

Nanofabrication by Scanning Probe Instruments: Methods, Potential Applications and Key Issues

Roland Wiesendanger
University of Basel, Dept. of Physics
Klingelbergstrasse 82, CH-4056 Basel, Switzerland

ABSTRACT

Scanning probe instruments can be applied to fabricate nanometer-scale structures by using a variety of different experimental methods. Depending on the physical mechanism involved in the modification process, the size of the written features may vary from submicron down to the atomic scale. The modification can be achieved by changing either the topographic structure or some physical property of the substrate locally. Potential applications of nanofabrication by scanning probe instruments as well as key issues, particularly with regard to nanometer-scale recording, are discussed.

1. INTRODUCTION

The increasing degree of miniaturization in microelectronics leads us to the major challenge to fabricate structures on a lateral length scale below 0.1 μm and to understand their physical properties. Besides already established tools for nanolithography, such as focussed high-energy electron beams, focussed ion beams and X-rays, the scanning probe instruments (SPI) have recently proven their potential for nanofabrication down to the atomic level [1-4].

The efforts towards SPI-based nanofabrication are primarily motivated by the smallness of the features which can directly be written onto various solid substrates. Conventional high-resolution electron beam lithography, being the most widely used and the most highly developed technique for the fabrication of nanometer-scale structures, is based on exposing an electron resist film (e.g. PMMA) by well-focussed high-energy electrons.

These high-energy electrons can penetrate several microns into the solid substrate of the resist film, thereby generating showers of secondary electrons which spread out into a volume having a size much larger than the diameter of the primary electron beam. Consequently, the resist also becomes exposed by these secondary electrons which are spatially much less confined and therefore set the limit to the size of the written structures [5]. In contrast, scanning tunneling microscope (STM) - based writing, for instance, involves the use of low-energy electrons, and an electron resist film is not necessarily required for the successful fabrication of small structures. Moreover, STM allows to address individual atomic sites, thereby offering the opportunity to manipulate matter down to the atomic level.

To be able to distinguish the written features from structural features of the substrate, its surface has to be sufficiently 'flat' with a local surface roughness considerably less than the size of the written features. Substrates which have most widely been used for SPI-based nanofabrication include:
- cleaved layered materials (e.g. graphite, transition metal dichalcogenides, mica, layered copper oxides etc.),
- single crystalline semiconductors prepared under ultra-high vacuum (UHV) conditions (e.g. Si, Ge),
- single crystalline metals prepared under UHV conditions (e.g. Au, Pt, Ni etc.),
- metal thin films (e.g. Au epitaxially grown on mica),
- noble metal balls prepared by melting a noble metal wire in the flame of an oxy-acetylene torch, and
- amorphous materials, either semiconductors or metals.

Various experimental methods have been developed for SPI-based nanofabrication, involving different microscopic mechanisms for the local modification process [1-4]. In the following, I shall review the experimental methods for the fabrication of nanometer-scale structures which we have exploited at the University of Basel since 1984.

2. FABRICATION OF NANOMETER-SCALE STRUCTURES

2.1. Chemical surface modifications

The finely focussed 'beam' of low-energy electrons, as offered by an STM, is ideally suited to induce local chemical reaction processes which leads to an important method for the fabrication of artificial surface structures from a submicrometer down to the atomic scale.

The first controlled nanometer-scale writing by STM was demonstrated by using amorphous metal substrates (Pd-81/Si-19) covered by a thin film of silicon dioxide and about two monolayers of hydrocarbons as well as carbon-oxygen species [6,7]. The experiment was performed in a vacuum of $10E(-8)$ Torr at room temperature. By scanning a tungsten tip at a speed of 100 nm/s with a constant tunneling current of 10 nA and a bias voltage of 100 mV, it was possible to draw lines on the surface which have subsequently been studied by high-resolution scanning electron microscopy (SEM). Fig. 1a shows a SEM image of a line pattern which was written by the STM on the glassy Pd-81/Si-19 substrate. The average spacing between the lines is about 16 nm, whereas the line-width is less than 2 nm. As an initial attempt to fabricate a nanojunction, crossed lines were generated as well (Fig. 1b).

Sample tilting experiments with the SEM and the absence of any deposits at the end points of the drawn lines indicated that the contrast, as observed in the SEM images, arises from differences in the yield of secondary electrons rather than from surface topographic features such as grooves or scratches. The different yield of secondary electrons along the written lines has been explained by a locally restricted chemical transformation of the adsorbate layer induced by the STM. Any chemical reaction, either polycondensation of the hydrocarbon film or reduction of the surface oxide layer, may well have been assisted by the catalytic activity of the tungsten tip.

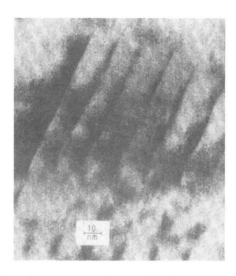

Fig. 1a: SEM image of parallel lines written by the STM on a glassy Pd-81/Si-19 substrate. (From [6,7]).

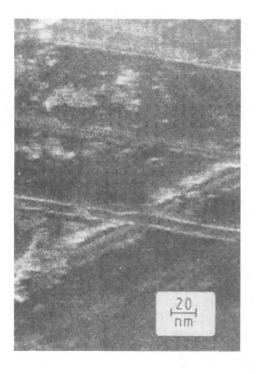

Fig. 1b: SEM image of a nanojunction fabricated with the STM on a glassy Pd-81/Si-19 substrate. (From [6]).

2.2 Thermal induced surface modifications

It is well known that effects caused by thermal heating are unimportant in STM experiments on crystalline substrates if a tunneling current I of about 1 nA and a bias voltage U on the order of 1 V are used [8-10]. The local temperature rise ΔT can roughly be estimated by: $\Delta T \approx (IU)/(4\pi\kappa\lambda)$, where κ and λ denote the thermal conductivity and electron mean free path of the substrate material. For STM experiments on a typical metal ($\kappa \approx 300$ W/Km, $\lambda \approx 30$ nm), ΔT is less than 1 mK and therefore entirely negligible. However, if the tunneling current and the bias voltage are significantly raised and, in addition, an amorphous metal is used as substrate, for which both the thermal conductivity and electron mean free path are lower by a factor of about one hundred compared with crystalline materials, a temperature rise up to several hundred Kelvin can be reached. Depending on the current and the electric field strength, enhanced diffusion, local crystallization of the glassy substrate or even local melting might occur.

Evidence for an STM-induced local melting process has indeed been found in surface modification experiments with amorphous metal substrates performed under UHV conditions [11-17]. At first, clean and sufficiently flat substrate surfaces were obtained by argon ion etching. For amorphous materials preferential etching at grain boundaries, as for polycrystalline substrates, or anisotropic etching, as for single crystals, does not occur. Therefore, ion-etched surfaces of glassy metals typically appear flat down to a one-nanometer scale, as directly confirmed by real-space STM images [18].

To modify the substrate surface locally, the STM tip was positioned above a preselected surface site and the bias voltage was increased from about 0.1 V, as used for non-destructive STM imaging, up to 1 V. Subsequently, the demanded current was slowly (time scale of several seconds) increased from 1 nA up to several hundred nA or even several μA, until oscillations in the current appeared, resulting from instabilities related with the locally molten surface. The molten substrate material is drawn towards the tip under the influence of the high electrostatic forces acting

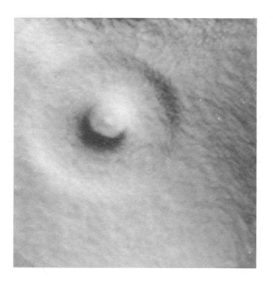

Fig. 2a: STM image (150 nm x 150 nm) of a cone written on a glassy Co-35/Tb-65 substrate. The FWHM of the cone is 29 nm whereas the topographic height is 23 nm. (From [15]).

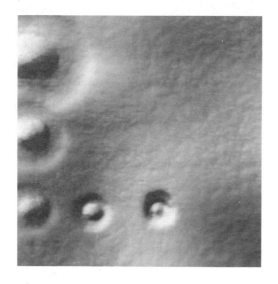

Fig. 2b: STM image (300 nm x 300 nm) of five cones written with different bias voltages (1.7 V, 1.2 V, 1.0 V, 0.9 V, and 0.8 V) on a glassy Co-35/Tb-65 substrate. The FWHM of the cones from the upper left to the lower right part of the image are 44 nm, 30 nm, 23 nm, 16 nm, and 10 nm. The corresponding topographic heights are 31 nm, 22 nm, 11 nm, 8 nm, and 7 nm. (From [15]).

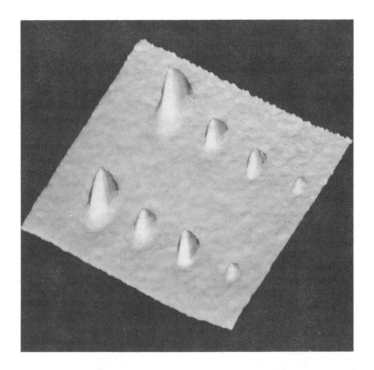

Fig. 2c: STM image (170 nm x 170 nm) of two sets of four cones written with different bias voltages (1.0 V, 0.9 V, 0.8 V, and 0.7 V from left to right) on a glassy Rh-25/Zr-75 substrate. The FWHM of the cones are 10.4 nm, 7.2 nm, 6.4 nm, 4.0 nm for the upper row and 11.5 nm, 8.2 nm, 8.2 nm, 5.7 nm for the lower row. The corresponding topographic heights are 4.3 nm, 2.0 nm, 1.7 nm, 0.7 nm and 3.5 nm, 2.0 nm, 2.4 nm, 1.0 nm. (From [15]).

between tip and sample surface, thereby forming a bridge between the two electrodes. Upon feedback-induced tip retraction, this liquid metal bridge breaks up and a cone of rapidly solidified material is left on the surface which can subsequently be imaged with the STM under normal operation conditions. Fig. 2a shows an STM image of such a cone written on a glassy Co-35/Tb-65 substrate. A depression is observed around the cone which can be explained by the material transport during cone formation.

The bias voltage applied during the surface modification process affects the power density as well as the electric field strength, which both have an influence on the size of the written cones. Therefore, by using different values of the applied bias voltage, the size of the written cones can directly be controlled. This has been demonstrated for several different amorphous metal substrates, including Co-35/Tb-65 (Fig. 2b) and Rh-25/Zr-75 (Fig. 2c). The dimensions of the cones, both their diameter as well as their height, were found to scale linearly with the applied bias voltage during the modification process [17]. On the other hand, a dependence on the polarity of the applied bias voltage was not observed. The smallest cones fabricated by this thermal-induced modification process had a base diameter of about 3 nm [15]. However, the actual size of the written cones may even be smaller because convolution with the tip shape is likely to increase the apparent size of the cones as imaged by STM.

Lines, letters and even messages (Figs. 3a and 3b) have been written by scanning the tip with a typical speed of 1-10 nm/s during local melting of the surface at elevated current and bias voltage. The line-width varies between 10-50 nm, depending on the sharpness of the STM tip. At the end of each line a cone is left, resulting from the break-up of the liquid metal bridge between tip and substrate upon tip retraction. By using seven different amorphous metal substrates (Co-35/Tb-65, Fe-28/Sc-72, Ni-64/Zr-36, Rh-25/Zr-75, Ni-60/Nb-40, Fe-80/B-20, and Ir-28/B-9/Mo-63), it has experimentally been verified that the power required to induce local melting of the surface scales, at least qualitatively, in a correct way with the product of the thermal conductivity and the melting temperature of these materials [17].

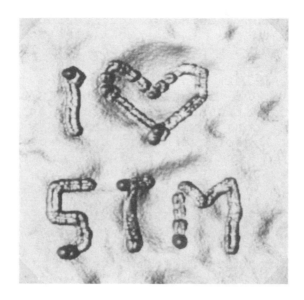

Fig. 3a: STM image (800 nm x 800 nm) of a message written on a glassy Rh-25/Zr-75 substrate at a bias voltage of 1.8 V and a demanded current of 630 nA. (From [15]).

Fig. 3b: STM image (650 nm x 650 nm) of another message written on a glassy Rh-25/Zr-75 substrate. (From [16]).

2.3. Mechanical surface modifications

Local surface modifications can also be achieved by using the force interaction between tip and sample, either in an STM-type or an AFM-type configuration. Depending on the amount of the applied force and the mode of SPI operation, different types of mechanical surface modifications have been demonstrated.

For instance, by increasing the tip-sample force interaction at a particular surface location, indentation marks can be produced with an STM or an AFM (atomic force microscope) tip. In STM experiments, the current and the applied bias voltage can be used as control parameters for the strength of the tip-sample interaction. By operating the STM beyond the point of tip-sample contact, indentation marks may be formed on the substrate surface as shown in Fig. 4.

Alternatively, the AFM can be used as a micromechanical tool, allowing for the direct measurement and control of the tip-sample force interaction during surface modification [20]. For instance, indentation marks of 50-nm diameter and 2.5-nm depth were produced by an AFM tip on a polycarbonate substrate of a commercial compact disc with an applied force of 8.2×10^{-8} N. The indentations are likely to cause plastic flow, well known for polymers, or even rupturing of covalent bonds.

To produce lines instead of single indentations, the tip can be scanned with an increased applied static force of about 10^{-7} N. At the end of each line a hillock is formed, resulting from the displaced substrate material (Fig. 5a). As an alternative, the applied force may be modulated during scanning with a modulation frequency of up to several kHz and an amplitude of up to 10^{-7} N. Cantilever torsion and slip-stick behaviour, as commonly observed during scanning at high static forces, are drastically reduced by using this modulation mode of operation. Therefore, the 'dynamic' AFM writing procedure proves to be more reliable. An example is presented in Fig. 5b.

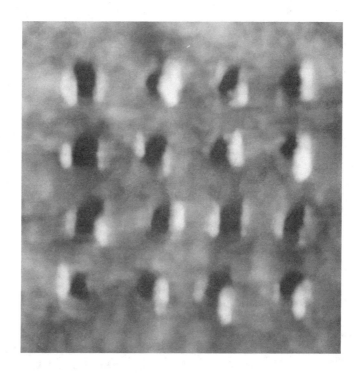

Fig. 4: STM image (300 nm x 300 nm) of a 4x4 array of indentation marks written on an amorphous metal substrate. (From [19]).

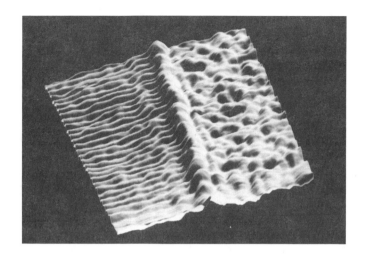

Fig. 5a: AFM image (1.35 μm x 1.35 μm) showing a partially modified surface region of a polycarbonate substrate. The individual lines, written with an applied force of 10E(-7) N, are 10 nm deep and 70 nm wide. The native polycarbonate surface, as visible on the right hand part of the image, exhibits a roughness of about 10 nm. (From [20]).

Fig. 5b: 'HEUREKA' written with an AFM between two information pits of a compact disc being separated by 1.6 μm. Writing was performed with an applied static force of 2x10E(-8) N and a superimposed force modulation having an amplitude of 6x10E(-8) N. The height of the letters is 700 nm and the indentation depth is 10 nm. (From [20]).

2.4. Electric-field induced surface modifications

By raising the bias voltage in an STM experiment up to several volts, an electric-field strength on the order of 1 V/Å can easily be reached. Electric fields of this magnitude are strong enough to cause modifications at the tip apex and on the substrate surface.

For instance, material deposition from a gold tip, caused by the application of voltage pulses, can be used to write several thousand gold mounds onto a substrate with no apparent degradation of the tip's quality [21,22]. The threshold voltage for a successful deposition of tip material was found to be between 3.5 V and 4.0 V for a gold tip and a gold substrate. The linear dependence of this threshold voltage on tip-surface separation indicated the existence of a critical electric field strength, which was derived to be on the order of 0.4 V/Å. This value is significantly lower than that required to cause field evaporation in FIM (field ion microscope) studies. However, it has been argued that the close proximity of tip and substrate in the STM experiment tends to lower the threshold field for atomic emission from the tip by field evaporation [21,22].

Similar surface modification experiments with gold substrates based on the voltage pulsing technique, have been repeated by using different tip materials (W, Pt, Pt-Ir, Mo, Fe, Ni) [23,24]. The experimental results, including the value obtained for the threshold pulse amplitude and the reliability of material transfer as a function of tip material, indicated that the large electrostatic forces acting between tip and substrate during a voltage pulse, rather than a field evaporation process, may be responsible for the material transfer. The tensile stress S acting on a tip of radius R due to the electrostatic force F can be estimated by the following expression:
$S = F/(\pi R^2) \approx (1/2)\varepsilon_o \varepsilon_r (U/d)^2 f(d/R)$, where U, d, ε_o and ε_r denote the applied bias voltage, the tip-substrate separation, the dielectric constant of vacuum and the dielectric constant of the gap material, respectively. The dimensionless function $f(d/R)$ has values between 0 and 2 [24]. For a given set of values for U, d and R, the tensile stress can be calculated and compared with the tensile strength of the selected tip materials

Material	Tensile strength in N/mm^2
Au	120 - 130
Pt	130 - 140
Fe	240 - 280
Ni	800 - 850

Tab. 1: Macroscopic tensile strengths of some metals

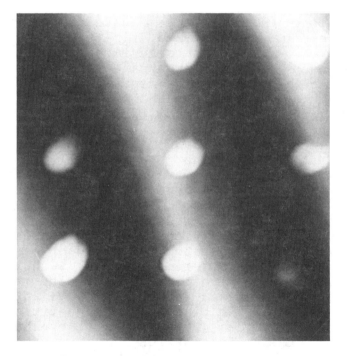

Fig. 6: STM image (270 nm x 270 nm) showing an array of hillocks that have been created on a gold substrate by applying 200-ns voltage pulses to a gold tip. The pulse height has been 3.0 V, 3.5 V, and 4.0 V for the hillocks in the upper, middle and lower row, respectively. (From [24]).

(Table 1). Indeed, it has been found experimentally that material deposition from a gold or platinum tip (Fig. 6) occurs much more reliable than from a nickel tip [24], which agrees reasonably well with the relative tensile strengths of these materials (Table 1). Furthermore, the threshold pulse amplitudes and therefore the threshold fields were generally found to be independent of pulse polarity. This lack of polarity dependence can easily be explained by assuming an electrostatic-force induced modification mechanism since the force is proportional to the field squared and therefore the same for either polarity of the applied voltage pulse.

2.5. Magnetic surface modifications

There exist several possiblities for magnetic surface modifications by using scanning probe microscopes, covering the submicron down to the atomic length scale. For instance, the magnetic stray field of a magnetized tip can cause local changes of the substrate surface magnetization at sufficiently small tip-substrate separations as a consequence of magnetic dipole interactions. By using this modification method, magnetic bits have been written on a surface of a hard disk by means of a tunneling-stabilized magnetic force microscope with a magnetized iron thin film tip, which also served to read the magnetically recorded information [25]. The size of the written bits were on the order of several hundred nanometers.

By using a spin-polarized scanning tunneling microscope, modifications of the surface spin structure down to the atomic level can be induced and subsequently observed [26]. Such localized changes of the magnetic surface structure are likely to be caused by the exchange interaction between the magnetic tip and the magnetic substrate at close distances ($\lesssim 3$ Å), rather than by the long-range dipole interaction.

Spin-polarized STM (SPSTM) is based on the spin-dependence of the tunneling current flowing between two magnetic electrodes, as theoretically predicted [27] and experimentally verified for the STM by the direct observation of vacuum tunneling of spin-polarized electrons between a ferromagnetic chromium dioxide tip and a Cr(001) test surface [28-30]. Recent improvements in the in-situ preparation of atomically sharp and clean magnetic sensor tips under ultra-high vacuum (UHV) conditions [31,32] have led to the capability of magnetic imaging at the atomic level on a magnetite (Fe_3O_4) (001)-surface [33-35].

Magnetite, the best-known natural magnetic material, has cubic inverse spinel structure and is ferrimagnetic with a Curie temperature of 860 K. There exist tetrahedrally coordinated Fe 'A-sites' as well as octahedrally coordinated Fe 'B-sites'. The A-sites are occupied by $Fe(3+)$, whereas the B-sites are occupied by $Fe(3+)$ and $Fe(2+)$ with spin configurations $5 \times (3d\uparrow)$ and $5 \times (3d\uparrow)/1 \times (3d\downarrow)$, respectively. At room temperature, fluctuations of the sixth electron (3d) among B-sites are rapid in the bulk, explaining the relatively high electrical conductivity of about $100/(\Omega\,cm)$ compared with other iron oxides. Below the so-called Verwey transition [36] at about 120 K, these charge fluctuations cease, and a static periodic arrangement of $Fe(3+)$ and $Fe(2+)$ ions sets in, leading to a decrease in conductivity by about two orders of magnitude. The driving force is interatomic Coulomb interaction, so the Verwey transition in magnetite is considered to be an example of a Wigner crystallization [37,38] in three dimensions, albeit modified by electron-phonon coupling.

For the SPSTM studies, magnetite (001) surfaces were prepared by mechanical polishing followed by in-situ annealing of the single crystals up to about 1000 K in UHV. Well-ordered and clean surfaces were obtained, as verified by low-energy electron diffraction (LEED) and Auger electron spectroscopy (AES) [34,39]. The correct surface stoichiometry was checked by X-ray photoelectron spectroscopy (XPS). Topographic STM images obtained with non-magnetic tungsten tips revealed terraces separated by either 4.2-Å or 2-Å high steps, depending on the sample surface location. On top of the terraces being separated by 2-Å high steps, parallel atomic rows with a spacing of 5.9 Å were observed which change their orientation by 90° from one terrace to the next. These atomic rows have been identified with the rows of octahedrally coordinated Fe B-sites in the Fe-O planes of magnetite (Fig. 7a inset), which are indeed

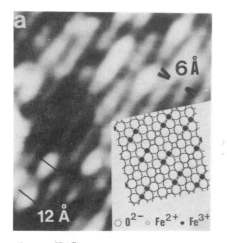 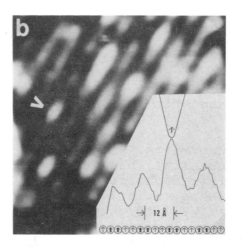

Fig. 7a: Magnetic STM image obtained with an iron tip showing the rows of Fe B-sites with the 5.9-Å spacing and a clear modulation along these rows. Inset: schematic structure of the Fe-O (001) plane of magnetite. (From [26]).

Fig. 7b: Magnetic STM image obtained after the one presented in Fig. 7a. A new feature (marked by an arrow) appears which fits into the predominantly observed 12-Å periodicity between Fe(3+) and Fe(2+) sites. Inset: single line section showing the dominant 12-Å periodicity due to the different spin configuration of Fe(3+) and Fe(2+) sites which are indicated by arrows of different size. (From [26]).

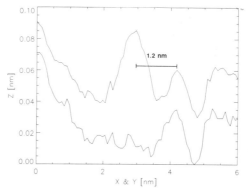

Fig. 8: Comparison of the line sections taken from Fig. 7a and Fig. 7b along the particular row of Fe B-sites which becomes modified in the magnetic STM image of Fig. 7b. The new maximum fits exactly into the 12-Å periodicity characteristic for the surface spin structure, in contrast to the 3-Å periodicity of the atomic structure. (From [26]).

separated by 2 Å. The oxygen sites remain invisible in the STM images because the corresponding O-1s and p-states lie well outside the energy window accessible with the STM [35]. No atomic-scale structure has been resolved along the rows of Fe B-sites by using non-magnetic tungsten tips, indicating a smooth spin-averaged density-of-states corrugation along these rows.

In contrast, a strong corrugation along the rows of Fe B-sites has been measured with ferromagnetic Fe probe tips in SPSTM experiments (Fig. 7). Several different periodicities were observed along these rows which are, however, always multiples of 3 Å, the nearest-neighbor distance of Fe-sites along the rows. A statistics of the observed periodicities [40] revealed a clear dominance of a 12-Å period (Fig. 7b inset) which is equal to the $Fe(3+)-Fe(2+)$ periodicity observed in the bulk for the low temperature phase of magnetite below the Verwey transition [41]. The observed static pattern with a dominant 12-Å periodicity at the (001) surface of magnetite at room temperature may be explained by an increase of the Verwey transition temperature at the surface due to a band narrowing as a result of the reduced coordination at the surface [38]. The lack of long-range order indicates that the Wigner crystallization of the 3d↓ electrons on the surface takes the form of a Wigner glass [40].

The static arrangement of $Fe(3+)$ and $Fe(2+)$ in the Wigner glass state with only short-range order can be imaged over extended time periods without any observable changes. However, occasionally local modifications between two successive SPSTM images were observed. For instance, an additional feature (marked by an arrow) appears in Fig. 7b, which is not present in Fig. 7a. Remarkably, the new feature fits exactly into the dominant 12-Å periodicity along the rows of Fe B-sites, as verified by a comparison of individual line sections along the particular modified row, taken from the SPSTM images of Figs. 7a and 7b. The corresponding two line sections are presented in Fig. 8. The observed 12-Å period clearly indicates a modification of the spin configuration of the magnetite surface because this is the expected repeat period between $Fe(3+)$ and $Fe(2+)$ with their different spin configurations, whereas the atomic periodicity, disregarding the difference between the two magnetic ions, would only be 3 Å.

Most remarkably, the modification appears to be spatially localized to a few ionic sites, leaving the magnetic surface structure in the vicinity unchanged. Several possibilities for such local modifications in the surface spin configuration exist. One involves charge transfer of one or more electrons to convert $Fe(3+)$ into $Fe(2+)$, or vice versa, at an adjacent B-site. Alternatively, a local modification of the spin configuration of an atom or group of atoms may be associated with non-collinear spin structures of the surface. Future experiments will focus on the influence of the tip-to-substrate spacing on the observed local magnetic surface modifications.

3. POTENTIAL APPLICATIONS

The primary application of nanofabrication by scanning probe instruments (SPI) will certainly be the investigation of the physical and chemical properties of well-defined nanometer-scale test structures. At a second stage, individual nanostructure devices based on the nanometer-scale physics may be developed. These devices will have dimensions smaller than the electron phase-coherence length and probably will be based on the properties of the quantum-mechanical wave functions. As an initial step towards this direction, a tunnel diode effect has been discovered in STM studies of B-rich Si(111) surfaces [42,43]. In contrast to the conventional Esaki tunnel diode, negative differential resistance (NDR) in this STM experiment arose as a result of tunneling between localized quasi-atomic states. The existence of such localized states gives rise to allowed and suppressed energies for tunneling, leading to NDR. A third possible application of nanofabrication by SPI is nanometer-scale recording, leading to ultra-high density data storage capabilities. In the following, the particular examples of nanofabrication discussed in the previous section will be reviewed with regard to their specific potential applications and possible future developments.

Chemical surface modifications, such as polymerization or dissociation and fragmentation of molecules, generally have a great potential for nanometer-scale lithography. Conventional electron beam lithography is already widely based on the local chemical modification of resist films. STM-based lithography may allow to extend the capabilities of nanometer-scale patterning to new types of resist films which require electrons of only a very small energy (0.1-5 eV) to induce a local chemical transformation. The chemical modification based on these low-energy electrons can spatially be confined to a significantly smaller scale than with high-energy electrons, for which backscattering and secondary electron generation is known to limit the size of the written structures.

Thermal surface modifications by STM appear to be promising because the possibility of inducing structural and phase transformations on a local scale can quite generally be exploited to modify the physical properties of small sample volumes. For instance, magnetic bits can be written thermally by either using phase-change or magneto-optical systems. In phase-change systems, localized heating causes a transition between crystalline and amorphous states which exhibit different magnetic properties. This is usually accomplished by means of a focussed laser beam with a spot size of typically 1 μm, which sets the size limit for the written bits. By using an STM as a local heating source, rather than a laser beam, a significantly smaller bit size on the nanometer scale may be achieved. However, effective heating by STM requires an amorphous substrate, as discussed in section 2.2, thereby limiting the application to local amorphous-to-crystalline or amorphous-to-amorphous transformations.

A particular promising application for STM used as a local heating source would be thermomagnetic recording on magneto-optical amorphous substrates, such as heavy rare earth - transition metal alloy films, as already suggested in 1989 [15]. By locally increasing the temperature of the film to the point, where an externally applied magnetic field can reverse the magnetization in the region being heated, a magnetic bit is obtained. In the conventional thermomagnetic recording process, a focussed laser beam is used for local heating, which again limits the size of the written bits to the micrometer length scale. In contrast, STM allows to confine the locally heated region of an amorphous metal substrate to nanometer-scale dimensions. Thermal recording by STM on amorphous CoTb-substrates was already demonstrated, as described in section 2.2, even in the presence of an externally applied magnetic field [15]. However, the magnetic properties of the written bits could not be characterized in these STM experiments. A UHV-compatible magnetic force microscope (MFM) appears to be most appropriate for this purpose.

Mechanical surface modifications by SPI, such as scratching, may become important for inducing structural anisotropies on a substrate surface at the nanometer level which could probably be useful for orientating molecular adsorption layers exhibiting specific properties [20]. This application would certainly require extending the surface modification to large surface areas, which could be achieved with scanning probe instruments having a scan range up to the millimeter scale.

Electric-field induced surface modifications by material transfer from an STM tip onto a substrate generally have a great potential for creating nanometer-scale structures made from a different material than the substrate itself. For instance, arrays of nanometer-scale metallic dots on semiconducting substrates, or magnetic dots on non-magnetic substrates can be produced by using this nanofabrication method. This could probably lead to ultra-high density storage devices as well as to tailored catalysts. However, the reproducibility of the material transfer process for non-gold tips has to be further improved.

Magnetic surface modifications by SPI offer a wide variety of potential applications, ranging from magnetic data storage to magneto-catalysis. While submicrometer-scale modifications induced by the stray field of a magnetic tip have already been studied to a relatively large extent, both experimentally and theoretically, atomic-scale modifications of the surface spin configuration, as described in section 2.5, have to be focussed on in more detail. In particular, the spin-polarized STM technique has to be further developed to allow for a routine characterization of surface spin configurations at the atomic level.

4. KEY ISSUES

The technological potential of a given nanofabrication method can be judged by considering the following key issues:
- the size of the written structures,
- the error rate in the writing process,
- the writing speed,
- the environmental conditions for the writing process, and
- the temporal stability of the written structures for a given environment.

As already mentioned in the introduction, the efforts towards SPI-based nanofabrication are primarily motivated by the smallness of the features which can directly be written onto various solid substrates. The error rate in the writing process depends critically on the particular surface modification method as well as on the choice of the tip and substrate material. In favourable cases, a reliability approaching 100% has been achieved [15,21,44]. For instance, the thermal induced surface modification process, described in section 2.2, has proven to be highly reliable for a wide variety of amorphous metal substrates. The electric-field induced material deposition process, described in section 2.4, has been found to be highly reliable for gold tips, but less reliable for other tip materials.

One of the most serious limitations of SPI-based nanofabrication with regard to nanometer-scale recording is given by the relatively slow writing speed being inherent in a serial writing process. Introducing parallelism by the microfabrication of arrays of simultaneously working miniaturized SPI units may partially solve this problem [45]. In addition, fast nanofabrication methods are favourable for recording applications, e.g. surface modifications based on ultra-short voltage pulses. For instance, the electric-field induced material deposition, initiated by a nanosecond voltage pulse (section 2.4), appears to be reasonably fast. On the other hand, the thermal induced surface modification, described in section 2.2, involves a relatively slow process.

In some cases, the application of SPI-based nanofabrication methods is restricted to special environmental conditions, such as UHV and/or low temperatures. In contrast to conventional electron beam lithography, which inherently requires vacuum conditions for the writing process, most SPI-based nanofabrication methods would, at least in principle, work in an ambient environment as well. However, the limited temporal stability of the written nanometer-scale structures under ambient conditions quite often determines the choice of the experimental environment. For instance, surface contamination under poor vacuum conditions or surface diffusion at finite temperatures can limit the stability of the artificially created surface structures to a time scale of minutes or even below. Temporal changes of the appearance of nanometer-wide lines, written with the STM on glassy metal substrates by using the chemical surface modification procedure (section 2.1), have been observed by SEM [6]. It was proposed that a segregation process may have widened the initially drawn lines. The nanometer-scale structures fabricated on amorphous metal substrates by using the thermal induced surface modification process, described in section 2.2, have proven to be stable over a time period of several hours. However, the evolution over longer time periods is unknown. Different time-dependent observations have been made for nanometer-scale structures obtained by the deposition of tip material using the electric-field induced surface modification procedure (section 2.4). In some STM experiments, performed with gold tips and gold substrates, a high degree of temporal and thermal stability of the written marks has been demonstrated [22], while in other STM experiments surface diffusion has been found to lead to a temporal change in the appearance of the nanometer-scale deposits [24].

Generally, to prevent the deterioration of molecular-scale structures and devices will perhaps constitute the most significant task which has to be attacked for the development of what we may call 'nanoelectronics'. Other key issues which have to be addressed are given by the problem of energy dissipation as well as the problem of effectively and reliably contacting the molecular-scale structures with the 'macroscopic world'. In particular, the development of appropriate architectures, which make use of the extremely small size of the fabricated structures, deserves increasing attention. Certainly, major efforts will be required to displace established technology

5. ACKNOWLEDGMENTS

I would like to thank my collaborators in the work on nanofabrication at the University of Basel: H.-J. Güntherodt, H. Hug, T. Jung, A. Moser, M. Ringger, L. Scandella, T. Schaub, I.V. Shvets, and U. Staufer. Financial support from the Swiss National Science Foundation and the Kommission zur Förderung der wissenschaftlichen Forschung is gratefully acknowledged.

6. REFERENCES

[1] G.M. Shedd and P.E. Russell, Nanotechnology 1, 67 (1990).
[2] C.F. Quate, in: Scanning Tunneling Microscopy and Related Methods, eds. R.J. Behm, N. Garcia and H. Rohrer, NATO ASI Ser. E: Appl.Sci. Vol. 184, p. 281. Kluwer, Dordrecht (1990).
[3] R. Wiesendanger, Appl. Surf. Sci. 54, 271 (1992).
[4] U. Staufer, in: Scanning Tunneling Microscopy Vol. II, eds. R. Wiesendanger and H.-J. Güntherodt, Springer Series in Surface Sciences Vol. 28, p. 273. Springer, Berlin/Heidelberg (1992).
[5] A.N. Broers, IBM J. Res. Develop. 32, 502 (1988).
[6] M. Ringger, H.R. Hidber, R. Schlögl, P. Oelhafen, and H.-J. Güntherodt, Appl. Phys. Lett. 46, 832 (1985).
[7] M. Ringger, B.W. Corb, H.R. Hidber, R. Schlögl, R. Wiesendanger, A. Stemmer, L. Rosenthaler, A.J. Brunner, P.C. Oelhafen, and H.-J. Güntherodt, IBM J. Res. Develop. 30, 500 (1986).
[8] B.N.J. Persson and J.E. Demuth, Solid State Commun. 57, 769 (1986).
[9] F. Flores, P.M. Echenique, and R.H. Ritchie, Phys. Rev. B 34, 2899 (1986).
[10] P.F. Marella and R.F. Pease, Appl. Phys. Lett. 55, 2366 (1989).
[11] R. Wiesendanger, Ph.D. Thesis, University of Basel, Basel, Switzerland (1987).
[12] U. Staufer, R. Wiesendanger, L. Eng, L. Rosenthaler, H.R. Hidber, H.-J. Güntherodt, and N. Garcia, Appl. Phys. Lett. 51, 244 (1987).
[13] U. Staufer, R. Wiesendanger, L. Eng, L. Rosenthaler, H.R. Hidber, H.-J. Güntherodt,

and N. Garcia, J. Vac. Sci. Technol. A 6, 537 (1988).
[14] R. Wiesendanger, L. Eng, H.R. Hidber, L. Rosenthaler, L. Scandella, U. Staufer, H.-J. Güntherodt, N. Koch, and M.A. von Allmen, in: The Structure of Surfaces Vol. II, eds. J.F. van der Veen and M.A. Van Hove, Springer Series in Surface Sciences Vol. 11, p. 595. Springer, Berlin/Heidelberg/New York (1988).
[15] U. Staufer, L. Scandella, and R. Wiesendanger, Z. Phys. B 77, 281 (1989).
[16] U. Staufer, Ph.D. Thesis, University of Basel, Basel, Switzerland (1990).
[17] U. Staufer, L. Scandella, H. Rudin, H.-J. Güntherodt, and N. Garcia, J. Vac. Sci. Technol. B 9, 1389 (1991).
[18] R. Wiesendanger, M. Ringger, L. Rosenthaler, H.R. Hidber, P. Oelhafen, H. Rudin, and H.-J. Güntherodt, Surf. Sci. 181, 46 (1987).
[19] R. Wiesendanger, in: EPS-8, Trends in Physics, Proc. part II, ed. J. Kaczér, p. 567, Prometheus, Prague (1991).
[20] T.A. Jung, A. Moser, H.J. Hug, D. Brodbeck, R. Hofer, H.R. Hidber, and U.D. Schwarz, Ultramicroscopy (1992).
[21] H.J. Mamin, P.H. Guethner, and D. Rugar, Phys. Rev. Lett. 65, 2418 (1990).
[22] H.J. Mamin, S. Chiang, H. Birk, P.H. Guethner, and D. Rugar, J. Vac. Sci. Technol. B 9, 1398 (1991).
[23] C.X. Guo and D.J. Thomson, Ultramicroscopy (1992).
[24] T. Schaub, R. Wiesendanger, and H.-J. Güntherodt, Nanotechnology (1992).
[25] J. Moreland and P.Rice, Appl. Phys. Lett. 57, 310 (1990).
[26] R. Wiesendanger, I.V. Shvets, J.M.D. Coey, D. Bürgler, G. Tarrach, and H.-J. Güntherodt, (submitted).
[27] J.C. Slonczewski, Phys. Rev. B 39, 6995 (1989).
[28] R. Wiesendanger, H.-J. Güntherodt, G. Güntherodt, R.J. Gambino, and R. Ruf, Phys. Rev. Lett. 65, 247 (1990).
[29] R. Wiesendanger, D. Bürgler, G. Tarrach, A. Wadas, D. Brodbeck, H.-J. Güntherodt, G. Güntherodt, R.J. Gambino, and R. Ruf, J. Vac. Sci. Technol. B 9, 519 (1991).
[30] R. Wiesendanger, D. Bürgler, G. Tarrach, H.-J. Güntherodt, and G. Güntherodt, in:

Scanned Probe Microscopy, ed. H.K. Wickramasinghe, AIP Conf. Proc. Vol. 241, p. 504. AIP, New York (1992).

[31] R. Wiesendanger, D. Bürgler, G. Tarrach, T. Schaub, U. Hartmann, H.-J. Güntherodt, I.V. Shvets, and J.M.D. Coey, Appl. Phys. A 53, 349 (1991).

[32] R. Wiesendanger, D. Bürgler, G. Tarrach, I.V. Shvets, and H.-J. Güntherodt, Mat. Res. Soc. Symp. Proc. Vol. 231, 37 (1992).

[33] R. Wiesendanger, I.V. Shvets, D. Bürgler, G. Tarrach, H.-J. Güntherodt, and J.M.D. Coey, Z. Phys. B 86, 1 (1992).

[34] R. Wiesendanger, I.V. Shvets, D. Bürgler, G. Tarrach, H.-J. Güntherodt, J.M.D. Coey, and S. Gräser, Science 255, 583 (1992).

[35] R. Wiesendanger, I.V. Shvets, D. Bürgler, G. Tarrach, H.-J. Güntherodt, and J.M.D. Coey, Europhys. Lett. 19, 141 (1992).

[36] E.J.W. Verwey, Nature 144, 327 (1939).

[37] E. Wigner, Trans. Far. Soc. 34, 678 (1938).

[38] N.F. Mott, Metal-Insulator Transitions, Taylor Francis Ltd., London (1974).

[39] I.V. Shvets, R. Wiesendanger, D. Bürgler, G. Tarrach, H.-J. Güntherodt, and J.M.D. Coey, J. Appl. Phys. 71, 5489 (1992).

[40] R. Wiesendanger, in: New Concepts for Low-Dimensional Electronic Systems, ed. G. Bauer, Springer Series in Solid State Sciences, Springer, Berlin/Heidelberg (1992).

[41] S. Iida, K. Mizushima, M. Mizoguchi, K. Kose, K. Kato, K. Yanai, N. Goto, and S. Yumoto, J. Appl. Phys. 53, 2164 (1982).

[42] P. Bedrossian, D.M. Chen, K. Mortensen, and J.A. Golovchenko, Nature 342, 258 (1989).

[43] I.-W. Lyo and Ph. Avouris, Science 245, 1369 (1989).

[44] T.R. Albrecht, M.M. Dovek, M.D. Kirk, C.A. Lang, C.F. Quate, and D.P.E. Smith, Appl. Phys. Lett. 55, 1727 (1989).

[45] T.R. Albrecht, S. Akamine, M.J. Zdeblick, and C.F. Quate, J. Vac. Sci. Technol. A 8, 317 (1990).

[46] R. Landauer, Physica A 168, 75 (1990).

DIRECT WRITING OF METALLIC NANOSTRUCTURES WITH THE SCANNING TUNNELING MICROSCOPE

A. L. de Lozanne, W. F. Smith, and E. E. Ehrichs

Department of Physics,
The University of Texas
Austin, Texas 78712-1081

Abstract

We review our work on the fabrication of structures with sizes below 100 nm by using the Scanning Tunneling Microscope (STM) to decompose gaseous precursor molecules, a technique we call STM-CVD (Chemical Vapor Deposition). We start with our first demonstration and review the development of the technique, including work by other authors, culminating in a sophisticated UHV (Ultra-High Vacuum) STM combined with a Scanning Electron Microscope (SEM) that allows us to build high purity metallic structures at a desired point on the substrate. We review the application of STM-CVD to the deposition of cadmium, aluminum, tungsten, iron, and gold, as well as the etching of silicon.

Introduction

Metallic structures with sizes below about 100 nm display a variety of interesting quantum phenomena, like electron localization, universal fluctuations, magnetofingerprints and flux quantization[1]. The size of these effects usually grows as the metal structures are made smaller and smaller. Most conventional microfabrication techniques have been used to make these structures with sizes below 1 μm, and down to the limit of the particular technology. Some clever extensions of conventional techniques have been able to produce straight metal wires with only a 30 nm x 30 nm cross section[2].

Soon after the STM was invented it demonstrated the capability to probe the structure of surfaces with resolution down to the atomic level. The natural question to ask is whether the STM can also modify surfaces down to that level. Very early on, Abrahams et al.[3] showed that trenches could be made with some control on the surface of gold, down to a few tens of nanometers in size. To this date, the most advanced demonstration of surface modification is the beautiful work of Eigler et. al.[4], where single atoms are manipulated to form patterns. A review of a variety of fabrication techniques with STM has appeared recently[5].

The most common method of microfabrication involves the use of a resist which is exposed and developed to form a pattern. This pattern is later transferred to the desired material by etching, or by deposition and lift-off. Direct writing, on the other hand, is a process whereby the pattern is made directly, without the need for a resist, thus reducing the number of process steps. This is important when making the smallest structures, since every process step introduces more danger of losing resolution. Focused beams of photons, electrons and ions have been used for direct writing. A very convenient way of adding/removing atoms to/from the surface is to use surface chemistry. In the case of deposition, for example, the semiconductor industry has developed high purity organometallic and other precursor gases for most elements of interest.   These gases are used in conventional Chemical Vapor Deposition (CVD) to deposit large areas of thin films by breaking down the organometallic precursor with heat and/or a plasma discharge. In the case of direct writing, the energy to break the precursor molecules comes from the focused beam. The power of this technique has been demonstrated by the fabrication of a complete transistor device by direct writing with focused lasers onto a silicon chip that was never taken out of the fabrication chamber [6].

Our technique was inspired by the previous work on direct writing with focused lasers and gas precursors, as well as the work of Bard's group with localized electrochemical deposition using the STM[7]. We replaced the photons from the laser with electrons from the STM tip, with the idea of breaking down the gas precursors on a much smaller scale. After we submitted the first paper demonstrating the feasibility of STM direct writing[8] we learned that the idea had already been patented by Binnig, Rohrer and coworkers[9], and that they did not succeed in making it work after a short effort[10].

In the remainder of this paper we review the work done in our laboratory on the first demonstration and further development of STM direct writing with organometallic gas precursors, which we call STM-CVD. Our interest in this technique is to fabricate the smallest lines possible and to measure their electronic transport properties in order to study interesting quantum effects.

First Results

In order to demonstrate that it is possible to break organometallic gas molecules with electrons from the STM tip we built a very simple STM inside a chamber evacuated by a turbomolecular pump, as shown in Fig. 1. The chamber was hung from the ceiling with elastic cords and the STM was supported by springs within the chamber. No damping was added. The STM itself was a simple tripod scanner, made with three piezoelectric tubes (0.25in diameter, 1.5in long) mounted on a solid aluminum cube, which is illustrated in Fig. 2. The tip-sample approach was just a 7:1 reduction lever pushed by a 00-90 screw. While this system is very simple, it

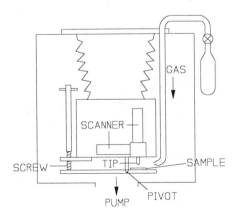

Figure 1. Sketch of the simple chamber and STM first used to demonstrate that it is possible to break organometallic molecules with the STM.

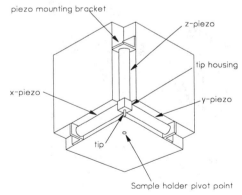

Figure 2. Schematic diagram of the high vacuum STM scanner included in Figure 1.

allowed us to demonstrate the principle of STM-CVD.

Our first depositions were made with dimethylcadmium and trimethylaluminum. We chose these gases because they are readily available and their photodissociation was known to occur at low energies (1-4 eV)[11-13]. This means that, in principle, it should be possible to operate in the tunneling regime of the STM. One of our first deposited features[8,14] is shown in Fig. 3. This STM image, 20 nm on a side, was taken immediately after depositing the mound by bringing the tip to the center of the field of view and putting a higher voltage (11.8 V, 500 nA) on the tip in the presence of 1 Torr of dimethylcadmium. Images of this silicon surface were also taken prior to the deposition to make sure that the surface was featureless on this scale.

As we tried to reduce the size of the deposits we reduced the pressure of the organometallic gas, until there was no gas added to the chamber, with the result that we were still able to write. In hindsight this was to be expected, since even a moderately good vacuum (10^{-8} Torr base pressure) and our simple ex-situ and in-situ cleaning of the silicon surface still left a few monolayers of organic contamination which was polymerized by the STM to form patterns. In fact, a similar technique called "contamination resist lithography" uses high energy (>50 keV) electron beams to make fine patterns on thin layers of organic contamination[15]. Our smallest patterns to date have been obtained with this contamination layer. The process is reproducible enough that we were able to make simple patterns under the control of a computer, as shown in Fig. 4.

Since our main goal is to produce metallic wires and not carbonaceous structures, we made larger deposits (1 μm lines) with dimethylcadmium and analyzed them, ex-situ, in a scanning Auger system. Smaller deposits would have been impossible to find due to the limited resolution of the Auger system. The results showed that we were indeed depositing cad-

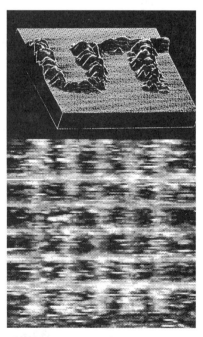

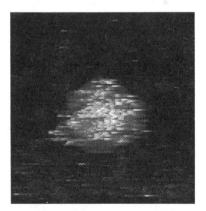

Figure 3. STM image of the deposits made with dimethyl-cadmium. The image is 200 nm on a side, and the deposit is about 80 nm in diam. and 20 nm high. As explained below, this deposit is likely to be mostly carbon.

Figure 4. Two sample patterns generated by polymerizing surface contamination with the STM tip. These STM images are 200 nm on a side and the smallest linewidth is 10 nm. The substrate is silicon.

mium only in the area of interest, but the deposits had large amounts of carbon, up to 60%[16].

We also attempted to measure the electrical properties of these deposits[17], which can give indirect information about purity of a metal. There are several complications in such measurements. The first is that in order to use the STM we need a conducting (or semiconducting) substrate, while the measurement of the fabricated wire requires an insulating substrate. The solution is simply to use a semiconductor substrate with doping just high enough to allow STM operation at room temperature. Since the measurements of the wire are most interesting at low temperature, where the semiconductor is effectively insulating, this provides an ideal solution. The second problem is to attach leads to the nanofabricated wire. Since the wire would be impossible to find after it is made, we opted to put the thin-film connection pads on the silicon surface prior to STM writing. The remaining problem is due to the limited range (20 µm in our case) of the STM. The pads must therefore be spaced by a few microns for the STM to be able to connect a line between them. Finally, we made an interdigitated pattern of electrodes, with 5 µm lines and spaces, so that the tip could easily find an electrode from each side. The net result of these experiments was that small lines were either too resistive or discontinuous to be measured, while micron-size deposits had a resistance lower than the thin-film electrodes used to measure them.

The problems with carbon contamination lead us to a short term search for precursor gases that do not contain carbon, and to a long term design and construction of a new UHV STM. The carbonless gas we tried was tungsten hexafluoride. As Fig. 5 shows, we were able to etch the silicon surface under some conditions, as well as deposit mounds[18]. The smallest pits made in silicon were 20 nm in diameter and about 10 nm deep. Tungsten hexafluoride is very corrosive for the vacuum pumps, therefore we moved on to the long term solution, which is described next.

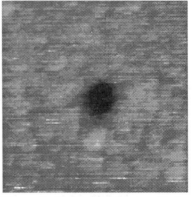

Figure 5. Etching of silicon with the STM and tungsten hexafluoride.

Ultra High Vacuum STM and SEM

Our first work with STM direct writing not only demonstrated the feasibility of the technique, but it also made it clear that a cleaner environment was necessary. Thus a UHV apparatus was designed an built. It was also evident that a four point measurement of the nanostructures was required for meaningful results. This mandated the design of a compact positioning mechanism to bring the tip over four prefabricated electrodes on the sample, and a way of guiding this process with both a large field of view and a high resolution. The positioning was implemented with homemade stick-slip motors, while the viewing was accomplished with a commercial Scanning Electron Microscope (SEM). This is a difficult combination, because the SEM has poor vacuum (10^{-5} Torr) and we required a UHV STM. The solution is to have the STM in a UHV chamber that plugs into the commercial SEM, and to have the SEM electron beam go through a small aperture which is the only opening between the UHV and non-UHV regions of the apparatus. The alternative would be to purchase a UHV SEM. This is not attractive because such instruments are very expensive, and they should not be exposed to the reactive gases that we use in our process. A further advantage of the commercial SEM is that it is conveniently available for other projects when it is not in use with the STM.

The implementation of the STM/SEM system is quite involved; we only give a few highlights here and refer to our publications for more details[19-23].

UHV STM

The design of our STM, sketched in Fig. 6, is fairly compact since it has to fit inside its own UHV vessel which in turn has to fit inside the SEM. We use a tube scanner to move the sample in three dimensions and we also move the tip in the z-direction (defined as the axis of the tip) with a concentric tube. The tube scanner is mounted on a commercial inchworm which provides long range z motion and fine tip-sample approach. The tip assembly is mounted on a polished ceramic substrate that rides on three ruby balls. The motion against friction and up/down a 45° incline is driven by homemade stick-slip motors. The friction between the ruby balls and the ceramic substrate, and against pressure clips, is necessary in order to keep the assembly from sliding downhill once the desired position is reached. The STM is supported by a stack of concentric stainless steel rings that have small viton spacers between them. Table 1 summarizes the motion of the eight degrees of freedom required to operate this (or any)

	X	Y	Z
STM scanner	20 μm	20 μm	3 μm
TIP-SAMPLE positioner	2 mm	2 mm	10 mm
STM-SEM positioner	5 mm	2 cm	not needed

Table 1. Range of the eight positioners of the STM/SEM.

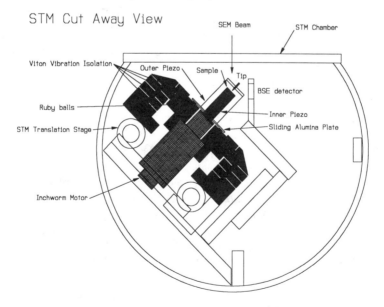

Figure 6. Schematic diagram of the STM. This view is from the left of Fig. 7, along the axis of the chamber. The SEM beam comes in vertically through a 300 μm aperture on the top.

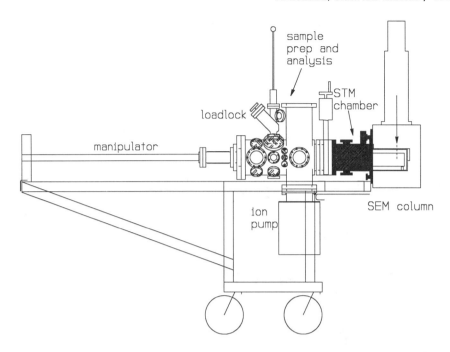

Figure 7. Schematic of the UHV STM chamber showing the sample/tip loadlock, manipulator and surface preparation area. The chamber is on wheels to move in and out of the SEM.

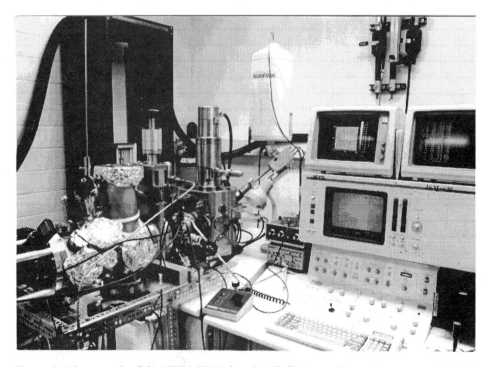

Figure 8. Photograph of the UHV STM chamber (left) plugged into the commercial SEM (right). The two monitors on top are for STM display and control.

STM/SEM. Figure 7 shows a sketch of the UHV chamber that houses the STM and provides a loadlock for tip and sample exchange, and in-situ surface preparation. This chamber is on wheels to enable it to be plugged into the commercial SEM, as shown in Fig. 8.

The best demonstration that all of the essential elements of this complex instrument work is to bring the tip to a unique location on the substrate under the guidance of the SEM, as shown in Fig. 9.

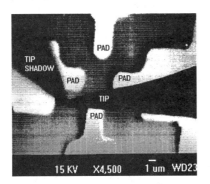

Figure 9. SEM image of a sharp tip hovering over a unique location on the surface. This location has four contact pads that were created by conventional techniques before loading the sample in the STM/SEM. These contacts are used for transport measurements presented below.

Experimental Results Obtained with the UHV STM/SEM

Figure 10 shows a line written using nickel carbonyl as the organic precursor. Auger measurements of larger deposits have shown that these deposits have at least 95% nickel. The pattern consists of a vertical wire, with voltage probes attached diagonally near the top and bottom. The left voltage probe appears discontinuous in the micrograph, however it was in fact electrically continuous, allowing a genuine four-probe measurement of the wire, as shown in fig. 11. The general behavior is qualitatively similar to that previously observed by a two-probe measurement [17] for larger and more contaminated aluminum deposits. As the temperature is lowered from room temperature, the resistance increases sharply. The observed characteristic temperature for the resistance rise is the same as for a blank test sample, indicating that the rise is due to freeze-out of conduction through the Si substrate. However, from the symmetry of the four-probe behavior with respect to the current path (fig.

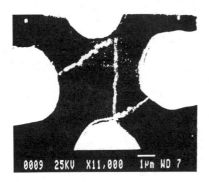

Figure 10. SEM images of lines connecting the contact pads of Fig. 9. The vertical line is the "sample", while the two diagonal lines are voltage probes. All lines are about 170 nm wide and 5 nm thick; the scale bar at the bottom is one micron long. The poor quality of the image is due to the fact that these lines are too thin for the sensitivity of the commercial SEM.

10), it would appear that much of the conduction through the substrate must be through the area directly under the nickel deposit. It may be that the process of writing creates a more highly doped region of Si directly under the nickel deposits. A more likely explanation is that there are one or more points in the sample where the nickel is very thin or narrow. At high temperatures, the current flows through the Si in these small regions. As the temperature is lowered, the Si freezes out, and the current flows through the Ni.

The inset to fig. 11 shows a magnified view of the behavior below 230 K. After the initial rise, the resistance decreases slowly as the temperature is lowered, with a temperature coefficient of resistance of 150 ppm/K. This is as expected for a very thin disordered metal film[24]. Below 50 K, there is a slight rise in the resistance. This may be due to freeze-out of small semiconducting regions in the sample or to the temperature dependence of the resistance of a parallel channel of nickel silicide (similar rises in resistance are observed in 6 nm-thick films of iron or cobalt silicide on silicon [25]). A more recent sample with similar dimensions shows the same qualitative behavior seen in Fig. 11.

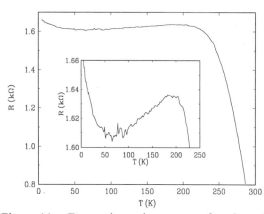

Figure 11. Four point resistance as a function of temperature for the sample shown in Fig. 2. The inset shows a close-up of the metallic-like region. The calculated low temperature resistivity is 35μΩ-cm.

The dimensions of the wire in Fig. 10 have been determined with an atomic force microscope: length=3.7μm, width=190 nm, and thickness=4nm. The resistance of the wire implies a resistivity of about 34 μΩ-cm at 300 K. Clearly, the resistivity of this wire is much less than the previous lowest figure obtained with this technique, 0.01Ω-cm [26]. Also, our estimate of the resistivity is consistent with Auger analysis [19] of larger deposits, which indicated very high (95%) Ni content.

The narrowest line that we have obtained in the STM/SEM system is 35 nm wide and 4 nm thick, as shown in Fig. 12. While this line was quite uniform and longer than 2 μm, unfortunately it did not reach the contact pads, so that a resistance measurement was not possible. In the future we will reduce the distance between the contact pads to have a higher probability of obtaining a continuous line between the pads. In our initial work demonstrating this technique we obtained lines as narrow as 10 nm [8] by just using the contamination present on a silicon surface as a "precursor". We therefore hope to make 10-nm lines with very high purity in our UHV STM/SEM and to measure their magnetoresistance in the near future.

Figure 12. AFM image of a nanoscopic nickel line, only 35 nm wide and 4 nm thick. The image is 1.5μm on a side.

Work on STM-CVD by other groups

One year after our first paper on STM-CVD [8] McCord et al. also succeeded in obtaining STM deposition from gas precursors [26]. They used tungsten hexacarbonyl and gold trifluoroacetate to make dots as small as 10 nm in diameter. They also deposited a large tungsten line (3.5μm long, 1μm wide and 50 nm thick) between two electrodes to estimate a resistivity of 0.01 Ohm-cm. This indicates a fair amount of impurities in the metal. Auger measurements on 10μm square deposit confirm this since the composition was 48% tungsten, 40% carbon and 12% oxygen.

The most interesting scientific use of STM CVD to date is the study of nanometer-scale magnets carried by Awschalom, McCord and Grinstein[27] . They decomposed iron pentacarbonyl with the STM and formed iron-rich particles with sizes ranging from 15nm x 38nm to 15nm x 70nm. The particles were fabricated inside the input coil of a dc SQUID (Superconducting QUantum Interference Device), which allowed the most sensitive measurement of the magnetic susceptibility of the particles as a function of frequency. They observed a clear size dependence of a resonance peak in the susceptibility, which is not understood, and they did not see clear evidence of the MQT (Macroscopic Quantum Tunneling) effects they were trying to find (such effects were later observed in a biomolecule [28]).

Nayfeh's group has added an extra degree of freedom by using a focused laser to help the writing process[29]. The laser can photodissociate the precursor gas in the vicinity of the tip, and metal atoms can be steered by the field of the tip. Clear mounds, 16nm x 16nm in size, were written in this fashion by dissociating trymethylaluminum on graphite. Smaller features, down to a few atoms in size, were also obtained. The chemical composition of any of these deposits is not known, however. Unknown composition is a universal problem for any technology that generates nanometer-size structures. Our measurements address this problem by measuring the electrical properties of the nanostructure. More recently[30,31], this group has used the residual contamination on a silicon surface to form patterns, without the use of a laser, in a similar fashion to our previous work. It is interesting to see that a double tip is able to write closely-spaced double lines[31], which is a prerequisite for implementing arrays of tips to enhance the throughput of STM lithography. Of course, a useful implementation would require, at the very least, independent control of the voltage on each tip.

Future Directions

The microelectronics revolution has taught us that smaller is faster, cheaper and altogether better. Therefore any technique, like STM-CVD, that has a capability for making structures as small as 10 nm is very attractive. Unfortunately STM-CVD suffers from the limitation on throughput shared by all "serial" techniques, namely that the tip has to visit all locations where something is to be built. This is in contrast with technologies like optical lithography which can process a whole wafer, or at least one die, in one step. One can imagine ways of circumventing this throughput limitation by, for example, using a one- or two-dimensional array of tips. As a simplest implementation, each tip would require a separate electrical lead and z-motion. Full, independent x-y-z motion of each tip would be useful, though not required. The feasibility of such an array of parallel tips remains to be demonstrated; the "STM on a chip" developed by Quate et al. [33,34] is an important step towards this goal.

A second trick to improve throughput is to use STM-CVD only for "seeding" the desired pattern with only a few monolayers. The rest of the material can be grown over the whole wafer by selective CVD, whereby the material grows only on the seeded areas. This is limited to a few combinations of substrate and deposited materials, however, so it may not be useful in most situations.

At this time the only practical applications of STM-CVD that can be envisioned for the next few years are those that require only a few simple patterns. Possibilities include the repair of submicron-linewidth masks for optical or x-ray lithography, custom wiring of programmable chips, and building prototype electronic devices. Of course, scientific applications will benefit greatly, since the behavior of matter in the 1-100 nm scale is just beginning to be explored. Perhaps the most important contribution of STM-CVD to Science and Technology will be to make it possible for many laboratories to own a tool capable of making structures smaller than 0.25 µm. Currently both electron and ion beam lithography systems are commercially available, but at a prohibitive cost.

One scientific, and perhaps some day commercial application of STM-CVD is the fabrication of a "quantum charge pump", a device invented by Niu in 1985[32]. The QCP uses four metallic electrodes to divide a 2D electron gas into four quadrants. At the center of the four quadrants a quantum dot is formed which can hold only one electron at a time. An appropriate sequence of voltages on the electrodes makes an electron from one quadrant enter the dot, be trapped there, and exit into any quadrant. One electron is thus pumped every cycle. Since the frequency of the driving voltages can be determined with high precision, the QCP becomes a standard for current. Since the performance of the QCP is better as it is made smaller, STM-CVD is an ideal tool for its implementation. Our current goal is to build a QCP with 100 nm minimum feature sizes, and to later scale it down to 10 nm. The smaller QCP's should operate in liquid nitrogen with higher precision than anything currently available.

Ultimate resolution

A final question to consider is whether there is a resolution limit for STM-CVD. As reviewed here, several groups have made controlled structures with sizes of 10 nm and larger. Thus, while there is no fundamental reason to expect a resolution limit, 10 nm seems to be a working limit. There is, however, evidence from the work of Lyo and Avouris[35] that single molecules on the surface of silicon can be dissociated, which should make near-atomic resolution possible.

Conclusion

STM-CVD is an inexpensive technique for making structures in the 10-100nm size range at present, and perhaps 1-100nm in the near future. Our work and that of other groups have demonstrated that this technique works with a variety of precursor gases to deposit a variety of elements on different surfaces. We have demonstrated that it is possible to make nanostructures with high purity.

Therefore we hope that this will encourage other groups to use this technique to fabricate prototype electronic devices and interesting nanostructures for basic measurements. The low cost of STM-CVD, compared to any technique capable of producing the same size structures, should make this an attractive technique for many laboratories, particularly those at universities.

Acknowledgements

Our work is supported by the National Science Foundation (DMR-8553305), the Defense Advanced Research Projects Agency, the Welch Foundation, and a Hertz Foundation fellowship (E. Ehrichs). We are grateful to Jesse Martinez for performing the contact-pad depositions. Some of the earlier results were obtained by R. Silver and S. Yoon, as indicated in the references.

REFERENCES

1. See, for example, *Nanostructures and Mesoscopic Systems*, edited by M.A. Reed and W.P. Kirk, (Academic Press, 1991).
2. D.E. Prober, "Microfabrication techniques for studies of percolation, localization and superconductivity, and recent experimental results," in *Percolation, Localization and Superconductivity*, edited by A.M. Goldman and S.A. Wolf, NATO ASI Series Vol.109 (Plenum Press, NY, 1984) 231-266.
3. D.W. Abraham, H.J. Mamin, E. Ganz, and J. Clarke, "Surface Modification with the STM," *IBM J. of Res. & Develop.* **30**(5), 492-499 (1986).
4. D. M. Eigler and E. K. Schweizer, Nature **344**, 524 (1990)
5. U. Staufer, "Surface Modification with a Scanning Proximity Probe Microscope" in *Scanning Tunneling Microscopy II: Further Applications and Related Scanning Techniques*, R. Wiesendanger and H.-J. Güntherodt, editors. (Springer Series in Surface Science, Vol. 28, pp 273-300; Springer-Verlag, New York, 1992).
6. B.M. McWilliams, I.P. Herman, F. Mitlitsky, R.A. Hyde, and L.L. Wood, "Wafer-scale laser pantography: Fabrication of n-metal-oxide-semiconductor transistors and small-scale integrated circuits by direct-write laser-induced pyrolytic reactions," *Appl. Phys. Let.* **43**, 946 (1983).
7. C. W. Lin, F. R. Fan and A. J. Bard, J. Electrochem. Soc. **134**, 1038 (1987)
8. R. M. Silver, E. E. Ehrichs, and A. L. de Lozanne, "Direct writing of metallic features with a scanning tunneling microscope," Appl. Phys. Lett. **51**, 247 (1987)
9. G. Binnig, R. M. Feenstra, R. T. Hodgson, H. Rohrer and J. M. Woodall, US patent No. 4,550,257.
10. H. Rohrer, private communication.
11. C. J. Chen and R. M. Osgood, J. Chem. Phys **81**, 327 (1984)
12. C. J. Chen and R. M. Osgood, J. Chem. Phys **81**, 318 (1984)
13. C.F. Yu, F. Youngs, K. Tsukiyama, R. Bersoh, and J. Presses, "Photodissociation dynamics of cadmium and zinc dimethyl," J. Chem. Phys. **85**, 1382 (1986).
14. E. E. Ehrichs, R. M. Silver, and A. L. de Lozanne, "Direct writing with the STM," J. Vac. Sci. Technol. **A6**, 540 (1988)
15. R.B. Laibowitz and C.P. Umbach, "Fabrication of sub-0.1 μm fine metal lines using high resolution electron beam techniques with contamination resist," in *Percolation, Localization and Superconductivity*, edited by A.M. Goldman and S.A. Wolf, NATO ASI Series Vol. **109** (Plenum Press, NY, 1984) 267-286.
16. E. E. Ehrichs, S. Yoon, and A. L. de Lozanne, "Direct Writing of 10 nm Features with the Scanning Tunneling Microscope," Appl. Phys. Lett. **53**, 2287 (1988)
17. E. E. Ehrichs and A. L. de Lozanne, "Fabrication of nanometer features with a scanning tunneling microscope," in *Nanostructure Physics and Fabrication*, M. A. Reed and W. P. Kirk, eds. (Academic Press 1989) pp. 441-446.
18. E. E. Ehrichs and A. L. de Lozanne, "Etching of Si(111) with the STM," J. Vac. Sci.

Technol. **A8**, 571 (1990)
19. E. E. Ehrichs, W. F. Smith, and A. L. de Lozanne, "An STM/SEM system for the fabrication of nanostructures," J. Vac. Sci. Technol. **B9** (2), 1381-1383 (1991).
20. Alex L. de Lozanne, Walter F. Smith, and Edward E. Ehrichs, "Synthesis and study of nanostructures," in <u>Scanned Probe Microscopy</u>, K. Wickramasinghe and F.A. McDonald, eds., AIP Proc. #214, 1992, pp 459-461.
21. W. F. Smith, E. E. Ehrichs, and A. L. de Lozanne, "Direct writing of nickel wires using a scanning tunneling microscope / scanning electron microscope system," in <u>Nanostructures and Mesoscopic Systems</u>, M.A. Reed and W.P. Kirk, eds. (Academic Press, 1991) pp85-94
22. E. E. Ehrichs, W. F. Smith, and A. L. de Lozanne, "Four probe resistance measurements of nickel wires written with a scanning tunneling microscope / scanning electron microscope system," *J. Ultramicroscopy* **42-44**, 1438-1442 (1992)
23. W. F. Smith, E. E. Ehrichs, and A. L. de Lozanne, "A novel ultrahigh vacuum manipulator using a shape-memory alloy actuator," J. Vac. Sci. Technol. **A10**, 576 (1992).
24. See, for example, <u>Experimental Techniques in Condensed Matter Physics at Low Temperatures</u>, edited by R. C. Richardson and E. N. Smith, Addison-Wesley, Redwood City CA, 1988, p. 201.
25. R. Miranda, private communication.
26. M. A. McCord, D. P. Kern, and T. H. P. Chang, J. Vac. Sci. Technol. **B6**, 1877 (1988)
27. D. D. Awschalom, M. A. McCord, and G. Grinstein, "Observation of macroscopic spin phenomena in nanometer-scale magnets," Phys. Rev. Lett. **65**, 783 (1990)
28. D. D. Awschalom, J.F. Smyth,, G. Grinstein, , D.P. DiVicenzo, and D. Loss, "Macroscopic quantum tunneling in magnetic proteins," *Phys. Rev. Lett.* **68** (20) 3092-3095 (1992).
29. S.T. Yau, D. Saltz, and M. H. Nayfeh, J. Vac. Sci. Technol. **B9** (2), 1371-1375 (1991)
30. S.-T. Yau, X. Zheng, and M. H. Nayfeh, "Nanolithography of chemically prepared Si with a scanning tunneling microscope," Appl. Phys. Let. **59** (19) 2457-2459 (1991).
31. X. Zheng, J. Hetrick, S.-T. Yau, and M.H. Nayfeh, "Parallel fabrication on chemically etched silicon using scanning tuneling microscopy," Ultramicroscopy **42-44**, 1303-1308 (1992).
32. Q. Niu, "Towards a quantum pump of electric charges," *Phys. Rev. Lett.* **64** (15) 1812-1815 (1990).
33. S. Akamine, T.R. Albrecht, M.J. Zdeblick, and C.F. Quate, "Microfabricated STM," IEEE Electron Device Lett. **10**(11) 490 (1989).
34. T.R. Albrecht, S. Akamine, M.J. Zdeblick, and C.F. Quate, "Microfabrication of integrated scanning tunneling microscope," J. Vac. Sci. Technol. **A8** (1) 317-318 (1990).
35. G. Dujardin, R.E. Walkup and Ph. Avouris, "Dissociation of individual molecules with electrons from the tip of a scanning tunneling microscope," Science **255**, 1232-1235 (1992)

Fabrication of Nanometer Scale Structures
Munir H. Nayfeh
Department of Physics
University of Illinois at Urbana-Champaign
1110 West Green Street
Urbana, Illinois 61801

Abstract

We have developed STM based techniques to fabricate nanometer scale structures at room temperature. We have made structures whose sizes range from a few hundred nanometers down to the size of individual atoms or molecules, on graphite, passivated silicon, photoresist coated silicon, and organometalic coated silicon surfaces. The technique utilizes processing with the biasing voltage / current of the tip of the microscope and tunable laser radiation coupled to the gap induces multiphoton excitation or ionization processes of precursor gasses thus providing material selectivity. On the other hand, at small enough tunneling gaps, the chemical potential collapses allowing the tip to suck material off the surface hence producing grooves. We have been able to fabricate continuos micrometer long lines of smallest widths ever (as small as 40 A). We have already used this capability to fabricate all sorts of two-dimensional patterns: triangular, rectangular, circular, parallel lines, grids, and others in the shape of some alphabets. In addition, we are in the process of integrating this capability (as a masking step) with novel molecular beam epitaxy (MBE) methods to fabricate and analyze two and three-dimensional nanometer scale structures such as quantum wires and dots, quantum gratings, arrays of quantum dots etc. The integration would provide structures of improved quality, purity, and of well controlled material composition which can be embedded in other devices . We are presently using these techniques to construct and test the quantum interference transistor, a micrometer size MOSFET with a nanometer scale grating or grid embedded in its gate area. These advances have important implications to mass storage of in formation which may lead to great reductions in the sizes of electronic circuits and devices and hence lead the microelectronics industry into a new regime of fineness--nanoelectronics [1].Eventually in this regime electronic equipment may operate on quantum mechanics so theywork faster and with less heat dissipated.

I. Introduction

Recently direct production of patterned features on a substrate was achieved using localized laser initiated chemical reactions. The process is called direct writing with lasers[2-3]. In direct writing, an ultraviolet or visible laser beam is focused onto the surface of a semiconductor wafer that is in a cell containing the appropriate reagent gases. Typically, argon-ion laser radiation at 514 nm is used directly; alternatively, the beam is first passed through a nonlinear crystal to obtain 257-nm radiation. If the laser wavelength is in the ultraviolet region, the laser light causes photochemical reactions to occur. Atoms that may either etch, deposit on, or dope a solid surface are produced. If the laser wavelength is in the visible region and the substrate absorbs light at the laser wavelength, surface heating occurs and the reactions are initiated by thermal chemistry.

The writing resolution is limited by the size of the laser spot. Focusing of visible laser beams can produce spots as small as .5 micrometer diameter; however it is not possible to get spots smaller than this because of limits imposed by diffraction. Because of their shorter wavelengths, ultraviolet light can be focused to submicrometer dimensions before the limits imposed by diffraction are reached, so photolithography with ultraviolet rather than visible light offers a decrease in the size of features that can be produced by the lithographic process.

Laser writing using visible lasers with resolution smaller than the spot size, i.e. less than .5 micrometer, has also been achieved by exploiting some nonlinear properties of the interaction. Ehrlich and Tsao[2,3] have written doped lines 0.2 micrometer wide in a silicon substrate using these nonlinear effects, thus pushing the technique to its utmost limit. While conventional techniques are satisfactory for most production operations, there are cases where it would be convenient to produce the patterned layer directly. These cases include the repair, design, and modification of circuits. Direct writing may even make it possible to monitor a device's performance during the fabrication.

More recently the characteristics of fabrication with electron or ion beams, have been refined by using the corresponding beams of the tunneling gap of a scanning tunneling microscope (STM), a device that was recently invented for observation of surfaces with atomic resolution [4]. For example, there have been experiments that used the scanning tunneling microscope (STM) to make nanometer scale features on chemically prepared surfaces[5,6]. In these experiments voltage pulses imposed across the tunneling gap were used to generate features. It was found that the process proceeds from ambient conditions to vacuum conditions. In the vacuum stud[5] it is believed that raised features on the surface are produced by chemical reactions in the contamination resist, however, quantitative characterization of the process is not available. On the other hand, the air study[6] indicates that the features are a result of differences in the electronic properties between the modified and unmodified regions brought about by oxidation than real topographic structures. Other experiments have utilized transplanting atoms or clusters of atoms from the tip to the surface[7] or from the surface to the tip and back to the surface[8]. These procedures succeeded in generating structures of nanometer scale.

II. Experiment And Results

We present schemes for deposition of nanometer structures4. The process utilizes the combined effects of the laser radiation, and the tunneling gap of a scanning tunneling microscope (see Fig. 1)[9,10]. The process is obviously highly nonlinear that

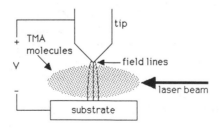

FIG. 1. Schematic of the STM tunneling junction bathed in laser radiation.

may involve several effects. These include excitation, ionization, heating ,melting and evaporation by laser radiation or electrons, field ionization, chemical interactions, and material transfer between the tip, the sample, and the precursor gas. One or several of these effects may play a role depending on the type of application sought , however there is no complete control over the effects at this time. We describe below some results when we we have fabricated nanometer scale deposits of dimensions as small as 1 x2.4 x 1 nm on graphite surfaces using laser induced fragmentation and ionization of Trimethylaluminum [TMA, Al(CH3)3] in the surface-tip field of a scanning tunneling microscope [9,10].

Radiation at 4300 A (2.9 eV) from a tunable pulsed dye laser pumped by a 308 nm eximer laser bathes the tip-surface in a grazing angle configuration, and focuses to a spot at the tip by a 15 cm focal length lens. The region is filled with 10-4 Torr of TMA gas. The idea of deposition relies on a multiphoton process for fragmentation and ionization of TMA.It is known that a three-photon process using radiation of wavelength in the range 3,900-4,600 A breaks the carbon-aluminum bonds, generating aluminum atoms and other radicals[11]. The wavelength of the radiation is short enough such that other subsequent multiphoton processes can occur such as ionization of the generated radicals, and Al atoms. Moreover, a competing process involves four photon ionizations of TMA, whose ionization potential is in the range 9-10 eV, yielding molecular ions.The ions were guided to the surface by the electric field between the STM tunneling tip and the surface. The nature and degree of the sticking depends on the materials of the deposits and the surface. Since the effective field of the tip has been confined laterally to a few nanometers, the deposition was also controlled with such resolution. The structures deposited were imaged subsequently with the STM.

The radiation of wavelength 4300A (photon energy of 2.9 eV) can photoionize TMA via a four photon process since the ionization potential of the molecule is in the range 9-10 eV. Moreover, it can dissociate TMA and happens to be near (by design) the multiphoton resonances of the Al atom. Therefore, the laser can be tuned to certain resonances while still being able to photodissociate the TMA molecule. The laser with wavelength at 430 nm excites the freed Al atom from the ground state 3p11PJ by the absorption of two photons. The excited Al atoms can further absorb a third photon, which ionizes them. It is seen that tunability can be used to enhance the deposition of Al atoms over other species, resulting in selective deposition. Moreover, TMA is chosen because at room temperature, the molecules are dimerised to form Al2(CH3)6, which is a liquid with a vapor pressure of 8.4 Torr. The high vapor pressure allows experiments to be performed at room temperature without major condensation problems. In addition, TMA has a relatively high pyrolytic decomposition temperature of 350C, which eliminates the possibility of pyrolytic deposition when slightly elevated temperature study is needed, or when the laser beam grazes the surface.

We used highly oriented pyrolytic graphite (HOPG). The surfaces were cleaved with a razor blade, and then a few layers were peeled off with scotch tape in order to prepare a flat sample. The sample is loaded and a vacuum of 10-9 Torr is established. This is followed by going through the alignment procedure of the laser beam. In the procedure the radiation is attenuated drastically in order to avoid any effect that might not be accounted for. The alignment on the tip-surface region is observed with an optical microscope that gives a magnification of 100, and monitored on a CRT screen via a TV camera. TMA is then introduced via a gas handling system which allows for static fills or a flow mode.

We made a number of deposits using single laser pulses with a tunneling current of 1 nA and different tip biasing voltages, ranging from 0.8 to 3 V. Figure 2 shows a deposit made with one laser pulse and a tip bias of 1 V. The deposit appears near the middle of the image, which also shows the individual graphite atoms. The deposit appears to be made of four components. The line profile in Fig. 2 is made through two of the components of the deposit and through a row of graphite atoms. The

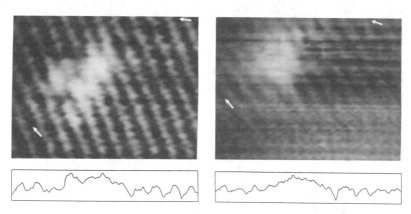

FIG. 2. STM image (27A x 27A) of a deposit. The line length between the arrows and the peak-to-valley height of the line profile are 28.5 and 1.4 A.
FIG. 3. STM image (43 A x 43 A) of a deposit. The line length between the arrows and the peak- to-valley height of the line prifile are 42 and 4 A.

profile clearly shows the corrugation of the equally spaced graphite atoms and the corrugation of the two components rising above them. Using 2.46 A for the spacing of the graphite atoms, we can measure the dimensions of the individual components of the deposit. Each of the components appears to have a size of approximately 5 x 4 A, and they do not have the shape of a TMA molecule. Therefore, they are likely to be a cluster of aluminum atoms rather than molecules. We believe the deposit is composed of four or five aluminum atoms, since the laser wavelength was tuned to an aluminum resonance line, where the yield of aluminum ions is about six times more than the yield of other species. It should be mentioned that direct identification of the deposit is not yet possible because of the lack of a theory which describes the features of small numbers of aluminum atoms on graphit surfaces. The line profile also shows some variation in the corrugation of the deposited atoms. This variation may be caused by the surface topography of graphite, which has three distinct atomic sites, namely, the A, and B, and the hollow (H) sites. Since the deposit consists of a small number of atoms, it may reveal to a certain degree the topography of the substrate surface.

Figure 3 shows another interesting deposit. It was made with one laser pulse and a tip bias of 1.1 V. Its area is approximately 14 A x13 A and the individual atoms in the deposit are not resolved. The line profile through the deposit gives a corrugation height of 2.0 A above the corrugation of the graphite atoms, compared to 0.6 A for the previous case (Fig. 2). Using 0.6 A for the corrugation height of a single layer, this corresponds to about three atomic layers, or a total of approximately 20 atoms in the deposit.

When the laser power is reduced, and the biasing voltage is increased, the field and current of the gap dominates and we tend not to break the molecules. Figure 4 shows one of our smallest deposits produced under these conditions namely a single

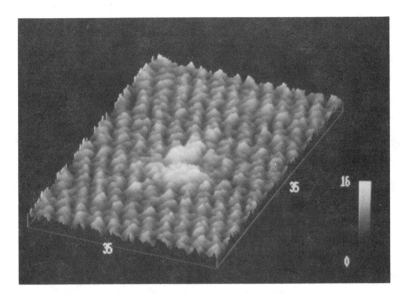

FIG. 4. One of our smallest deposits on a graphite surface, a TMA dimer.

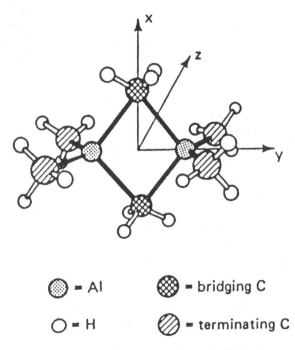

◯ = Al   ◉ = bridging C

○ = H   ⊘ = terminating C

FIG. 5. Schematic of the molecular structure of a TMA dimer.

molecule of TMA. The image (35A x 35 A) was taken after the application of a .02s, 4V pulse to the tip. A deposit, presented in light color, appears near the middle of the image, which also resolves the individual graphite atoms. The deposit appears to have two resolved components. Thus we believe our scan indeed resolves the substructure of the deposit, and the measured size agrees with the known size of the molecule as discussed below.

It is worthwhile to discuss qualitatively the doublet like shape of the deposit. For reference, a schematic of the structure of a TMA dimer is shown in Fig. 5. The shape is consistent with the fact that all Al-Al and C-Al interactions are bonding and hence electron pairs go into the bonding orbitals, distributing the electron density evenly to the atoms involved. Furthermore, the shape agrees with the fact that all interactions between methyl groups are nonbonding (the orbitals are empty). Therefore, the dimer effectively consists of two semilocalized three-center $CAl2$ bonds. An estimate of the dimensions of a TMA dimer can be made for comparison. Using 1.9 A for Al-C bonds, 1.0 A for C-H bonds, and 122 degrees for the bridging carbon bonds from the molecular axis, one gets satisfactory values of 5, 5, and 10 A for the width, corrugation, and length. Other deposits we made also show the doublet structure of the molecule.

We now examine the possibility of erasing and its addressability. Figure 6a gives a graphite surface with several nanometer structures. The tip was positioned directly above one of the structures (labeled by an arrow). The surface was then irradiated by a single pulse that bathes the gap. Figure 6b, taken right after, shows that this structure has disappeared while all of the other structures have not been affected.

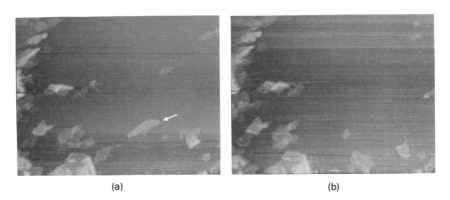

FIG. 6. (a) STM image (255 nm x 255 nm) of a graphite surface, showing several deposits. The tip was placed above the deposit indicated by the arrow, and a laser pulse was activated. (b) STM image of the same area showing that the indicated structure has disappeared.

Since all of the structures got exposed to the laser radiation as the beam cross section was larger than the area of the sample, then it is clear that the laser in this technique can only affect the structure placed right under the tip. This spatial selectivity of the erasing scheme (addressability) constitutes the best resolution achieved so far and we believe it will have tremendous implications.

In another development, we have used our technique for depositing material in grooves and holes that have been prepared by other means[10]. We drill holes in a graphite surface by a high voltage pulse across the gap. The STM-laser technique is then used to fill the holes. To illustrate the power of our technique, we show in Fig. 7 a series of images of the growth of the deposit that eventually fills the hole. Moreover, our measurements show that such fillings require higher laser intensities to erase than the deposits made on flat surfaces. This might add another dimension and feature to the versatility of the technique.

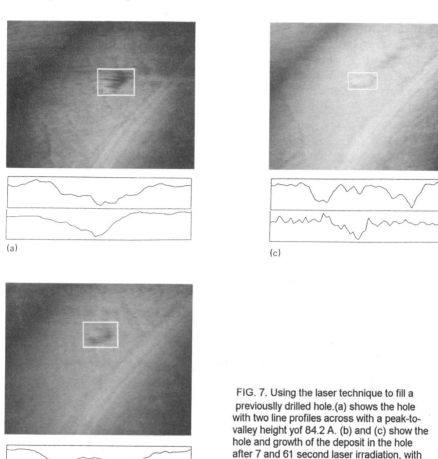

FIG. 7. Using the laser technique to fill a previouslly drilled hole.(a) shows the hole with two line profiles across with a peak-to-valley height yof 84.2 A. (b) and (c) show the hole and growth of the deposit in the hole after 7 and 61 second laser irradiation, with peak-to-valley of 46 and 37A respectively.

We also fabricated nanometer scale lithographic structures on silicon. Silicon is reactive and forms "contamination" layers of the percursor gas. In this case the process proceeds by electron polymerization and activation of the reaction of the gas as well as of the absorbed layer with the silicon surface. The substrates used were n-type silicon (111) samples with a conductivity of 0.1 Ohm-cm. The samples were chemically prepared with the method of Ishizake and Shiraki [12] in which 49% HF solution is used in the finishing step to remove any oxide on the sample surface. An as-prepared sample was loaded into the STM, which was immediately loaded into a UHV chamber. The chamber was then pumped down to a pressure of 8 x 10-9 Torr. The sample surface was first imaged with the STM operated in the constant-current mode. Images of the surface show typical topography where residual surface adsorbates appear as clusters with an average size of 20nm. The sample is then exposed to TMA at a pressure of 10-4 Torr for a certain period of time to form an adlayer, and then the gas is pumped out or changed to a lower value. After positioning the tip over a certain point on the surface, and selecting the tip-surface spacing by establishing the appropriate tunneling current and biasing voltage, a certain voltage pulse is applied at the tip while holding the tip in place. The surface is then immediately scanned (no more than 20 seconds delay). As an example we show in Figure 8 a three-dimensional view of a mound that was made using a single pulse of 5 volt amplitude, .2 second duration, and with 1nA current.

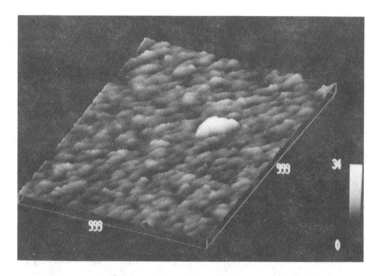

FIG. 8. A three dimensional STM image of a 12 x 14 x 16 nm mound on H - passivated silicon. Hydrocarbon species appear as clusters with an average size of 10 nm.

We believe that the features generated with this process are real topographic structures. Concerning the mechanism of the process, we share the same light as the authors in ref. 5 that the organic species adsorbed on the surface are polymerized upon electron exposure as first discovered by Laibowitz et al.[13] In the process, the tunneling electrons are used as electron beam to expose the carbonaceous species adsorbed on the sample surface. The resultant raised features, therefore, are real topographic structures due to the build-up of the polymerized species. In recent papers[11] we reported on some aspects of this phenomenon and presented current- voltage measurements to show that the fabrication results in a local transformation from a Schottky junction behavior to a MIS junction behavior, indicating the formation of topographic structures.

It is also interesting to note that the process can be reversible. Our results indicate that complete or partial erasure of mounds could be made with additional processing. In the procedure we position the tip right over a structure that has been previously made, and apply a series of pulses. Figures 9 (a) and 9 (b) show that a single

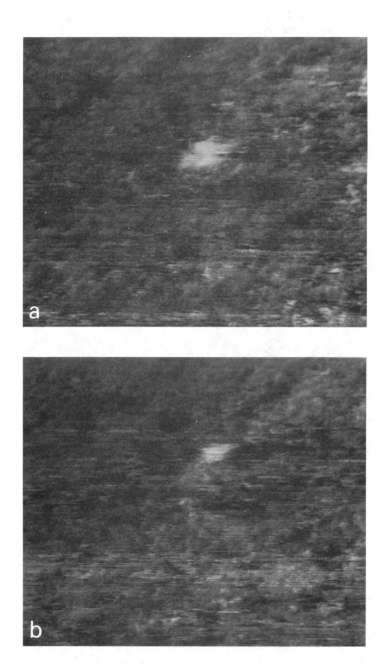

FIG. 9. A demonstration of the erasing cabability ot the STM by placing the tip over a section of the structure in (a) and firing a voltage pulse to remove the addressed section as shown in (b).

pulse erased the lower part of the structure with the upper part essentially preserved, thus effectively reducing the size of the structure. We find that the polymerization and adhesion to the substrate are strong enough as not to be erased by single pulses. Usually multiple pulses can achieve complete erasure of the structures Additional fabrication processes at nearby sites, with distances longer than 10nm, however, hardly affect the integrity of a given structure.

The next series of measurements involve continuous fabrication. For this purpose the tunneling current and bias are set to the appropriate values. The voltage across the gap is now increased to the required level for fabrication while simultaneously the tip is externally set in motion at a given speed. The motion and the high voltage are then simultaneously terminated at the end of the fabrication. Figure 10 gives a line fabricated using this procedure. In Figure 11 , we present several patterns made with this process. These patterns show the ability of making arbitrary shapes provided by the process.

FIG. 10. A line fabricated using a TMA contamination layer on H - passivated silicon.

Another procedure we employed for fabrication might be best described by the terms "shading". In this case a certain area of a surface is chosen and its coorindates (or boundary) is fed into the computer. The process then proceeds by fabricating 256 close lines inside the area, effectively, shading the region uniformly. Figure 12 is an example of such fabrication on silicon where three boxes of 500 x 500 A each were shaded.

We refer the reader to Reference 11 for more details on the basic principles of the fabrication process on silicon . These include several issues such as efficiency, resolution, repeatability, reproducibility, dependence on the voltage, current, and duration of the pulses, tip-surface distance, conditions of the surface, adjacent processing, repeated processing and reversibility, etc.

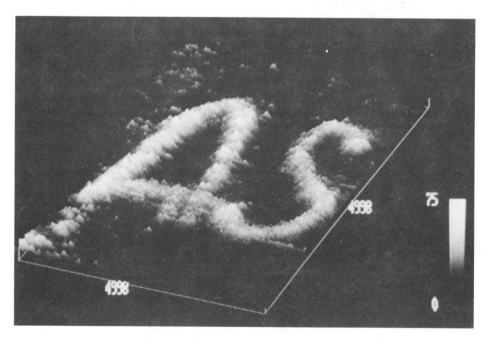

FIG. 11. Several two dimensional patterns fabricated on H - passivated silicon.

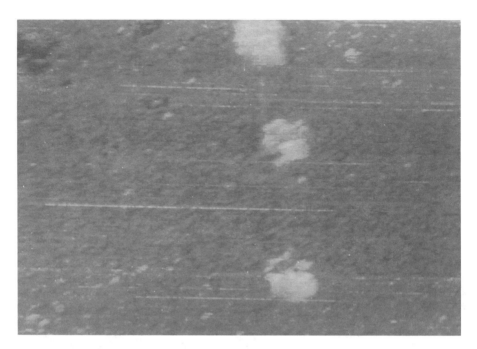

FIG. 12. The three structures are fabricated by fabricating 256 close lines within the areas of each of the resulting structures.

More recently, the high degree of control on interdistances of STM has revealed another promising feature that is useful for surface modification, namely control of the chemical potential between the tip and the surface[14]. At small enough interseparations, the chemical forces acting on the surface become very strong, and even comparable to the binding forces involved. This chemical effect combined with the strong field effect were recently used to demonstrate the ability of the STM to remove atoms with atomic resolution, off a clean silicon surface, a surface that involves strong and covalently bonded atoms[15]. Specifically, by combining the strong field effect and the chemical force effect, silicon atoms and silicon clusters up to tens of atoms were reproducibly transferred from the surface to the tip using negative biasing at the tip. Moreover, the clusters removed were subsequently redeposited on other locations on the surface using opposite biasing.

Here we report on successful fabrication of nanoscale grooves on a silicon surface that has been prepared by chemical etching (see ref 1). Micrometer long continuous grooves that are as narrow as 50 A FWHM were produced with good uniformity on H- passivated silicon under ultrahigh vacuum conditions. Figure 13 gives a three-dimensional image of silicon surface showing a silicon surface showing a groove pattern etched using a tip bias of 400 mV and 1 nA tunneling current. The pattern consists of four straight segments meeting at right angles, 50 A for the FWHM of the groove and 15 A for its depth. Figure 14 gives several line profiles across different parts of the groove to illustrate its quality. A closer look at the channels show some pile up of material at the sides. If one assumes near "local" redistribution of the material after it gets picked up by the tip, i.e. the material gets shaken off locally, then one concludes that the actual depth of the material that has been excavated is 9 A. This

FIG. 13. A groove pattern etched on H - passivated silicon.

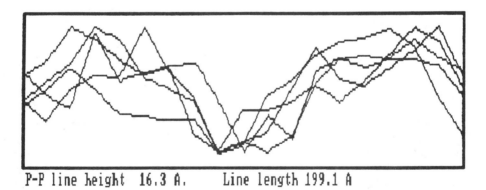

P-P line height 16.3 Å.   Line length 199.1 Å

FIG. 14. Several line profiles across different sections of the groove .

points to the conclusion that there is no appreciable pile up of material on the tip, and as such there is no need for reversal of voltage biasing to shake it off as was required in the fabrication on clean silicon samples [14].

III. Parallel Fabrication

The objective for commercial application is the development of a high speed of access and/or writing while maintaining high precision. We would also like to use the system for the mass production of a few thousand devices at the same time. The multitip linear array arrangement or "comb" will be able to make the concept of writing with atoms and molecules a device of wide commercial application. The linear array writer is a closely packed, equally spaced set of needles with spacing as small as a few micrometers.

We tested this concept by preparing an array of two and three tips. These multi-peak tips were produced in situ by zapping a single tip with a large current pulse. The resulting peak configuration depends on the magnitude and duration of the curren 11. We used such multi-peak tips to fabricate structures on silicon . It was very interesting to find that, after this procedure, more than one mound were made by a single fabrication process. Figure 15 shows several deposits of this kind each showing a

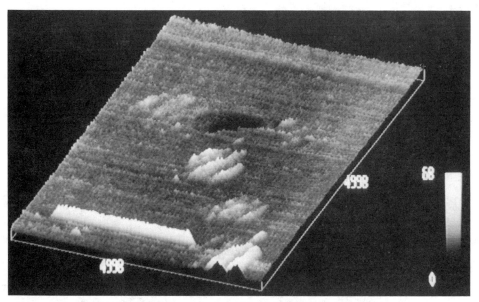

FIG. 15. An image of a silicon surface showing triplets of structures made in single trials

substructure of three mounds of the same interspacing. It appears that these were made by three sharp intrusion on the tip, separated by approximately 100 A from each other. We made other sets of depositions with a tip that has two sharp peaks separated by 500A. Thus one can see the structure of deposit might reflect the sharp structure of the tip, effectively providing means for recording the "footprint" of the tip.

By taking advantage of the multi-peak tip, we can develop a procedure for parallel fabrication; we just apply high voltage to the sample and move the tip. Figure 16 shows two parallel lines made on the surface by moving the tip diagonally. Figure 17

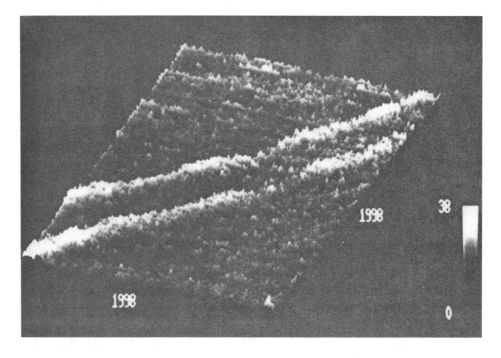

FIG. 16. Two parallel lines made using a two - peak tip.

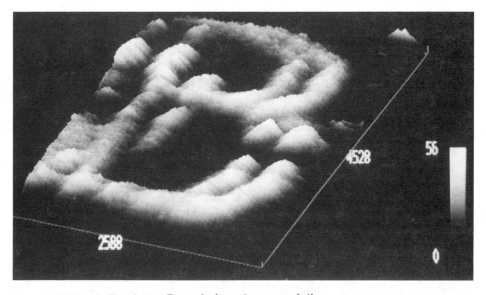

FIG. 17. The letter B made by a two - peak tip.

shows an example of a circuit fabricated with a two-peak tip for further illustration. The design is in the form of letter B. In the vertical direction we see double lines, while in the horizontal direction,we see a single line. One can also see that the double lines merge into one when the tip motion changes from the vertical to the horizontal direction. This exercise also shows that the structure made by the front peak does not get compromised or erased by the back peak as it cuts through it.

IV. Stability at Higher Temperature

The process described above has been demonstrated at room temperature, with the structures remaining stable for a long time at this temperature. We also tested the stability of the structures at higher temperatures by heating the substrate and the nano structure itself by a cw laser beam. The laser beam was directed at the structure obliquely with an incident angle of 70 degrees. Figures 12 showed an image of silicon with structures fabricated on it just before laser heating (room temperature). Figure 18 shows the same surface after laser irradiation of 25 minutes. It appears that for up to 25 minutes of heating, which we believe might have raised the temperature to 100C -150C, the structure remained stable except for loss of some periphery. On the other hand, the continuous heating started to affect the contamination layer in the unprocessed region of the silicon sample long before it did anything to the structure itself.

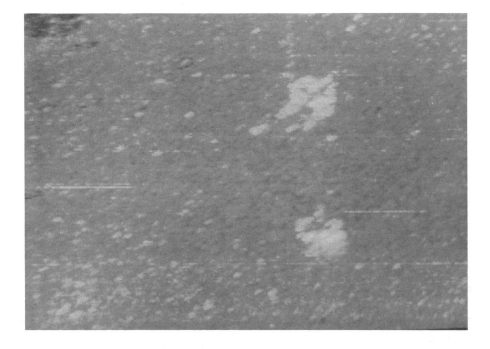

FIG. 18. An image of the surface of figure 12 after heating the surface to about 150 degrees Celsius using laser radiation.

V. Pattern Transfer and Replication of Nano Structures

The basic limitation of STM-based lithography techniques is the fact that they are eventually slow. Fabricating in parallel as suggested and demonstrated in this work is a step in the right direction but it is not enough. There is now agressive thinking to integrate the STM-based techniques with novel molecular beam epitaxy (MBE) growth techniques, reactive ion etching (RIE), and chemically-dominant etching, to transfer nano scale device features into a variety of substrates, and to develop mechanisms for pattern replication. There is however some progress in this direction in the research efforts of groups at NIST[16] and NRL[17] in Washington D.C. Their achievements have already stimulated similar research activities and thinking by other groups. If this goal is realized then one can imagine a nano electronic industry combining the precision of nanofabrication, with the speed of mass production a mask offers.

VI. Nanotechnology

The ability to draw very fine lines with an STM tip may make possible new types of devices, e.g. a quantum inerference transistor that operate on the wave-like behavior of electrons such as diffraction, refraction and interference. This device is a micrometer size MOSFET with a nanometer scale grating or grid embedded in its gate area. At temperatures in the neibourhood of liquid nitrogen temperatures (-76 C) electrons exported along the transister channel will behave as waves with a wavelength of about 10 nanometers ; hence a grating of about 10 nanometer line spacing could prvide a means for control of the flow of these electrons via the adjustment of the gate biasing of the device. Such quantum interference transistors would be faster than conventional transistors due to the fact that electrons need much less time to pass through the much smaller channels of these devices. In addition wave-like propagation of electrons use less energy because of the great reduction in scattering by the atoms of the material, a process that accounts for a large part of the energy expanded in conventional devices. We are part of an effort to carry out this project at the university of Illinois with support from the Office Of Naval Research under the University Research Initiative (URI) on nanolithography.

VII. Acknowlegments

This work was supported by the Naval Research Laboratory under contract NOS. N00014 - 87 - K - 0354, N00014 - 90 - J -1004 and N00014 - 92 - J - 1519.

References

1. See for example the news report by Daniel Clery, New Scientist vol 133 No 1811, 42 (1992).

2. See for example, R. M. Osgood and T. F. Deutsch, Science 227, 709 (1985).

3. D. J. Ehrlich and J. Y. Tsao, Appl. Phys. Lett. 44, 267 (1984).

4. See various articles in IBMJ Res. Develop. 30 (1985).

5. E. E. Ehrichs, S. Yoon, and A. L. de Lozanne, Appl. Phys. Lett., 53, 2287 (1988).

6. J. A. Dagata, J. Schneir, H. H. Harary, C. J. Evans, M. T. Postek, and J. Bennett, Appl. Phys. Lett., 56, 2001 (1990).

7. D. M. Eigler and E. K. Schweizer, Nature 344, 524 (1990).

8. H. J. Mamin, R. J. Hamers, and D. Rugar, Phys. Rev. Lett. 65, 2418 (1990); I.-W. Lyo and P. Avouris, Science 253, 173 (1991).

9. S. T. Yau, D. Saltz, and M. H. Nayfeh, Appl. Phys. Lett., 57, 2913 (1990); S. T. Yau, D. Saltz, A. Wriekat, and M. H. Nayfeh, J. Appl. Phys. 69, 2970 (1991); S. T. Yau, D. Saltz, and M. H. Nayfeh, J. Vac. Sci. Technol. B9, 1371 (1991).

10. S. T. Yau and M. H. Nayfeh, Appl. Phys. Lett. 59, 2457 (1991); X. Zheng, S. T. Yau, and M. H. Nayfeh, Ultramicroscopy 42-44, 1303 (1992); J Hetrick, X. Zheng and M.H. Nayfeh, J. Appl. Phys. May (1993).

11. S. A. Mitchell and P. A. Hackett, J. Chem. Phys. 79, 4815 (1983).

12. A. Ishizaka and Y. Shiraki, J. Electroshem. Soc. 133, 666 (1986).

13. R. B. Laibowitz and C. P. Umbach inPercolation, Localization and Superconductivity, edited by M. A. Goldman and S. A. Wolf, (Plenum, New York, 1984), pp. 267-286.

14. N. D. Lang, Phys. Rev. B 37, 1 10395 (1988); S. Ciraci and E. Tekman, Phys. Rev. B 40, 11969 (1989); J. K. Gimzewski and R. M-ller, Phys. Rev. B 36, 1284 (1987).

15. I. W. Lyo and P. Avouris, Science 253, 173 (1991).

16. Dagata, J. A., Schneir, J. Harary, H. H. Evans, C. J. Postek, M. T., and Bennett, J., Appl. Phys. Lett. 56, 2001, (1990) ; Dagata, J. A., Tseng, W., Bennett, J., Evans, C. J., Schneir, J., and Harary, H. H. Appl. Phys. Lett. 57, 2437 (1990) ; Dagata, J. A., Schneir, J., Harary, H. H. Bennett, J., Tseng, W., J. Vac. Sci. Technol., B 9 (1991) in press Dagata, J. A., Tseng, W., Bennett, J., Schneir, J., and Harary, H. H. J. Appl. Phys. (1991) in press ; Dagata, J. A., Tseng, W., Bennett, J., Schneir, J., and Harary, H. H., Appl. Phys. Lett. (1991) submitted.

17. Marrian, C. R. K., Colton, R. J. Snow, A., and Taylor, C. J. MRS Symp. Proc. 76, 353 (1987) ; Marrian, C. R. K. and Colton, R. J. Appl. Phys. Lett. 56, 755 (1990) ; Marrian, C. R. K., Dobisz, E. A., and Colton, R. J. J. Vac. Sci. Technol A8, 3563 (1990) ; Marrian, C. R. K., Dobisz, E. A., and Colton, R. J. J. Vac. Sci. Technol B9, 1367 (1991) ; Marrian, C. R. K., Bobisz, E. A., and Colton, R. J., Proc. of "Scanned Probe Microscopies: STM & Beyond", Santa Barbara, CA, American Institute of Physics(January 1991).

Modification and Manipulation of Layered Materials Using Scanned Probe Microscopies

Charles M. Lieber
Department of Chemistry and Division of Applied Sciences
Harvard University, Cambridge, Massachusetts 02138

ABSTRACT

Scanned probe microscopes, including the scanning tunneling microscope and the atomic force microscope (AFM), are promising techniques for manipulating matter on the atomic to nanometer length scales. Herein, recent work from the author's laboratory that has utilized the AFM as a technique to affect the direct modification and manipulation of layered materials on the nanometer scale is reviewed. The AFM has been used to machine linear and more complex patterns in thin layers of MoO_3 grown on the surface of MoS_2. The pattern lines have been formed with less than 10 nm resolution using conventional Si_3N_4 tips. These structures can also be inspected with out perturbation by imaging the surface with a small applied load. Nanometer scale MoO_3 structural objects, which were defined by AFM nanomachining, have also been manipulated over distances of several hundred nanometers using the AFM tip. The immediate and future applications of these results are discussed.

1. INTRODUCTION AND BACKGROUND

The controlled assembly of matter into structures on the atomic to nanometer scales is an important objective of researchers in the field of nanotechnology.[1,2] Underlying the successful achievement of this goal is the need to develop methods that can be used to modify and manipulate matter reproducibly on nanometer and shorter length scales. The scanned probe microscopies, including scanning tunneling microscopy (STM) and atomic force microscopy (AFM) are general techniques that have shown considerable promise for manipulating materials on the nanometer scale. Utilization of STM and AFM thus represents an attractive approach for assembling matter into nanometer scale structures.

To date, a number of interesting examples of STM- and AFM-based manipulation have been reported. On the atomic scale STM has been used to remove single atoms and clusters of atoms from the surfaces of several distinct materials,[3-6] to deposit and position atoms on surfaces[3,4] and to create an atomic switch.[7] On the nanometer scale STM has also been used to create structures by field-assisted diffusion,[7,8] to develop organic resists,[9,10] to expose passivated semiconductor surfaces,[11] and to deposit material from the probe tip.[12,13] STM has also been used to dissociate a single molecule on the surface of silicon.[14] The AFM has also been used with success to manipulate matter on the nanometer length scales. For example, several groups have used AFM to modify organic layers and transition metal dichalcogenide surfaces on a ≥ 50 nm scale.[15-18] More recently we have shown that AFM can be used to machine features in thin oxide layers with 5-10 nm resolution, and that distinct structural objects with nanometer dimensions can be manipulated on the surface.[19,20] These examples of materials modification and manipulation using STM and AFM serve to illustrate some of the unique capabilities of probe microscopies for nanotechnology. Several of these areas are discussed in greater depth in other chapters within this volume. Herein, we review the recent efforts in our research program that focus on the modification and manipulation of layered materials using AFM.

To date, our research program has focused on the modification and manipulation of layered materials, including transition metal dichalcogenide (MX_2) solids and aniso-tropic thin films grown on these solids.[16,17,19,20] A schematic view of these systems are illustrated in Fig. 1.

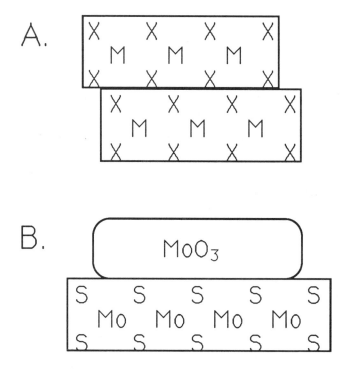

Figure 1. (A) Schematic side-view of a transition metal dichalcogenide (MX_2) material. (B) Illustration of a specific MX_2 material (MoS_2) with an oxide (MoO_3) film grown on the surface.

The MX_2 materials have a number of features that make them attractive for controlled studies of atomic to nanometer scale surface modifications; these features include: (1) single crystals can be cleaved between adjacent X/X sheets to yield atomically flat and ordered chalcogenide surfaces; (2) atomically flat thin films can be grown or deposited onto the surfaces of these materials; and (3) the electronic properties of the solid (surface) can be varied from metallic (e.g., $NbSe_2$) to semiconducting (e.g., MoS_2) through changes in M and X. On the one hand, the ability to prepare atomically flat surfaces facilitates controlled high-resolution experiments on these solids. The ability to vary the fundamental nature of the surface through the growth of atomically flat thin layers or through changes in M and X also provides a flexible approach to systems with unique electronic and optical properties. Below we review the status of our research program directed towards the modification and manipulation of these interfaces (section 2) and also suggest potentially exciting directions in which these investigations may proceed (section 3).

2. MODIFICATION AND MANIPULATION OF LAYERED MATERIALS

2.1. Nanomachining with the AFM

In force microscopy, a surface is imaged by scanning a surface below a probe that is attached to a cantilever; deflections of the cantilever are related to surface features. The force(s) that cause the cantilever deflections define the mode of imaging. For example, when the tip is in contact with the surface repulsive electrostatic forces usually are the dominant interaction; this contact regime is termed repulsive mode imaging. Alternatively, if the tip is removed from the surface, a van der Waals attractive interaction may be the dominant force between the sample and tip; this case is termed attractive mode imaging. All of our AFM studies reviewed below were carried out in the repulsive mode using a modified commercial instrument.[21] Si_3N_4 cantilever-tip assemblies were used for both imaging and modification studies. An illustration of the experimental apparatus is shown in Fig. 2.

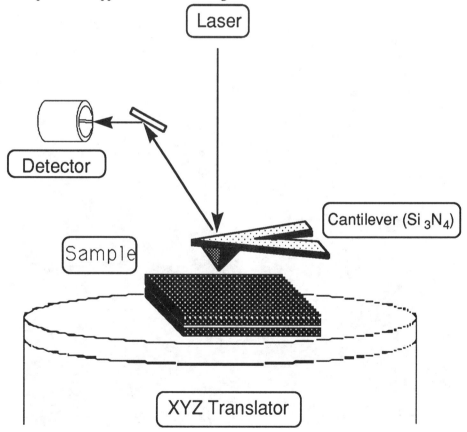

Figure 2. Schematic view of the atomic force microscope.

In general, there are two regimes of repulsive mode imaging: (1) if the forces between the tip and sample are sufficiently small then tip-sample sliding can occur without wear; alternatively, (2) at higher loads tip-sample motion leads to wear. It is this latter regime that we seek to exploit in our studies of layered materials.

Previous AFM studies of MX_2 materials carried out in our laboratory and elsewhere have shown that the tip can be used to modify the surfaces of these materials in air.[16-18] In Fig. 3 we show the effect of continuous scanning of the AFM probe over the surface of $NbSe_2$.

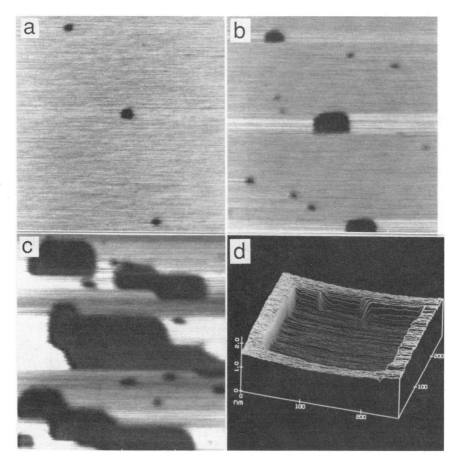

Figure 3. Removal of material from the surface of $NbSe_2$ while imaging in the repulsive mode with a force of $\approx 10^{-8}$ N. Images a, b, c and d were recorded after ≈ 0, 490, 1720 and 3350's of scanning. The top-view images a-c are 200 x 200 nm^2.

This series of images illustrates that the AFM can be used to pattern a structure in the surface of $NbSe_2$ (and other MX_2 solids), although the structures have sizes greater than 10 nm. It is also apparent from this series of images that the modification process is not well controlled. That is, scanning in air continuously removes material from the surface regardless of the imaging force, and thus it is not possible to modify and then non-destructively image and probe resulting nanostructures. Similar results have also been observed during AFM imaging of MoS_2 in air (Fig. 4).[16,20]

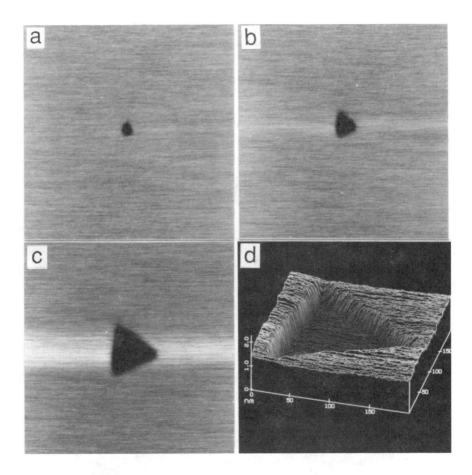

Figure 4. Removal of material from the surface of MoS_2 while imaging in air with a force of $\approx 10^{-8}$ N. Images a, b, c and d were recorded after $\approx$ 0, 590, 1760 and 8500 s of scanning. The top-view images a-c are 200 x 200 nm^2.

Scanning the MoS_2 surface also results in removal of material from the surface, although on this compound surface modification occurs preferentially along three crystallographic directions. Hence, we believe that the MX_2 materials themselves are not suitable systems to explore direct AFM-based surface modification. STM can, however, be used to reproducibly modify and probe the surfaces of MX_2 materials in an ultrahigh vacuum environment.[5]

To extend significantly the applicability of AFM to controlled, high-resolution surface modification we have therefore developed a new system that consists of a thin metal oxide film (MoO_3) on the surface of the metal dichalcogenide (MoS_2).[19,20] This thin film/substrate system has several unique features in comparison to materials studied previously, including: (1) the MoO_3 thin film is rigid and non-deformable; (2) MoO_3 can selectively modified (machined) or imaged depending on the applied load; and (3) the MoS_2 substrate, which is a good solid lubricant, acts as an integral stop layer that automatically sets the depth of features. Important results obtained with this system are described below.

2.2 Nanomachining MoO_3/MoS_2.

An image of a MoO_3 crystallite grown on the surface of MoS_2 is shown in Fig. 5.

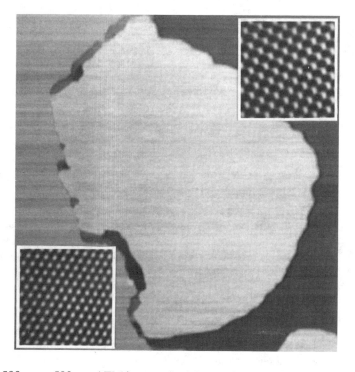

Figure 5. 500 nm x 500 nm AFM image of a 1.5 nm thick MoO_3 crystallite on the surface of MoS_2. The upper and lower insets show the atomic structure of MoO_3 and MoS_2, respectively.

The MoO$_3$ is oriented with the a-c axes parallel to the (0001) MoS$_2$ surface[19] and has a thickness of approximately 1.5 nm (corresponding to one unit all along the b-axis). The regions around the edges of this image are MoS$_2$. An essential point concerning this thin-oxide layer and MoS$_2$ substrate is that they are completely stable to repetitive scanning in an inert atmosphere when the imaging force is $\leq 10^{-8}$ N. Hence, it is possible to explore controlled surface modification with this system.

Notably, when the applied load is increased to $> 5 \times 10^{-8}$ N, we have shown that the MoO$_3$ layer (and only this material) can be machined with nanometer resolution in a controlled manner.[19,20] A line 150 nm long machined in the MoO$_3$ crystallite of Fig. 5 is shown in Fig. 6. This line has width of 10 nm at the MoO$_3$ surface and a width of only 5 nm at the MoO$_3$/MoS$_2$ interface.

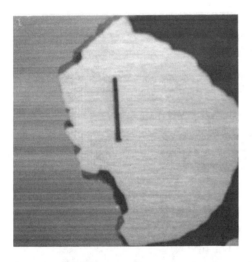 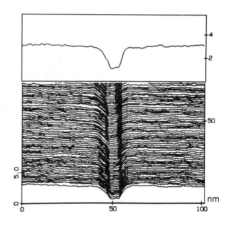

Figure 6 500 nm x 500 nm top-view image illustrating the line machined in the MoO$_3$ layer using a force of 5×10^{-8} N, and a line-scan view that illustrates the three-dimensional characteristics of this structure.

The line-scan data also show that modification results in a microscopically smooth structure in the MoO$_3$. Since the rate at which structures such as this line are formed is proportional to applied load ($> 5 \times 10^{-8}$ N) and scan rate we have termed the modification process, by analogy to macroscopic processes, nanomachining.

To explore the versatility of nanomachining with this system and its potential applications to nanotechnology we have created a number of different structures. For example, we find that a series of high aspect ratio line-structures can be readily created in the MoO$_3$ layers as shown in Fig. 7.

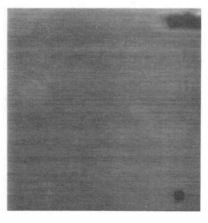
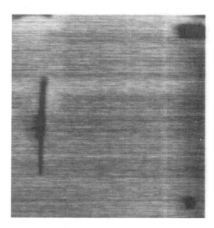

Figure 7. Series of 800 x 800 nm^2 AFM images illustrating the nanomachining of three lines in a large MoO$_3$ crystallite. The lines were created using an applied load of 1 x 10^{-7} N. The images were recorded with a load of 10^{-8} N.

These line structures are 400 nm long, ≈ 20 nm wide and 30 nm deep. While these features are typical of the linear structures we have created in the MoO$_3$ layers, we do not believe that they represent the limit of this technology. Recent advances in the fabrication of integrated tip-cantilever assemblies suggest that it may be possible to achieve < 5 nm resolution in the future. Notably, the ability to create readily arrays of closely spaced linear structures suggests that these techniques may be useful for producing nanometer resolution gratings.

In addition, it is possible to produce more complex structures in these MoO$_3$ layers without a loss of resolution. A series of nonparallel lines that were nanomachined to define the initials "H U", which stands for Harvard University, are shown in Fig. 8.

Figure 8. A series of 500 nm x 500 nm AFM images that illustrate nanomachining operations to define the pattern HU.

This sequence of images shows that it is possible to define patterns with intersecting lines, in addition to the parallel features shown in Fig. 7. The resolution (width) of all of these features is approximately 10 nm, although we believe that better resolution will be attainable using higher aspect ratio tips. In the future it may be possible to utilize complex patterning of MoO_3 (or other thin films) to prepare masks for high-resolution X-ray lithography (see below).

2.3 Manipulation of MoO$_3$ Structures.

We have also shown that it is possible to go beyond the level of nanomachining a series of lines or a complex pattern in the MoO$_3$ thin layers, and make distinct structures that can be manipulated on the surface.[19] The underlying basis for structure manipulation in the MoO$_3$/MoS$_2$ system is that the MoO$_3$ material is not strongly bound to the MoS$_2$ substrate. Hence, it is possible to machine a MoO$_3$ structure, separate this object from the MoS$_2$ substrate, and then manipulate the structure on the surface. This series of operations is illustrated in Fig. 9.

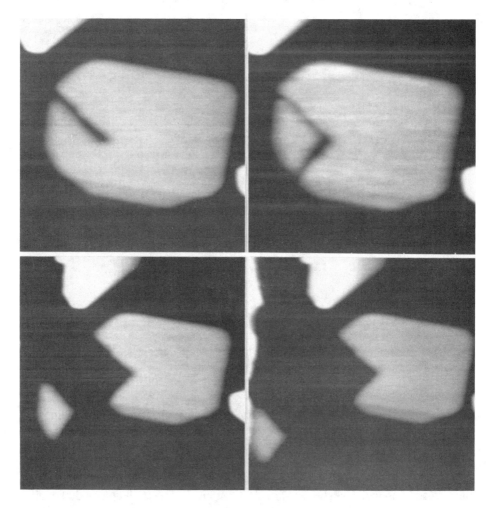

Figure 9. Sequence of AFM images that illustrate (1) the formation of two lines at the edge of a MoO$_3$ crystallite to define a triangular structure, and (2) the manipulation of the triangular structure across the MoS$_2$ surface. The triangular MoO$_3$ object is approximately 60 nm on edge. The two upper images are 450 x 450 nm^2, and the two lower images are 600 x 600 nm^2.

An exciting feature illustrated in Fig. 9 is the ability to manipulate nanometer scale objects on surface with the AFM. We have found that it is possible to move small structures across the surface by applying a high load ($\approx 1 \times 10^{-7}$ N) with the tip and then translating the object using the tip. It is also possible to image the manipulated structures without perturbation using a small applied load ($\leq 10^{-8}$ N). Since the electronic and optical properties of MoO_3 can be readily varied by doping we believe that this system and the manipulation techniques described above represent a promising approach for the fabrication of nanostructures with novel properties.

3. FUTURE DIRECTIONS

It is also interesting to consider potential applications of this work to nanotechnology and future directions in which this research may evolve. Below we discuss areas in which we believe these studies could have an immediate impact, as well as longer range and more speculative ideas.

3.1. Direct Applications of MoO_3/MoS_2 to Nanotechnology.

We believe that the ability to pattern the MoO_3 thin layers with < 10 nm resolution could have an immediate impact in several areas, including the fabrication of nanometer scale diffraction gratings, and X-ray lithography masks. To date, the longest line structures that we have fabricated are 400 nm. Practical gratings and masks for conventional technologies will, however, require longer features. We believe that 100 µm long high-resolution patterns can be obtained through several straight-forward modifications of our procedures. First, larger area MoO_3 layers will have to be grown on the MoS_2 substrates. At the present time the MoO_3 crystallites that we grow by thermal oxidation are usually < 1 µm. Large area MoO_3 layers can, however, be grown using thin film techniques such as pulsed laser ablation. Secondly, larger area scanners are needed for tip control over the 1 nm to 100 µm length scale. Recent advances in both commercial and custom instruments indicate that such scanning requirements will be met in the near future, and thus the scan range should not limit future applications. In addition, it will be important to extend further the complexity of the patterns that we machine with the AFM. For example, it would be interesting and useful to nanomachine structures such as a Fresnel zone-plate lens (Fig. 10).

Figure 10. Schematic illustration of a Fresnel zone-plate lens consisting of a pattern of concentric rings.

The zone-plate lens could have interesting application to both X-ray and atom optics. It is unlikely, however, that conventional sample manipulation will lead to successful nanomachining of this complex structure. We believe that an attractive approach will involve using the new sample translation method reported by Mamin and Rugar.[22] Application of this method to our system would involve spinning the sample on a precision drive, and then machining the MoO_3 at specified radial distances from the rotation axis.

Other approaches may also enable our system to be used for the fabrication of complex high-resolution masks. A current limitation to the direct application of our techniques to this problem is the need to maintain registry over very large distances (e.g., mm). It may be possible, however, to combine conventional photolithography with AFM to produce unique structures. Conventional photolithography could be used to define the basic mask pattern on the 10 μm to 1 mm scale. AFM could then be used to nanomachine intricate details and interconnects in this pattern, and thereby yield a novel mask structure.

3.2. Other directions.

We believe that it also should be possible to generalize the attractive features of the MoO_3/MoS_2 system and our nanomachining ideas to other materials. For example, the major requirements to achieve controlled and reproducible nanomachining are (1) that the thin layer wears more rapidly than the substrate, and (2) that the AFM tip is stable during the machining operation. There are many choices of systems that meet these requirements, although we believe two are particularly interesting. These systems are SiO_x/Si and copper oxide superconductor/$SrTiO_3$. The SiO_x/Si system would be interesting due to the central role these materials play in microelectronics; however, cantilevers with diamond probes may be needed for reliable modification. Preliminary work also indicates that the layered high-T_c copper oxide materials can be machined on the nanometer scale. Hence, it may be possible to make novel junction structures and devices by using AFM to nanomachine these superconductor materials.

In addition, we believe it will be important in the future to exploit our ability to manipulate MoO_3 structures (and perhaps other materials on the nanometer scale). For example, it will be interesting to consider whether it is possible to construct a nanomotor. An armature and gears could be readily machined on the < 100 nm scale and then used to assemble a motor on the surface. A motor on this scale would represent several orders of magnitude size reduction compared to the structures that can be obtained using micromachining techniques.[23] One obstacle that would have to be overcome in fabricating such a device would be the development of methods to "lift" parts, such as the gears, during assembly. We believe that it may be possible to lift small objects with the tip using electrostatic clamping. We must also consider how a nanomotor would be powered and/or observed. Finally, we believe that it will be possible to explore many other types of structures, such as those that exploit quantum confinement and/or the photochromic properties of MoO_3, as we continue to explore the applications of scanned probe microscopies to nanotechnology.

4. ACKNOWLEDGMENTS

The work described in this chapter was carried out by Drs. Yun Kim and Jin-Lin Huang. C.M.L. acknowledges support of this research by the Air Force Office of Scientific Research (contract AFOSR 90-0029) and the David and Lucile Packard Foundation.

5. REFERENCES

1. For recent reviews see: *Science* **254**, 1300-1311 (1991).
2. P. Ball and L. Garwin, "Science at the Atomic Scale," *Nature* **355**, 761 (1992).
3. I.-W. Lyo and Ph. Avouris, "Field-Induced Nanometer- to Atomic-Scale Manipulation of Silicon Surfaces with the STM," *Science* **253**, 173 (1991).
4. D. M. Eigler and E. D. Schweizer, "Positioning Single Atoms with a Scanning Tunneling Microscope," *Nature* **344**, 524 (1990).

5. J. L. Huang, Y. E. Sung, and C. M. Lieber, "Field-Induced Surface Modification on the Atomic Scale by Scanning Tunneling Microscopy," *Appl. Phys. Lett.* **61**, 1528 (1992).

6. J. A. Stroscio and D. M. Eigler, "Atomic and Molecular Manipulation with the Scanning Tunneling Microscope," *Science* **254**, 1319 (1991).

7. D. M. Eigler, C. P. Lutz, and W. E. Rudge, "An Atomic Switch Realized with the Scanning Tunneling Microscope," *Nature* **352**, 600 (1991).

8. L. J. Whitman, J. A. Stroscio, and R. A. Dragoset, R. J. Celotta, "Manipulation of Adsorbed Atoms and Creation of New Structures on Room-Temperature Surfaces with a Scanning Tunneling Microscope," *Science* **251**, 1206 (1991).

9. C. R. K. Marrian, E. A. Dobisz, and R. J. Colton, "Lithographic Studies of and e-Beam Resist in a Vacuum Scanning Tunneling Microscope," *J. Vac. Sci. Technol. A* **8**, 3563 (1990).

10. E. A. Dobisz and C. R. K. Marrian, "Sub-30 nm Lithography in a Negative Electron Beam Resist with a Vacuum Scanning Tunneling Microscope," *Appl. Phys. Lett.* **58**, 2526 (1991).

11. J. A. Dagata, J. Schneir, H. H. Harary, C. J. Evans, M. T. Postek, and J. Bennett, "Modification of Hydrogen-Passivated Silicon by a Scanning Tunneling Microscope Operating in Air," *Appl. Phys. Lett.* **56**, 2001 (1990).

12. H. J. Mamin, S. Chiang, H. Birk, P. H. Guethner, and D. Rugar, "Gold Deposition from a Scanning Tunneling Microscope Tip," *J. Vac. Sci. Technol. B* **9**, 1398 (1991).

13. R. M. Silver, E. E. Ehrichs, and A. L. deLozanne, "Direct Writing of Submicron Metallic Features with a Scanning Tunneling Microscope," *Appl. Phys. Lett.* **51**, 247 (1987).

14. G. Dujardin, R. E. Walkup, and Ph. Avouris, "Dissociation of Individual Molecules with Electrons from the Tip of a Scanning Tunneling Microscope," *Science* **255**, 1232 (1992).

15. G. S. Blackman, C. M. Mate, and M. R. Philpott, "Atomic Force Microscope Studies of Lubricant Films on Solid Surfaces," *Vacuum* **41**, 1283 (1990).

16. Y. Kim, J.-L. Huang, and C. M. Lieber, "Characterization of Nanometer Scale Wear and Oxidation of Transition Metal Dichalcogenide Lubricants by Atomic Force Microscopy," *Appl. Phys. Lett.* **59**, 3404 (1991).

17. C. M. Lieber and Y. Kim, "Characterization of the Structural, Electronic, and Tribological Properties of Metal Dichalcogenides by Scanning Probe Microscopes," *Thin Solid Films* **206**, 355 (1991).

18. E. Delawski and B. A. Parkinson, "Layer by Layer Etching of Two-Dimensional Metal Chalcogenides with the Atomic Force Microscope," *J. Am. Chem. Soc.* **114**, 1661 (1992).

19. Y. Kim and C. M. Lieber, "Machining Oxide Thin Films with an Atomic Force Microscope: Pattern and Object Formation on the Nanometer Scale," *Science* **257**, 375 (1992).
20. Y. Kim, Ph.D. Thesis, Harvard University (1992).
21. Nanoscope, Digital Instruments, Inc., Goleta, CA.
22. H. J. Mamin and D. Rugar, "Thermomechanical Writing with an Atomic Force Microscope Tip," *Appl. Phys. Lett.* **61**, 1003 (1992).
23. K. D. Wise and K. Najafi, "Microfabrication Techniques for Integrated Sensors and Microsystems," *Science* **254**, 1335 (1991).

Atomic force microscopy experimentation at surfaces: Hardness, wear and lithographic applications

*†‡ T. A. Jung, †A. Moser, ‡M. T. Gale,
†H. J. Hug, and †U. D. Schwarz

†Institute of Physics, University of Basel, Klingelbergstrasse 82,
CH-4056 Basel, Switzerland

‡Paul Scherrer Institute Zurich, Badenerstrasse 569,
CH-8048 Zurich, Switzerland

*On leave to IBM T. J. Watson Research Center, P. O. Box 218
Yorktown Heights, New York 10598, USA

March 21, 1993

Abstract

This chapter reviews the experimental techniques for surface modification using an atomic force microscope. The numerous examples presented demonstrate the ability of these methods to contribute to our knowledge of interface physics in various fields. Fundamental structural and dynamical behaviour of surfaces and adsorbates can be studied in detail, with implications to material science and technology. In the conclusion the future impact of even more sophisticated atomic force microscopy techniques is discussed as related to this rapidly growing field of research.

1 Introduction

The invention of scanning tunneling microscopy (STM) [1] and atomic force microscopy (AFM) [2] has triggered numerous research activities in the field of nanometer and subnanometer–scale modification with these new and versatile tools. Impressive examples are the controlled manipulation of atoms by a scanning tunneling microscope as reviewed by Stroscio [3], Avouris [4], Staufer [5] and Shedd [6]. Although these manipulations have all been performed in environments that are not currently cost effective for technical applications, there has been valuable insight gained into the physics and chemistry of inter-atomic interactions from these experiments as discussed in the above mentioned reviews and references [7, 8]. All these studies have in common that a scanning tunneling microscope is operated as the tool for locally manipulating surfaces, using different physical phenomena such as local heating, electron injection and mechanical interaction. A major limitation of the STM technique, however, is that it is only applicable to conductive samples.

The atomic force microscope, on the other hand, overcomes this impediment and introduces important new capabilities and parameters that widen the range of experimentally accessible information. This chapter is intended to give an overview of techniques that have been developed to locally modify surfaces using an atomic force microscope.

In previous reviews on surface modification techniques using scanning probe microscopes [6, 9, 10] there have been only a few reports on surface modifications observed with an atomic force microscope [11, 12, 13]. Recently, there has been a steady increase in publications on nanometer scale manipulation using AFM and applications thereof [14, 15, 16, 17]. First in this chapter the experimental background and the basic modification techniques are described. In the following section, the application of these techniques in various fields will be reported. Examples to be discussed include:

> The controlled production of persistent modifications in polymers and first results on their use in advanced lithographic techniques.

> The determination of surface elastic and plastic properties from defined modification experiments while monitoring the applied forces and the deformations.

> Different tribological mechanisms of surface wear studied locally with an atomic force microscope used as an ultra precision pin-on-disk tribometer, and molecular manipulations of Langmuir-Blodgett films.

After a discussion of the actual and possible impact of these results on current technologies we will give an outlook on future developments and technologies in this field.

2 Experimental background

An atomic force microscope consists of the sample, a cantilever with an integrated tip as a force-sensor, and a unit for monitoring the cantilever deflection, for example a scanning tunneling microscope [1], capacitive [18, 19], interferometric [20], piezoelectric [11, 21], and error-detection type optical heads [22], or a beam-deflection sensor [23]. In general, the sample is scanned with the cantilever–tip, while the interaction forces between the cantilever-tip and the sample are measured. AFM hardware and conductive force sensors can be used to map forces while simultaneously performing STM experiments gaining spectroscopic and electronic information [24]. Additionally, various force interactions, for example van der Waals [25], electrostatic [26, 27] and magnetic [28, 29], can be probed. This comprises the most exciting feature of AFM techniques, namely that different forces at surfaces can be mapped in order to study changes induced by the controlled variation of one or more of the many possible parameters in an atomic force microscope. The surface modification studies covered in this review will use the loading of the cantilever tip on the sample as the main parameter.

The atomic force microscope can be operated in both dynamical and static modes that can be combined in one setup [29]. The dynamical modes are most important for non–contact imaging [28, 30, 31, 32] and are not included in this review.

In the (static) variable–deflection mode, the sensor monitors the deflection of the cantilever beam during the scan. For images acquired in this mode the cantilever loading depends on the elevation of the sample. This has to be taken into account, most especially when imaging large ($\sim \mu m^2$) scan areas. If the sample tilt can be compensated for on line, this measuring mode is useful for high resolution imaging of flat samples, especially for magnetic field images [29]. This measuring mode is also important for the aquisition of force vs. distance curves and the calibration of the cantilever loading as will be detailed in the next section.

More important for the aquisition of the topographic images is the constant force operation of the atomic force microscope. The sample is scanned laterally while a feedback circuit controls the z position such that the cantilever deflection is kept constant. A more general overview of atomic force microscopy can be found in earlier reviews [33, 34, 35] and books [36].

Most types of atomic force microscopes are suited to provide topographic images as well as to perform some sort of surface modification. For controlled experimentation, however, it is important to adjust and measure forces reproducibly in the wide range of $\sim 10^{-9} - 10^{-6}$ N. Loadings below 10^{-9} N are required to image surfaces of bulk and thin film polymers without modification in the contact regime. For undistorted imaging of biological samples, even lower forces ($\sim 10^{-11}$ N) are recommended [37, 38]. These can be adjusted using AFM techniques in liquids (see for example [39]) or noncontact techniques. Much higher forces ($\sim 10^{-4}$ N) are required to create micrometer sized indentations in engineering materials such as metals. Force sensitivity and the maximum force range for detectable forces depend on three important factors: the sensitivity and range of the deflection detection mechanism, the force constant of the cantilever and the stability of the experimental setup. These will be adressed below in the context of their importance for surface modification and lithography.

Deflection sensing system: All sensors have inherent limits of sensitivity and detection range.It is advantageous in order to mount the sensor on a piezo tube to increase the detection range. This is a particular advantage for interferometric sensors, since their sensitivity generally depends sinusoidally on displacement. Another important drawback of some sensor principles is the convolution of the measured cantilever deflection with detection irregularities. This is most pronounced with the early tunneling detection setups [2, 40], as detailed in reference [41]. The major drawback of some of the techniques for measuring the cantilever deflections is that they are difficult to place on a piezo.

These difficulties might be resolved with the force balance techniques that have been recently introduced in AFM [42]. Microfabricated force balances allow the continous control of the effective (electronic) spring constant of a mechanically stiff cantilever sensor, virtually independently of the deflection sensing and force balancing interaction. This provides elegant ways to measure force vs. distance curves, and to adjust forces over a wide range.The resolution limit demonstrated so far (10^{-8} N for capacitance detection) is not yet comparable to the resolution of conventional AFM techniques (10^{-12} N, [29]).

Cantilevers: The cantilevers used for atomic force microscopy have significantly improved during recent years. Some manufacturers supply sets of cantilevers produced with different materials, with various cantilever and tip shapes, and force constants from ~ 0.03 to 30 N/m [43, 44]. Force constants are generally estimated from the geometrical shape of the cantilevers and are thus dependent on process control, most especially the thickness ($\pm$ 20% for Si_3N_4 cantilevers). For higher accuracy the force constants can be calibrated in the force microscope by measuring resonance frequencies or thermal noise amplitudes.[1] For modification experiments, it is important that the cantilever force constant is chosen appropriately, e.g. such that maximum range of deflection detection allows reproducible adjustment of the cantilever loading within the desired range ($\sim 5 \cdot 10^{-9}$ to $\sim 10^{-6}$ N respectively for imaging and modification of the samples discussed later in this review). For most of those applications using high probing forces or lateral force detection, it is very important to have cantilevers with defined (high and low) force constants for lateral deflection of the probing cantilever tip. As will be demonstrated later, the shape of the probing tip also influences performance and results of surface modification. Si_3N_4 cantilevers with pyramidal shape have proven to be very stable due to their low aspect ratio, but do not have a well–defined apex in many cases [45, 46].

---

[1] Some manufacturers supply exact specifications on demand.

The silicon cantilevers [44, 47] with microfabricated conical tips have a very high aspect ratio, and are well suited for precise modification, but are prone to damage at applied forces higher than 10^{-7} N. Controlling the tip geometry down to the nanometer scale is still one of the major difficulties in AFM cantilever production, although considerable progress has been achieved recently [44, 48, 49, 50]. For a review see also [45]. Undercut tips have been developed for improved imaging of steep structures [51].

Experimental hardware and software: High stability against mechanical excitation and thermal drift is a requirement for all scanning probe microscopy setups. Especially for the more developed experimental techniques using high and modulated cantilever loadings, this, and reliable control of all experimental parameters, is essential. Instrumentation and software control should therefore allow operation in many of the operation modes of a force microscope. Special emphasis has to be taken on the switching between one and another of these modes without movement of the cantilever tip. In our setup [52], this has been achieved by using variable gain amplifiers instead of mechanical switches. The AFM data acquisition and control software has also been customized to make it possible to program arbitrary movements and record more than one channel simultaneously, as well as obtain force vs. distance curves [53]. Another important requirement for some of the experiments is that the sample areas imaged and/or modified using AFM can be repositioned for characterization with other methods. This is usally achieved by having a reasonably large scanning area of the Atomic Force Microscopy (typically $100\mu m \times 100\mu m$), together with good optical/electron optical visibility of the cantilever tip positioned above the surface, and with some alignment patterns on the sample to allow repositioning if required.

3 Force calibration

Since the force on the cantilever is the most important parameter of AFM modification experiments, a serious effort must be undertaken to obtain an accurate force calibration. Force vs. distance curves have been introduced by Meyer [25] Weisenhorn [54] and Burnham [40] to study the tip–sample interaction. Fig. 1. a) illustrates a force vs. distance curve experiment. Typically, the cantilever is approached to the sample (1), jumps to the surface as soon as the sample force gradient df/dz exceeds the force constant of the cantilever(2–3), and the cantilever load is increased to a chosen peak value (4) before retraction starts. To separate the cantilever from the surface, a certain amount of retractive force has to be applied (5). Carefully performing this simple experiment provides insight into all kinds of interactions between the tip and the sample surface. Whereas long distance interactions such as electrostatic or magnetic forces alter the slope in (1), short ranged interactions such as van der Waals and adhesion forces alter the acquired curve at (2–3), and non–conservative adhesive forces, for example due to surface adsorbates, influence the height of the jump (5). The measured force vs. distance curve for graphite can be fitted to get reasonable parameters for a Lennard–Jones potential [25]. Nevertheless, the observed force vs. distance curves are highly sensitive to the materials chosen, tip geometry, surface topography, and adsorbate layers. The principles of these forces are discussed in [55]. In ambient air, in which most of the AFM experiments are performed, tip- and sample surfaces are generally contaminated by adsorbate layers. Cohen [56] reviews previously published experimental data to compare adhesion values for different systems and finds rather poor agreement. Therefore, conclusions on interatomic interactions cannot be drawn from such experiments, which strongly motivates better defined experiments using better defined tips and samples in the UHV.

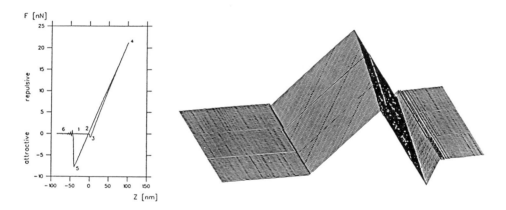

Figure 1: a) Cantilever deflection (z_{Tip}) plotted against sample position (z_{Sample}) recorded on a magneto–optical disc sample. The scale measures net deflection of the cantilever, i.e. $\vec{F}_{Lever}$. The maximum force required to separate probing tip and sample is indicated at position (5). Therefore the relevant force in succeeding AFM experiments is shifted by ~ -8 nN, to coincide with point (5). The loading and unloading portion of the curve do not match due to creeping of the domain walls in the piezo actuator, that has not been corrected for in this setup. b) An array of force versus distance curves as acquired on the same sample one after another. The reproducibility of the single curves is essential for reliable force calibration.

Fig. 1. b) shows an array of force vs. distance curves acquired automatically by computer. In spite of the apparent excellent reproducibility for successive acquisitions, reproducibility is usually observed to be poor over longer time periods (∼ 1/2 day). This can be attributed to changes in the amount and type of adsorbates that condense from ambient air. This assumption is confirmed by the fact that interaction forces required to separate tip and sample in ultra high vacuum [57] are up to an order of magnitude smaller than forces measured in force vs. distance experiments performed in air.

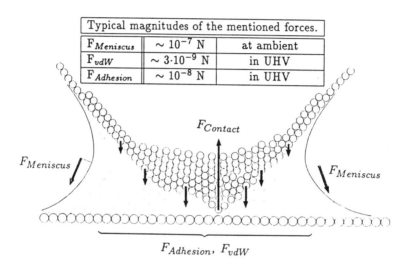

Figure 2: Schematic representation of the forces acting in the tip region of an atomic force microscope. $\vec{F}_{Meniscus}$: meniscus force; $\vec{F}_{vdW}$: Van der Waals force; $\vec{F}_{Adhesion}$: adhesion force; $\vec{F}_{Contact}$: repulsive force acting on the topmost atomic cluster ot the cantilever tip (compare [41, 58]). The typical magnitudes of these forces are summarized in the table above.

As shown schematically in Fig. 2., the force sensing cantilever measures $\vec{F}_{Lever}$, the sum of all forces acting on the cantilever tip. Neglecting electrostatic and magnetic forces, $\vec{F}_{Lever}$ can be separated into a short ranged part, dominated by the repulsive force due to the overlapping electron shells of the exposed atoms in the junction ($\vec{F}_{Contact}$), and more slowly varying components (the meniscus force $\vec{F}_{Meniscus}$ and adhesion $\vec{F}_{Adhesion}$ and van der Waals force: $\vec{F}_{vdW}$) summing up to a net attractive force. In this context local is meant to be in a few atomic dimensions, the relevant lengthscale for the interaction forces during contact mode AFM imaging.

$$\vec{F}_{Lever} = \overbrace{\vec{F}_{Meniscus} + \vec{F}_{Adhesion} + \vec{F}_{vdW}}^{\sum \text{nonlocal,}} + \overbrace{\vec{F}_{Contact}}^{\text{local}}$$

The force $\vec{F}_{Contact}$, acting in the frontmost tip region and not the overall force measured by the cantilever deflection sensor $\vec{F}_{Lever}$ is the relevant force for AFM experiments. $\vec{F}_{Contact}$ is minimal when $\vec{F}_{Lever}$ balances the other acting forces, just before tip and sample are separated at position (5) in Fig. 1. a). Therefore, this point is the most reasonable definition of a zero-force calibration for AFM experiments in air. The calibration accuracy for contact mode experiments performed in air is often limited by the reproducibility of force vs. distance curves acquired on one sample, not by the force sensitivity. The force calibration should be verified prior to taking AFM images or performing AFM experiments. Due to thermal drift and because adhesive properties may depend on the lateral position on the sample, the force calibration should be repeated during a series of AFM experiments. This is equally important not only for modification experiments but also for AFM imaging of soft samples, where modification might occur due to scanning with unintended high forces.

4 Modification experiments using an atomic force microscope as the tool

This section will give an overview of the modifications performed on various polymer substrates namely polycarbonate (PC), nylon (NY) and polyvinyl alcohol (PVA). These samples have been chosen to demonstrate the principal capabilities of surface modifications using AFM. In a recent paper [14] three methods of modifying surfaces using AFM have been described. The first and simplest is indentation of the sample surface by a controlled increase of the force exerted by the tip on the surface. The second is to produce modifications by performing lateral movements with an increased cantilever loading and the third is to modulate the cantilever loading.[2] Surface modifications were first produced by increasing the force [11] and by driving the cantilever into a feedback oscillation [12, 13].

By increasing the load of the cantilever on the sample surface, indentations can be reproducibly created with depths in the nanometer to micrometer range, as can be seen from Figs. 5 and 7. The lower left indentation in Fig. 7, for example, was created by applying a loading of $\sim 4.1 \cdot 10^{-8}$ N to the sample surface. Because of the depth of ~ 2 nm and the width of ~ 40 nm, the indentation is hardly visible on the polycarbonate surface, which has a corrugation of ~ 5 nm.

Larger surface areas can be patterned using the ploughing technique. With a typical force of $\sim 10^{-7}$ N, parallel lines in Fig. 3 a) with a period of 70 nm and a depth of 10 nm have been created. In spite of the high

---

[2] These methods have been referred to as "nanoindenting", "nanoploughing" and "nanohammering".

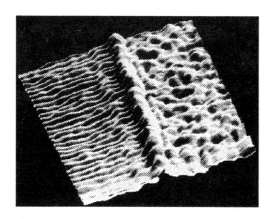 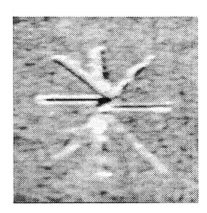

Figure 3: Two typical surface modifications produced by laterally moving the cantilever–tip across a polycarbonate surface with a loading increased to $\sim 10^{-7}$ N. a) This image, 1.35μm × 1.35μm in size, shows a line pattern with 70 nm period and 10 nm depth. On the right hand side the uniform corrugation ($\sim 10\ nm$) of the unmodified polycarbonate surface is imaged. b) ($\sim$3μm × $\sim$3μm) demonstrates the direction dependence of this modification method: Frictional forces due to the increased loading lead to torsion of the cantilever and therefore cause the observed poor reproducibility of the tip position. In addition the apparent asymmetry of the created pattern is caused by asymmetries in the frontmost tip region, as discussed in the next section (Sect. 5.1).

reproducibility for creating line patterns, this method is not ideal for creating more complex patterns, as can be seen from Fig. 3 b): The increased loading of the cantilever required to overcome the elastic limit of the sample leads to friction and torsion of the cantilever. For this reason the star shaped lines in Fig. 3 b) do not intersect at one center. This inaccuracy has been observed to be up to $\sim$0.5 μm, thus complicating the creation of smaller structures. For some experimental parameters the line depth varies periodically as expected from stick–slip movement.

These drawbacks have been resolved with the development of a dynamic technique. The cantilever loading has been modulated by adding an ac (4 kHz) voltage to the z-piezo electrode while performing lateral movements. This technique reduces stick–slip and the torsion of the cantilever and therefore overcomes the difficulties in positioning the cantilever–tip. To correct for the other observed anisotropies, the cantilever position has also been modulated laterally. Using a modulated force of $2.5 \cdot 10^{-8}$ N, the letters in Fig. 4 have been written between the tracks of a compact disc with a letter size of $\sim$700 nm.

Figure 4: "HEUREKA" scribed with a modulated cantilever loading, as detailed in the text. The oval pits are information coding in the compact disc. The letter height is 700 nm and the indentation depth is 10 nm.

All modifications created on polymer surfaces in air as shown here have proved to be stable for at least 24 hours. Some of the structures described below have been imaged many weeks after their creation using another instrument. The use of the cantilever as a stylus for lithography is thus a versatile method for reproducibly creating structures and patterns on the nanometer scale. As demonstrated by the development of a storage device using laser–assisted AFM modification, recently introduced by Mamin [16], there is plenty of room for new developments. The next section will give an introduction to applications of these techniques in various fields of research and technology.

5 Applications

From the many phenomena of interface science that can be studied locally using AFM, we have chosen the following topics to demonstrate the application of surface modification techniques. The first part of this section introduces methods for the probing of surface mechanical properties, e.g. surface hardness and wear. The second part reports on techniques for creating sub-100 nm grid structures due to a combination of AFM lithography, chromium (Cr) shadowing and reactive ion etching technique, and the last part reviews recent results on direct manipulation of molecular adsorbate layers.

5.1 Indentation hardness and wear experiments

5.1.1 Introduction

Tribology is concerned with the science of improving and modeling the frictional behaviour of solid–solid interfaces under load. To investigate applied problems (mainly the performance of different sorts of bearings and bushings), different ways for performing macroscopic experiments with optimized control of all relevant parameters have been developed (pin-on-disk tribometer [59], indentors for hardness measurements). The experimental results have been used to make predictions about other material pairings, and to verify phenomenological models (e.g. Amonton 1699, Coulomb 1781, Desagulier, 1734, Hertz 1881). Dowson reviews this in his book [60].

The application of these macroscopic tribological test systems to bearings represents an extrapolation over powers of ten. High loads have to be applied in order to be able to detect the effects of wear with the limited resolution of optical and electron optical methods. Therefore, experimentation has been improved in order to investigate the so called molecular tribology, e.g. the study of initial processes of wear and the modelling of extended surface properties from this knowledge. Highly sensitive experimental tools such as the "surface force apparatus"[61], the "quartz crystal microbalance" as well as the scanning probe methods (SPM) have been developed. Increasing theoretical (continuum elasticity theory for layered structures, Tomanek [62]; molecular dynamics, Landman [63]) and experimental (Krim [64], Blackman [65], Mate [66]) effort has been undertaken to better understand plastic and elastic properties at surfaces in the context of physics on a molecular scale. Very similar to these classical techniques AFM can be applied to investigate this subject with increased lateral resolution, as we will demonstrate in the following.

5.1.2 Indentation techniques

Conventional indentation techniques using a rod with diamond cones and inductive motion sensors [67] have been improved such that forces as low as 10^{-6} N can reproducibly be applied. Using shadowing techniques and transmission electron microscopy, indentations less than 10 nm in depth have been resolved [68, 69]. Pethica [70] and Weihs [71] developed a variety of experimental techniques to measure other material parameters, such as Young's modulus, during an indentation experiment. Most of these techniques are applicable to a specially configured atomic force microscope. The major advantages of such a setup are the combination of imaging and indentation capabilities, together with the higher resolution imaging and force detection.

An indentation resulting from applying a force pulse with a peak force of $\sim 10^{-6}$ N to a polycarbonate surface is shown in Fig. 5. The AFM image has

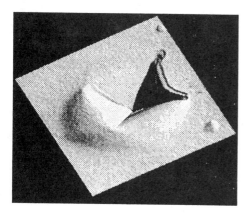 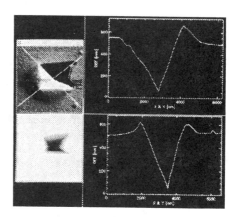

Figure 5: a) Indentation generated in polycarbonate by applying a peak force of 10^{-6} N with the atomic force microscope. Structural details, as for example the convexity of the edges or the decreasing corrugation can clearly be observed. This AFM image has been acquired with a cantilever–loading of $5 \cdot 10^{-9}$ N. From the crossections in b), the size (1900 nm×2300 nm×600 nm) as well as structural details can clearly be recognized.

been acquired after creation of the indent using the same cantilever and a loading adjusted to $\sim 10^{-9}$ N. The hole punched by the pyramidal Si_3N_4 tip is clearly imaged. Structural details of the indentations, e.g. the convexity of the edges and the ridges of excessive material due to elastic recovery of the sample, are imaged and can be measured. The asymmetry of the created indentation is probably caused by the axial symmetry of the cantilever–beam and possible non–orthogonality of the piezo movement, which can be corrected with our current setup. The depth of the created indentation as well as the height of the ridges can be measured from the cross sections in Fig. 5 b)

The schematic drawing in Fig. 6 illustrates that the elastic recovery of the sample alters the opening angle of the created indents. In our case, the angle between the 111 faces of the of Si_3N_4 tip, $\alpha=70°$ creates an indentation with $\alpha'=120°$. This can be explained by the simple assumption that the material next to the indenting tip is plastically deformed, whereas for a larger distance from the penetrating tip the deformation of the material is below the plasticity limit. From such assumptions the penetration of the cantilever tip $h(L)$ can be calculated from the applied load L [72]. As shown in Fig. 6, the observed rim, as well as the convexity of the edges shown in Fig. 5, follow from these basic assumptions. Most of the techniques that have been developed for the analysis of indentations in metals are useful for the understanding of indentations generated with an atomic force microscope. For a collection of articles on indentation techniques see ref. [73]. However, in this size range

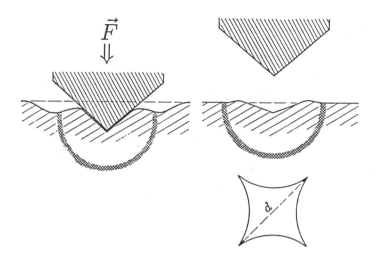

Figure 6: Schematic drawing of an indentation: a) with the load applied, the surface is deformed plastically (close to the sample surface) as well as elastically. b) After removing the load, the elastically deformed material recovers. Assuming a hemispherical boundary of the material being deformed plastically, the experimentally observed structural details (e.g. the convexity of the edges and the rims at the edge of the indentation) can be modelled. Comparison of more accurate models with the experimental data sheds light on the materials properties.

and below, the phenomenological theories tend to fail, and surface effects such as adhesion, diffusion and surface tension have to be taken into account.

There are therefore ongoing discussions whether surface hardness values will increase or decrease for low loads [68]. From AFM indentation experiments, as shown in Fig. 7, these effects can be studied in detail. An array of indentations has been created while the applied forces have been monitored by acquiring force vs. distance curves. The force vs. distance curves that have been acquired during the generation of the leftmost indentations in Fig. 7 are shown in Fig. 8. While the forces ranged from 26 – 82 nN from the bottom to the top, these have been modulated with an amplitude of 0 – 9 nN increasing by 1.8 nN from the left to the right. The reproducibility of the technique is demonstrated from the size of the created indentation increasing from the left to the right and from the bottom to the top. In this experiment we observe a linear relationship between the applied loads L and the measured diameters of the indentations. Therefore these experiments would support an increase of hardness with smaller load. However this effect is most probably due to the limited tip radius of the Si_3N_4 tip used (typically 20–50 nm [45]), and not to material properties. Future experiments

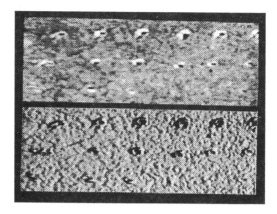

Figure 7: Array of indentations with 2–20 nm depths, generated in polycarbonate by increasing the cantilever load and measuring a force vs. distance curve simultaneously. Both images represent the same AFM data, displayed as a topview at the top, and shaded for improved visibility at the bottom. From the bottom to the top of each array, the applied maximum force has been increased (26 nN, 53 nN, 82 nN); from the left to the right the force modulation amplitude has been increased in 1.8 nN steps starting from 0–9 nN. Due to the high modulation amplitude the force calibration has been shifted in the third row as visible from the size of the indentations.

shall investigate other effects, for example the effect of microstructure and mechanical properties of the contacting materials at the boundary.

To conclude, we have shown how reproducible AFM indentation experiments can be performed using the cantilever as the stylus and the force vs. distance curve technique as a measure of the applied load and deformation. This is a major achievement compared to recent studies using STM and AFM for high resolution mapping of conventionally created indentations [74]. Microscopic experiments can be readily performed using AFM for both modification and characterization, thus gaining new experimental possibilities (for example on time dependance and dynamics) and saving time and labour. For more conclusive quantitative experimentation on the sub–50 nm scale however, better control of the cantilever tip shape is essential. Further developments of the AFM technique to study surfaces will include the dynamical methods that have been developed for conventional indentation [70]. In general, these AFM characterization techniques have the potential to gain insight into mechanical properties at small scales. For example, established methods like thermomechanical probing or the study of glass transitions and mechanical spectroscopy can be applied to very small volumes of material using AFM. One step in this direction, although not making use of the full

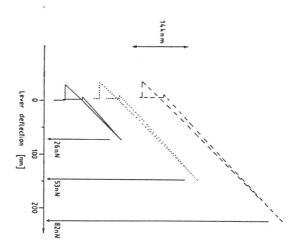

Figure 8: Force vs distance curves as they have been acquired during the generation of the leftmost indentations in Fig. 7. The maximal forces of 26 nN, 53 nN and 82 nN can be determined.

AFM capabilities as described in this review, is the study of Thibeaudau [75] who detects the local changes of the mechanical properites of mica due to the irradiation with Krypton ions. There are many more interesting mechanical properties to be measured with the high resolution of the AFM, e.g. plastic flow, defect induction and mobility in the vicinity of surfaces, and mechanical properties at grain boundaries and other inhomogeneities.

5.1.3 Probing of wear properties using AFM

In addition to the indentation techniques described above, AFM can also be used as a versatile tool for probing wear properties. First studies in this context have applied the atomic force microscope as a high resolution tool to image surface roughness in relation to surface wear, for example on hard disc materials [76, 77, 78]. Whereas this method now has been applied to a variety of problems, methods have been developed using AFM to directly probe wear properties. In analogy to the experiments with pin–on–disk tribometers [59], or point contact microscopes [79], defined loadings can be applied to the sample while laterally moving the cantilever tip over the surface. In Fig. 3, the result of such a ploughing experiment performed at a loading of 10^{-7} N has already been shown. In the discussion we have mentioned that the observed modification is mainly due to "ploughing" of material with high plasticity. Besides this, more complex experiments can be performed to probe the surface's wear resistance. Simultaneously with the lateral movement across the sample, the cantilever loading in Fig. 9 has been increased.

The parameter changed between the lines from bottom to the top is the force

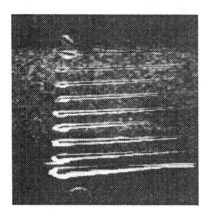

Figure 9: Wear experiment performed on a polycarbonate surface. Simultaneosly with the lateral movement of the cantilever tip across the surface, the loading of the cantilever tip on the surface is increased from 10^{-8} N to $1-5 \cdot 10^{-7}$ N continuosly. The protruding edges due to the penetration of the modifying tip through the surface are clearly visible.

modulation amplitude. The result of these experiments, imaged at $5 \cdot 10^{-9}$ N, implies that, in contrast to the previous experiment at high forces and modulation amplitudes, the surface is penetrated, leading to sharp protrusions and edges. Experiments such as that shown might be interesting for wear studies of systems in emergency running conditions.

In contrast to the surface penetration as shown in these studies, Fig. 10 demonstrates another wear effect: The lower left area of the micrograph has been modified using the ploughing technique as discussed before, leaving the typical parallel traces. The experiment has been continued, while changing the direction of the scanning movement, during ~10 min. This created the rectangular area with the apparently high corrugation at the lowest left. Different changes have occured: The sample no longer shows the direction of the last scanning movement with the high force, the corrugation of the surface has increased significantly and the instabilities in the upper part of the modified area could be attributed to the formation of wear debris. All these are different effects of surface wear.

To draw a conclusion on the physical background of these effects, the energy dissipation mechanisms involved in these processes have to be studied. In general for polymers, thermal and mechanical effects as well as structure changes have to be considered. In this context, also the volume increase of the polymer surface as observed by Grütter [80] and Hamada [15] has to be mentioned. From our experiments a slight volume increase could be deduced,

Figure 10: Surface wear probed on polycarbonate: With the force increased to a value of 10^{-7} N, the larger rectangular area in the lower left has been modified in an experiment similar to that shown in Fig. 3 a). In the smaller area in the lowest left, continuous modification while changing the scan direction for about 10 min. led to structural changes such as increase of the corrugation and the formation of wear debris as discussed in the text.

but further experimentation is needed to confirm this due to the difficulties in separating this effect from experimental artifacts such as changing mechanical properties and the convolution of tip geometry with the highly corrugated topography after the modification.

These studies have clearly shown that AFM can provide many parameters for a detailed study of wear properties as they are well known from more macroscopic experiments, e.g. microploughing, microcutting and microfatigue. The most exciting aspect of the AFM technique is, however, that the effects can be studied down to their molecular or atomic origins. For example Bosbach has been able to correlate the forces required to produce modifications on layered molecular crystals with the interlayer interaction and the macroscopic elastic constants [81]. More detailed studies require the acquisition of additional data and will focus on the initiation and distribution of wear on different tribological model systems. Friction and wear properties of tribological systems including lubricants can also be studied using the demonstrated techniques. For example Mate [82] deduced significantly different wear resistances for liquid polymers with and without hydroxylated end groups from force vs. distance curves and "friction loops": lateral force vs. distance curves measured with a bidirectional force microscope.

To summarize, we have demonstrated the principal experimentation techniques for performing micromechanical experiments using a specially designed atomic force microscope. The cantilever and the sample form a well defined tribological system that can be varied to gain insight into many different problems. The combination of the techniques discussed in this review with bidirectional force microscopes [19, 83, 84] and the imaging of electric charges allows even more detailed studies, such as the study of dissipation energies, vibration and tribocharging. The ongoing development of these experimental techniques by different researchers in various fields will certainly provide more and more insight into applied and fundamental problems of materials science.

5.2 Sub–100 nm lithography

Lithographic techniques have played a major role in scientific and technical development from the historic past to the present. Patterning is required in a wide variety of solid substrate and film materials, and the patterns vary over many decades in size depending on the application. With the demand for increasingly higher integration of devices, the necessity arises to further investigate and better understand the physical properties of structures of smaller and smaller size. This is the motivation of the following study, namely not only to produce very fine structures in polymeric films, but also to investigate methods to transfer these patterns into other materials, as required in the majority of technical applications.

The experiments described in this section are intended to further investigate the question of whether AFM modification techniques as described in the previous sections can be extended to produce sub-100 nm structures. As a first approach towards the patterning of substrates of various materials, an established chromium (Cr) slope evaporation technique (shadowing) for the fabrication of submicron grating structures [85] has been applied to produce a hard etch mask on topographic structures created with the atomic force microscope. The 'decorated' patterns were then processed using reactive ion etching to transfer the structure into the underlying substrate. The resolution of decoration and shadowing techniques is known to be sufficient to image single atomic steps on various crystalline surfaces [86, 87]. Lithographic Cr shadowing techniques in photoresist and polymer materials generally require deeper relief structures, with typical amplitudes of at least 10-20 nm. An important step towards reproducible patterning is thus to optimize AFM techniques for producing structures with small lateral dimensions and largest possible relief amplitudes.

The crucial parameter in achieving this aim is the aspect ratio of the cantilever–tips used. Experiments with the very high aspect ratio silicon tips [44, 47] have shown that they tend to break after a few modification experiments, usually leaving a double tip with low aspect ratio. This was most pronounced with the dynamic modification techniques as described in Section 4. The experiments presented here have therefore been performed using Si_3N_4 cantilevers with integrated pyramid-shaped tips. Their aspect ratio is less than unity, due to the production process which uses anisotropic etching of silicon 111 faces. The evolution of techniques to produce higher aspect ratio tips in a hard material would significantly benefit the fabrication of higher amplitude relief patterns using the atomic force microscope.

From exploratory studies using the different modification techniques on various polymer films, including nylon, polyvinylalcohol, PMMA and polycarbonate, we have concluded that polycarbonate is best suited to produce the highly corrugated patterns. We did not observe significant differences in the mechanical properties of 170 and 520 nm thick polycarbonate films on glass compared to bulk polycarbonate as used in previous studies. This is probably due to the small amplitude of the indentations created here ($\sim$30 nm) compared to the thickness of these films. The thinner polymer films are, however, favoured for the subsequent etching steps.

The modification technique best suited to our specific requirements was the static ploughing technique with forces of $\sim$7 μN. Torsion of the cantilever under this high loading, as well as creep of the piezo actuator, led to a certain distortion of the created line patterns (chirp and non-orthogonality). This did not impose any real problem in these preliminary, investigative tests, and can be corrected for future experiments.

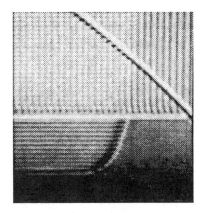

Figure 11: Crossed grating structure produced by successive AFM scanning with a force of 0.35 μN in two orthogonal directions. The diagonal line results from the start of the scanning movement and can be suppressed in a software upgrade. The well-formed crossed grating structure was imaged using the same cantilever and a loading of $\sim 10^{-9}$ N.

Fig. 11 demonstrates the expertise and control of AFM modification that has been acquired during this study. Two orthogonal, linear arrays of lines were fabricated, producing a very regular, crossed grating nanostructure with a period of about 80 nm. Together with the pattern shown in Fig. 4, this demonstrates the basic feasibility of fabricating fairly complex patterns using the AFM lithographic technique. In order to simplify the optimization of the writing parameters in this investigation, the experiments were limited to the fabrication of linear arrays of lines (grating structure) in a polycarbonate thin film on glass.

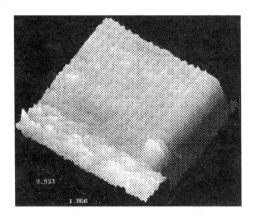 

Figure 12: The result of a Cr shadow masking and O_2 reactive ion etching step applied to AFM nanolithographic patterns, as revealed by subsequent AFM studies. a) The grating area on the left, with an amplitude of about 8 nm, has been fully covered with Cr and was therefore resistant to the etching step. The grating relief created in the polycarbonate thin film is clearly visible. The etched area on the right corresponded to a grating amplitude of about 15 nm, which was sufficient to form a patterned, shadow Cr mask. Etching then produced the depression shown by a combination of pattern transfer and underetching. b) Detail on the bottom of the etched depression, showing residual grating structure.

Fig. 12 shows a result of initial experiments on producing an effective etch mask from the modified polymer film surface. For this investigation, the polycarbonate film was patterned with the forces of 3.5 μN and 7 μN, producing two rectangular grating areas of about 300 nm period and different relief amplitudes of about 8 nm and 15 nm respectively. After AFM writing, the modified polycarbonate surface was subjected to a chromium (Cr) shadow evaporation step at an angle of 5° to the surface, followed by reactive ion etching in O_2 (20 mTorr, 60W). The figure shows the resulting surface as imaged in a commercially available atomic force microscope [88], which was used for the characterization of the patterns produced.

Fig. 12 a) shows the 2 grating areas. On the left, the grating relief of 8 nm amplitude has been fully covered by Cr and is thus resistant to etching

— the AFM image shows the residual grating relief. On the right, the deeper (~15 nm) grating amplitude has been successfully shadowed and the pattern heavily etched, leaving a rectangular depression of about 50 nm depth in the film. As shown in Fig. 12 b), the surface in this depression still shows the periodicity of the initial pattern, so that a combination of grating transfer and underetching has taken place. In any case, the processing has produced a film patterned with an etched relief in excess of 50 nm and with a resolution, as judged by the sharpness of the step, of better then 100 nm. Further work is required to improve the etching of the actual grating structure and to produce a mask of the corresponding linewidth.

We would like to point out that, to our knowledge, this is the first time a pattern created by AFM lithography has been produced in one instrument, further processed and then imaged using another instrument. The time between the production and final characterization of these structures was more than one month, demonstrating the stability of the developed techniques.

In conclusion, our investigative study clearly shows that AFM lithography can be combined with the classical lithographic tools of masking and etching. With smaller size structures, however, major improvements have to be made regarding the etching techniques. Further downscaling is clearly feasible, and the limits have to be determined in future studies. The above results have also to be put into perspective with respect to other techniques for sub-100 nm lithography, for example the well-developed direct e-beam writing, other methods that use stylus techniques to create different patterns (e.g. the nanoindenter [89] or the nanoruler [90]), or the widely used electron beam lithographies [91] as well as the newer lithographies using the STM as the source of either electron beams [92, 93, 94] or even other matter (e.g. local electrochemical deposition, local chemical reactions or field deposition of patterns [95]).

This study also most clearly exemplifies the versatility of AFM, in that it can be used for both characterizing and modifying samples in different states of the lithographic process, without irradiation damage problems that can be encountered with electron beam and optical methods. Very important insight into nanostructured surfaces will be gained from AFM experiments in the growing field of nanometer lithography.

5.3 Molecular scale modification

There have been many examples of AFM studies showing defects generated by various mechanisms like mechanical stress, irradiation damage, lattice misfit or other growth parameters. As another demonstration of the versatility of the AFM technique, this section reviews AFM modification experiments that give insight into molecular or atomic processes at surfaces. There have been many impressive examples of AFM imaging and manipulation of molecular layers on surfaces [17, 11, 12]. Weisenhorn [12], for example, has been able to probe the interaction between adsorbate actin filaments and the supporting mica as a function of the changing ionic conditions of the liquid solution. In the following part of this review we will mainly focus on modification experiments on AgBr and Langmuir–Blodgett films that illustrate the ongoing research.

Using similar techniques to those shown above, Meyer has been able to create indentations with depths of ~ 10 nm in AgBr (100) thin films [96]. In contrast to the long term stability of indentations in polymers, these have been observed to fill in on a time scale of ~ 12 min. The filling in was observed to be virtually independent of the scanning movement at adjusted low loadings, but could be accelerated by scanning with higher forces. This has been attributed to the high surface diffusion of AgBr, and diffusion rates could be calculated from the AFM experiments that confirmed the order of magnitude of diffusion measurements using radioactive bromine isotopes.

Figure 13: AFM image of an artificially created hole on a four-layer Cd-arachidate film (3000 nm×3000 nm). The density of the pores is decreased in the vicinity of the hole (courtesy of E. Meyer).

In the same paper Meyer has studied friction and topography of Cd-arachidate Langmuir–Blodgett films. Applying high loadings (10^{-8}–10^{-7} N) at high scan speeds (0.5–5 μm/s), they report erosion of molecular monolayers starting preferentially at inhomogeneities such as steps and pores of the molecular coverage. Fig. 13 shows the result of one of these experiments, and demonstrates the tendency of the removed molecules to retain the Langmuir Blodgett structure, i.e. to fill preexistent pores and form new layers. This and the observation of slow rearragements of the molecular layers exemplify the AFM capabilities to study structure and dynamics of molecular layers on surfaces.

The use of the more developed friction force microscope for even more advanced studies on various mixtures has made possible an impressive number of experiments: Meyer [97] reports on a detailed study of boundary lubrication on Cd-arachidate films, whereas Overney [98] and Meyer [99] clearly demonstrate that phase separated thin films can be classified for the type of chain (hydrocarbon and fluorocarbon respectively) using the measured frictional forces. With this knowledge and even more refined experimentation techniques they have been able to use the AFM as the tool to shape the domain structure of the studied Langmuir–Blodgett film mixture [99]. These studies clearly illustrate the power of the AFM techniques to modify not only topography but also the chemical composition of the surface layer.

6 Summary and outlook

To conclude, we have given an extensive overview of the experimental techniques using AFM for probing and modifying surface properties. The broad range of applications demonstrates that the techniques are of immediate significance to scientific and technological applications. However, the examples given here are pioneering experiments in a rapidly growing field of knowledge. We have only chosen examples how the atomic force microscope can be applied to gain control of and insight into the physics of small pieces of matter, and have tried to demonstrate the width of the field opened up by these experimental techniques. Together with the improved theoretical modelling, much more physics and chemistry at boundaries and interfaces will be discovered and explained.

In addition to this, the many experimental modes making AFM sensitive to weak electric and magnetic fields on surfaces will allow even more controlled experimentation. First examples are the detection of charges dissipating on a surface by Schoenenberger [27] and the control of surface magnetization as demonstrated by Moreland [100] and Hug [101]. In combination with the surface mechanical techniques mainly covered by this review, triboelectric and elecrochemical effects such as the formation of latent images in photosensitive materials, can be approached. Another exciting topic is the study of static and dynamic electromagnetic properties of nanofabricated structures, for example quantum dot and quantum wire arrays. This research might gain insight into possible elementary processes in working "devices" such as enzymes and biosensors as well as concepts or ideas for more developed nanotechnological devices as already proposed by Feynman [102] more than 30 years ago. The particular importance of the AFM techniques for this kind of study is the applicability to both conductors and non–conductors and the versatility for producing structures as well as sensing different forces at surfaces.

Acknowledgements

E. Meyer, R. Lüthi, R. Overney and L. Howald are gratefully acknowledged for providing Fig. 13. We are indebted to H.-J. Güntherodt, J. Frommer, H. Haefke, K. Knop, B. Curtis, J. E. Epler, Ph. Avouris, R. Walkup, Y. W. Mo, I.-W. Lyo, J. Söchtig, M. Rossi and L. Baraldi for interesting and helpful discussions. The help of D. Brodbeck, H. R. Hidber, R. Hofer and A. Tonin in developing the soft- and hardware is appreciated. We also thank I. Parashikov, H. Breitenstein, S. Messmer, H. Schütz, C. Appassito and J. Pedersen for technical assistance.

The authors gratefully acknowledge grants received from Swiss National Science foundation. One of the authors (T.J.) has received Swiss National Science Foundation and IBM postdoctoral fellowships while writing this review.

References

[1] G. Binnig, H. Rohrer, Ch. Gerber, E. Weibel. *Surface studies by scanning tunneling microscopy.* Physical Review Letters **49**, 57–61 (1982).
G. Binnig, H. Rohrer. *Scanning tunneling microscopy.* Helvetica Physica Acta **55**, 726–735 (1982).
G. Binnig, H. Rohrer, Ch. Gerber and E. Weibel. *Tunneling through a controllable vacuum gap.* Applied Physics Letters **40**, 178–180 (1982).

[2] G. Binnig, C. F. Quate and Ch. Gerber. *Atomic force microscopy.* Physical Review Letters **56**, 930–933 (1986).

[3] J. A. Stroscio and D. M. Eigler. *Atomic and molecular manipulation with the scanning tunneling microscope.* Science **254**, 1319–1326 (1992).

[4] Ph. Avouris and R. E. Walkup. *Atomic and nanometer scale modification of materials with the scanning tunneling microscope.* In: Encyclopedia of Advanced Materials. Eds. D. Bloor, R. J. Brook, M. C. Femings, S. Mahajan. Pergamon Press, Oxford (1993).

[5] U. Staufer. *Surface modification with a scanning proximity probe microscope.* Springer Series in Surface Sciences Vol. 28. Scanning tunneling microscopy II. Eds. R. Wiesendanger und H.-J. Güntherodt. Springer (1992).

[6] G. M. Shedd and P. E. Russell. *The scanning tunneling microscope as a tool for nanofabrication.* Nanotechnology **1**, 67–80 (1990).

[7] R. E. Walkup, D. M. Newns and Ph. Avouris. *Vibrational heating and atom transfer with the STM.* In: Atomic & nanoscale modification of materials: Fundamentals & applications. Ed. Ph. Avouris. Kluwer Academic Publishers, Dordrecht (1993).

[8] Y. W. Mo. *Precursor states in the adsorption of Sb_4 on $Si(001)$.* Physical Review Letters **69**, 3643–3646 (1992).

[9] C. F. Quate. *Manipulation and modification of nanometer scale objects with the scanning tunneling microscope.* In: Highlights in condensed matter physics and future prospects, NATO ASI Series B, **285**, Plenum Press, New York, 573-630 (1991).
C. F. Quate, *Surface modifications with the scanning tunneling microscope and the atomic force microscope.* In: Scanning tunneling microscopy and related methods. Eds. R. J. Behm et. al., 281–297, Kluwer Academic Publishers (1990).

[10] R. Wiesendanger. *Fabrication of nanometer structures using scanning tunneling microscopy.* Applied Surface Science **54**, 271–277 (1992).

[11] T. R. Albrecht. *Advances in atomic force microscopy and scanning tunneling microscopy.* Ph. D. Thesis, Stanford University (1989).

[12] A. L. Weisenhorn, B. Drake, C. B. Prater, S. A. C. Gould, P. K. Hansma, F. Ohnesorge, M. Egger, S.-P. Heyn and H. E. Gaub. *Immobilized proteins in buffer imaged at molecular resolution by atomic force microscopy.* Biophysics Journal **58**, 1251–1258 (1990).

[13] A. L. Weisenhorn, J. E. MacDougall, S. A. C. Gould, S. D. Cox, W. S. Wise, J. Massie, P. Maivald, V. B. Elings, G. D. Stucky and P. K. Hansma. *Imaging and manipulating molecules on a zeolite surface with an atomic force microscope.* Science **247**, 1330–1333 (1990).

[14] T. A. Jung, A. Moser, H. J. Hug, D. Brodbeck, R. Hofer, H. R. Hidber and U. D. Schwarz. *The atomic force microscope used as powerful tool for machining surfaces.* Ultramicroscopy **42**, 1446-1451 (1992).

[15] E. Hamada, R. Kaneko. *Microtribological evaluations of a polymer surface by atomic force microscopy.* Ultramicroscopy **42–44**,184–190 (1992).

[16] H. J. Mamin and D. Rugar. *Thermomechanical writing with an atomic force microscope tip.* Applied Physics Letters **61**, 1003–1005 (1992).

[17] Y. Kim and C. M. Lieber. *Machining oxide thin films with an atomic force microscope.* Science **67**, 375–377 (1992).
Y. Kim, J. L. Huang, and C. M. Lieber. *Characterization of nanometer scale wear and oxydation of transition metal dichalcogenide lubricants by atomic force microscopy.* Applied Physics Letters **59**, 3404–3406 (1991).

[18] T. Göddenhenrich, H. Lemke, U. Hartmann, C. Heiden. *Force microscope with capacitive displacement detection.* Journal of Vacuum Science and Technology **A 8** 383–385 (1990).

[19] G. Neubauer, S. R. Cohen, G. M. McClelland, D. Horne and C. M. Mate. *Force microscopy with a bidirectional capacitance sensor.* Review of Scientific Instruments **61**, 2296–2308 (1990).

[20] G. McClelland, R. Erlandsson and S. Chiang. In: Review of progress in quantitatative nondestructive evaluation. Eds. D. O. Thomson and D. E. Chimenti. Plenum New York **6**, 307 (1987).

[21] T. Goeddenhenrich, U. Hartmann, M. Anders and C. Heiden. *Investigations of bloch wall fine structures by magnetic force microscopy.* Journal of Microscopy **152**, 527–536 (1988).

[22] Focusing error detection type optical head, Olympus, ZP 01.
T. Kohono, N. Ozawa, K. Migamoto and T. Musha. *High precision optical surface sensor.* Applied Optics **27**, 103–108 (1988).

[23] G. Meyer and N. M. Amer. *Novel optical approach to atomic force microscopy.* Applied Physics Letters **53**, 1045–1047, (1988).
S. Alexander, L. Hellemans, O. Marti, J. Schneir, V. Elings, P. K. Hansma, M. Longmire and J. Gurley. *An atomic-resolution atomic force microscope implemented using an optical lever.* Journal of Applied Physics **65**, 164–167 (1989).

[24] D. Anselmetti, Ch. Gerber, B. Michel, H. Rohrer and H.-J. Güntherodt. *A compact, combined scanning tunneling/force microscope.* Review of Scientific Instruments **63**, 3003–3006 (1992).

[25] E. Meyer, H. Heinzelmann, P. Grütter, T. A. Jung, Th. Weisskopf, H.-R. Hidber, R. Lapka, H. Rudin and H.-J. Güntherodt. *Comparative study of lithium fluoride and graphite by atomic force microscope.* Journal of Microscopy **152**, 269–280 (1988).

[26] B. D. Terris, J. E. Stern, D. Rugar and H. J. Mamin. *Contact electrification using force microscopy.* Physical Review Letters **63**, 2669–2672 (1989).
J. E. Stern, B. D. Terris, H. J. Mamin and D. Rugar. *Deposition and imaging of localized charge on insulator surfaces using a force microscope.* Applied Physics Letters **53**, 2717–2719 (1988).

[27] C. Schönenberger and S. F. Alvarado. *Observation of single charge carriers by force microscopy.* Physical Review Letters **65**, 3162–3164 (1990).

[28] Y. Martin and H. K. Wickramasinghe. *Magnetic imaging by "force microscopy" with 1000 Å resolution.* Applied Physics Letters **50**, 1455–1457 (1987).
Y. Martin, D. Rugar and H. K. Wickramasinghe. *High resolution magnetic imaging of domains in TbFe by force microscopy.* Applied Physics Letters **52**, 244–246 (1988).

[29] H. J. Hug, A. Moser, T. A. Jung, A. Wadas, O. Fritz, I. Parashikov and H.-J. Güntherodt. *Low temperature magnetic force microscopy.* Submitted.

[30] D. Rugar, H.-J. Mamin and P. Guethner. *Improved fiberoptic interferometer for atomic force microscopy.* Applied Physics Letters **55**, 2588 (1989).

[31] Chr. Schönenberger, S. F Alvarado. *Understanding magnetic force microscopy.* Zeitschrift für Physik B **80**, 373–383 (1990).

[32] A. J. den Boef. *Scanning force microscopy using optical interferometry.* Ph. D. Thesis, University Twente, 1991.

[33] E. Meyer and H. Heinzelmann. *Scanning force microscopy (SFM).* In: Scanning tunneling microscopy I. Eds. H.-J. Güntherodt and R. Wiesendanger, Springer, Berlin, Heidelberg (1992).

[34] J. E. Frommer. *The emerging presence of scanning tunneling microscopy (AFM) in organic chemistry.* Invited review in: Angewandte Chemie. English edition **31** 1298–1328 (1992).

[35] L. Scandella. *Atomic force microscopy.* Proceedings of International School of Electron Microscopy, Mesagne, Italy (1992).

[36] D. Sarid *Scanning force microscopy.* Oxford University Press, New York 1991.
J. Chen *Introduction to scanning tunneling microscopy.* Series in Optical and Imaging Sciences 4, Oxford University Press, in press, 1993.
R. Celotta, T. Lucatorta, J. Stroscio and W. Kaiser, eds. *Methods of experimental physics: Scanning tunneling microscopy.* **27**, Academic Press (1992).
N. A. Burnham and R. J. Colton. *Scanning tunneling microscopy and spectroscopy.* In: Fundamental concepts and applications. Ed. D. Bonnell, VCH, New York, 191–249 (1993).

[37] B. N. J. Persson. *The atomic force microscope: Can it be used to study biological molecules?* Chemical Physics Letters **141**, 366–368 (1987).

[38] T. P. Weihs, Z. Nawaz, S. P. Jarvis, and J. B. Pethica. *Limits of imaging resolution for atomic force microscopy of molecules.* Applied Physics Letters **59**, 3536–3538 (1991).

[39] A. L. Weisenhorn, P. Maivald, H. -J. Butt and P. K. Hansma. *Measuring adhesion, attraction, and repulsion between surfaces in liquids with an atomic force microscope.* Physical Review B **45**, 11226–11232 (1992).

[40] N. A. Burnham and R. J. Colton. *Measuring the nanomechanical properties and surface forces of materials using an atomic force microscope.* Journal of Vacuum Science and Technology **A 7**, 2906–2913 (1989).

[41] H. J. Hug, T. A. Jung and H.-J. Güntherodt. *A high stability and low drift AFM.* Review of Scientific Instruments **63**, 3900–3904 (1992).

[42] S. A. Joyce and J. E. Houston, *A new force sensor incorporating force-feedback control for interfacial force microscopy.* Review of Scientific Instruments **62**, 710–715 (1992).
G. L. Miller, J. E. Griffith E. R. Wagner and D. A. Grigg. *A rocking*

beam electrostatic balance for the measurement of small forces. Review of Scientific Instruments **62**, 705–709 (1992).

[43] Si$_3$N$_4$ Cantilevers from Park Scientific Instruments, Mountain View, CA 94043.

[44] O. Wolter, Th. Bayer and J. Greschner. *Micromachined silicon sensors for scanning force microscopy.* Journal of Vacuum Science and Technology **B 9**, 1353–1357 (1991).
O. Wolter, Nanoprobe, D–7042 Aidlingen 3, Germany.

[45] U. D. Schwarz, H. Haefke, P. Reimann and H.-J. Güntherodt. *Tip artifacts in scanning force microscopy.* To be submitted to Journal of Microscopy.

[46] T. R. Albrecht, S. Akamine, T. E. Carver and C. F. Quate. *Microfabrication of cantilever styli for the atomic force microscope.* Journal of Vacuum Science and Technology **A 8**, 3386–3396 (1990).
P. Grütter, W. Zimmermann–Edling and D. Brodbeck. *Tip artifacts of microfabricated force sensors for atomic force microscopy.* Applied Physics Letters **60**, 2741–2743 (1992).

[47] Ultralevers by Park Scientific Instruments as released recently seem to have similar tip geometries; Mountain View, CA 94043.

[48] E. P. Visser, J. W. Gerritsen, W. J. P. Enckevort and H. van Kempen. *Tip for scanning tunneling microscopy made of monocrystalline, semiconducting, chemical vapor deposited diamond.* Applied Physics Letters **60**, 3232–3234 (1992).

[49] G. J. Germann, G. M. McClelland, Y. Mitsuda, M. Buck and H. Seki. *Diamond force microscope tips fabricated by chemical vapor deposition.* Review of Scientific Instruments **63**, 4053–4055 (1992).

[50] D. J. Keller and C. Chih–Chung. *Imaging steep, high structures by scanning force microscopy with electron beam deposited tips.* Surface Science **268**, 333–339 (1992).

[51] K. L. Lee, D. W. Abraham, F. Secord and L. Landstein. *Submicron Si trench profiling with an electron–beam fabricated atomic force microscope tip.* Journal of Vacuum Science and Technology **6**, 477–481 (1991).
IBM SXM Tip Series, CD-Tips, Manufacturing Technology Center.

[52] A. Moser, H. J. Hug, T. A. Jung, U. Schwarz and H.-J. Güntherodt. *A miniature fiber optic force microscope scanhead.* Submitted.

[53] D. Brodbeck, L. Howald, Roland Lüthi, E. Meyer and R. Overney. *Scan control and data acquisition for bidirectional force microscopy.* Ultramicroscopy **42–44**, 1580–1584 (1992).

[54] A. L. Weisenhorn, P. K. Hansma, T. R. Albrecht and C. F. Quate. *Forces in atomic force microscopy in air and water.* Applied Physics Letters **54**, 2651–2653 (1989).

[55] J. N. Israelachvili *Intermolecular and surface forces.* 2nd Edition Academic Press (1991).

[56] S. R. Cohen. *An evaluation of the use of the atomic force microscope for studies in nanomechanics.* Ultramicroscopy **42–44**, 66–72 (1992).

[57] U. Dürig, J. K. Gimzewski and D. W. Pohl. *Experimental observation of forces acting during scanning tunneling microscopy.* Physical Review Letters **57**, 2403–2406 (1986).
U. Dürig, O. Züger, D. W. Pohl. *Observation of metallic adhesion using the scanning tunneling microscope.* Physical Review Letters **65**, 349–352 (1990).

[58] L. Howald, H. Haefke, E. Meyer, R. Overney, G. Gerth and H.-J. Güntherodt. *The surface of AgBr crystals studied with the AFM.* L'actualité chimique, 200–202 (1992).

[59] T. E. Karis and V. J. Novotny. *Pin-on-disk tribology of thin-film magnetic recording disks.* Journal of Applied Physics **66**, 2706–2711 (1989).

[60] D. Dowson. *History of tribology.* Longman, London (1979).

[61] D. Tabor and R. H. S. Winterton. *The direct measurement of normal and retarded van der Waals forces.* Proceedings Royal Society London **A 312**, 435 (1969).

[62] D. Tomanek, G. Overney, H. Miyazaki, S. D. Mahanti and H.-J. Güntherodt. *Theory for the atomic force microscopy of deformable surfaces.* Physical Review Letters **63**, 876 (1989); ibid **63**, 1896 (E) (1989).

[63] U. Landman and W. D. Luedtke. *Nanomechanics and dynamics of tip-substrate interactions.* Journal of Vacuum Science and Technology **B 9**, 414–423 (1991).

[64] J. Krim, D. H. Solina, R. Chiarello. *Nanotribology of a Kr monolayer: A quartz-crystal microbalance study of atomic-scale friction.* Physical Review Letters **66**, 181–184 (1991).

[65] G. S. Blackman, C. M. Mate and M. R. Philpott. *Interaction forces of a sharp tungsten tip with molecular films on silicon surfaces.* Physical Review Letters **65**, 2270–2273 (1990).

[66] C. M. Mate, G. M. McClelland, R. Erlandsson and S. Chiang. *Atomic-scale friction of a tungsten tip on a graphite surface.* Physical Review Letters **59**, 1942–1945 (1987).

[67] See for example: The nanoindenter II. The mechanical properties microprobe. Nano Instruments, Inc. P. O. Box 14211 Knoxville TN 37914.

[68] J. B. Pethica, R. Hutchings and W. C. Oliver. *Hardness measurements at penetration depths as small as 20 nm.* Philosophical Magazine A **48**, 593–606 (1983).

[69] W. C. Oliver, R. Hutchings and J. B. Pethica. *Measurement of hardness at indentation depths as low as 20 nanometers.* Microindentation Techniques in Materials Science and Engineering, ASTM STP 889. Eds. P. J. Blau and B. R. Lawn. American Society for Testing and Materials, Philadelphia, 90–108 (1986).

[70] J. B. Pethica and W. C. Oliver. *Mechanical properties of nanometer volumes of material: Use of the elastic response of small area indentation.* Materials Research Society Symposium Proceedings, **130**, 13–23, Materials Research Society, Boston, MA (1989).

[71] T. P. Weihs, S. Hong, J. C. Bravman and W. D. Nix. *Mechanical deflection of cantilever microbeams: A new technique for testing the mechanical properties of thin films.* Journal of Material Science **3**, 931–942 (1988).

[72] J. L. Loubet, J. M. Georges and G. Meille. *Vickers indentation curves of elastoplastic materials.* Microindentation Techniques in Materials Science and Engineering, ASTM STP. Eds. P. J. Blau and B. R. Lawn. American Society for Testing and Materials, Philadelphia, 72–89 (1986).

[73] P. J. Blau and B. R. Lawn, Eds. *Microindentation techniques in materials science and engineering, ASTM STP.* American Society for Testing and Materials, Philadelphia, 72–89 (1986).

[74] M. R. Castell, M. G. Walls and A. Howie. *Imaging of low load indentations into Si and GaAs by scanning tunneling microscopy.* Ultramicroscopy **42–44**, 1490–1497 (1992).
Y. Miyazaki, Y. Koga and H. Hayashi. *Observation of vickers imprints by scanning tunneling microscopy.* Journal of Vacuum Science and Technology **A 8**, 628–630 (1990).

B. M. DeKoven and G. F. Meyers. *Friction studies in ultrahigh vacuum of Fe surfaces with thin films from exposure to perfluordiethylether.* Journal of Vacuum Science and Technology **A 9**, 2570–2577 (1991).

[75] F. Thibaudau and J. Cousty. *Atomic force microscopy observations of tracks induced by swift Kr ions in mica.* Physical Review Letters **67**, 1582–1585 (1991).

[76] E. Meyer, H. Heinzelmann, P. Grütter, T. A. Jung, L. Scandella, H.-R. Hidber, H. Rudin, H.-J. Güntherodt and C. Schmidt. *Investigation of hydrogenated amorphous carbon coatings for magnetic data storage by atomic force microscopy.* Applied Physics Letters **55**, 1624–1626 (1989).

[77] E. Meyer, H. Heinzelmann, P. Grütter, T. A. Jung, H.-R. Hidber, H. Rudin and H.-J. Güntherodt. *Atomic force microscopy for the study of tribology and adhesion.* Thin Solid Films **181**, 527–544 (1989).

[78] H. Heinzelmann, E. Meyer, L. Scandella, P. Grütter, T. A. Jung, H. J. Hug, H.-R. Hidber, H.-J. Güntherodt and C. Schmidt. *Topography and correlation to wear of hydrogenated amorphouscarbon coatings: An atomic force microscopy study.* Wear **135**, 109–117 (1989).

[79] T. Miyamoto and R. Kaneko. *Tribological characteristics of amorphous carbon films investigated by point contact microscopy.* Journal of Vacuum Science and Technology **B 9**, 86–89 (1991).

[80] P. Grütter, personal communication.
R. Kaneko, personal communication.

[81] D. Bosbach and W. Rammensee. *Surface manipulation on layered organic crystals by scanning force microscopy.* Ultramicroscopy **42–44**, 973–976 (1992).

[82] C. M. Mate. *Atomic force microsope study of polymer lubricants on silicon surfaces.* Physical Review Letters **68**, 3323–3326 (1992).

[83] G. Meyer and N. M. Amer. *Simultaneous measurement of lateral and normal forces with an optical-beam- deflection AFM.* Applied Physics Letters **57**, 2089–2091 (1990).

[84] O. Marti, J. Colchero and J. Mlynek. *Combined scanning force and friction microscopy of mica.* Nanotechnology **2**, 141–144 (1990).

[85] M. T. Gale, B. J. Curtis, H. Kiess and R. Morf. *Design and fabrication of submicron grating structures for light trapping in silicon solar cells.* Proc. SPIE **1272**, 60–66 (1990).

[86] G. A. Bassett. *A new technique for decoration of cleavage and slip steps on ionic crystal surfaces.* Philosophical Magazine **8** (**3**), 1958.

[87] for a review see for example: H. Bethge, M. Krohn and H. Stenzel. *Indirect imaging of surfaces by replica and decoration techniques.* In: Materials Science Monograph **40**. Electron microscopy in solid state physics. Eds. H. Bethge and J. Heydenreich. Elsevier (1987).

[88] Digital Instruments, Inc., 6780 Cortona Drive, Santa Barbara, CA 93117. *Nanoscope III.*

[89] J. Gobrecht and J. B. Pethica. *The potential of mechanical microlithography for submicron patterning.* Microelectric Engineering **5**, 471–474 (1986).

[90] T. Kita and T. Harada. *Ruling engine using a piezoelectric device for large and high groove density gratings.* Applied Optics **31**, 1399–1406 (1992).

[91] see for example: A. Classen, S. Kuhn, J. Straka and A. Forchel. *High voltage electron beam lithography of the resolution limits of SAL 601 negative resist.* Microelectronic Engineering **17**, 21–24 (1992).
A. Forchel, B. E. Maile, H. Leier, G. Mayer and R. Germann. *Optical emission from quantum wires.* In: Science and engineering of one and zero dimensional semiconductors. Proceedings of a NATO advanced research workshop. Plenum, New York (1990).

[92] M. A. McCord and R. F. W. Pease. *Exposure of calcium fluoride resist with the scanning tunneling microscope.* Journal of Vacuum Science and Technology **B 5**, 430–433 (1988).

[93] C. R. K. Marrian and E. A. Dobisz. *High resolution lithography with a vacuum scanning tunneling microscope.* Ultramicroscopy **42–44**, 1309–1316 (1992).

[94] A. Majumdar, P. I. Oden, J. P. Carrejo, L. A. Nagahara, J. J. Graham and J. Alexander. *Nanometer-scale lithography using the atomic force microscope.* Applied Physics Letters **61**, 2293–2295 (1992).

[95] H. J. Mamin, S. Chiang, H. Birk, P. H. Guethner and D. Rugar. *Gold deposition from a scanning tunneling microscope tip.* Journal of Vacuum Science and Technology **B 9**, 1398–1402 (1991).

[96] E. Meyer, L. Howald, R. Overney, D. Brodbeck, R. Lüthi, H. Haefke, J. Frommer and H.-J. Güntherodt. *Structure and dynamics of solid surfaces observed by atomic force microscopy.* Ultramicroscopy **42-44**, 1580–1584 (1992).

[97] E. Meyer, R. Overney, D. Brodbeck, L. Howald, R. Lüthi, J. Frommer and H.-J. Güntherodt. *Friction and wear of Langmuir–Blodgett films observed by friction force microscopy.* Physical Review Letters **69**, 1777–1780 (1992).

[98] R. M. Overney, E. Meyer, J. Frommer, D. Brodbeck, R. Lüthi, L. Howald, H.-J. Güntherodt, M. Fujihira, H. Takano and Y. Gotoh. *Friction measurements on phase separated thin films with a modified force microscope.* Nature **359**, 133–135 (1992).

[99] E. Meyer, R. Overney, R. Lüthi, D. Brodbeck, L. Howald, J. Frommer and H.-J. Güntherodt. *Friction force microscopy of mixed Langmuir–Blodgett films.* Thin Solid Films **220**, 132–137 (1992).

[100] J. Moreland and P. Rice. *High resolution tunneling stabilized magnetic imaging and recording.* Applied Physics Letters **57**, 310–312 (1990).

[101] H. J. Hug, A. Moser, O. Fritz, I. Parashikov, H.-J. Güntherodt and Th. Wolf. *Low temperature magnetic force microscopy on High T_c superconductors.* Submitted to LT 20, 1993, Eugene, Oregon, USA.

[102] R. P. Feynman. *There's plenty of room at the bottom.* In: Miniaturization. Ed. H. D. Gilbert. Reinhold Publishing Group, New York (1961).

Fabrication and Characterization Using Scanned Nanoprobes

G.C. Wetsel, Jr., S.E. McBride, and H.M. Marchman
Erik Jonsson School of Engineering and Computer Sciences,
The University of Texas at Dallas, Richardson, TX 75083

I. Introduction

Microelectronics manufacturing employing methods such as molecular-beam epitaxy to form layers of material and electron-beam lithography to create lateral features are presently capable of producing electronic devices with feature sizes diminishing to about 100 nm, such as quantum-effect devices[1]. Estimates of improvements in current techniques indicate that feature sizes may decrease below 100 nm in the future. New measurement techniques are required to characterize such small devices. The higher-speed performance and higher device densities of the new generation of electronics—*nanoelectronics*—will involve devices with feature sizes of the order of 10 nm or less. A new technology is required to fabricate nanometer-scale devices. At the University of Texas at Dallas (UTD), we have developed scanned nanoprobe instruments (SNIs) that have been flexibly designed for topographical, electrical, and optical characterization as well as for fabrication, modification, and repair of nanostructures[2,3].

An example of the type of task to be performed by an SNI is the topographical and electrical characterization of an identifiable, individual, nano-scale device, such as a quantum-dot resonant-tunneling diode[4]. In this case the device to be characterized is about 150 nm in diameter, with its top extending several hundred nanometers above the surface of the substrate; the geography is similar to a mesa in the desert. The dot, of course, cannot be observed optically; however, its location relative to an optically-observable reference mark, which serves as the origin of coordinates, is known from the device design plan. The distance of the dot from the reference mark is of the order of 10μm. If the top of the dot can be located without damaging the structure, topographical imaging and electrical characterization can be accomplished using a scanning-tunneling-microscope (STM) probe. The coarse-positioning capabilities of translation stages and STM-tip scanners are more than sufficient for positioning the tip relative to the origin. Thus, the problem reduces to the precise initial positioning of the probe to the origin of coordinates. We describe in Section IIA how we have solved[3,5] this problem to obtain the first high-resolution images of identifiable quantum dots[6]; images of dots are presented in Section III.

Another serious difficulty is encountered when an STM probe is used to measure in air the current-voltage characteristics of a quantum heterostructure. Electrical contact with the device electrode must be established by carefully controlling the approach of the tip without damaging the tip or the device. Achievement of subnanometer precision tip extension and the transition from tunneling to electrical contact are described in Section IIB[3,7]. Results of electrical characterization of devices are presented in Section III.

Promising techniques for nanostructure fabrication involve interactions between sharp metallic tips and surfaces to be modified. The tips must be precisely positioned over distances of the order of Å and the tip-sample voltage (V_t) and current (I) must be carefully controlled; this can be accomplished using a scanned nanoprobe. Nanometer-scale features (<10 nm) can be created on sample surfaces in air by pulsing the tip/sample voltage usingscanned probes based on scanning-tunneling-microscope technology. Mounds on Au samples using Au tips have been formed at IBM (Almaden)[8]. Features ranging from craters to mounds have been formed at UTD[2,9-11] using various tip/sample combinations (W/Au, Au/Au, Au/Pt, W/Pt, and W/C). The process by which these features are formed is characterized by a threshold value of V_t[2,8-11]. The results of various experiments collectively indicate that the features were formed by direct transfer of tip material to sample or vice versa when a critical electric field characteristic of the junction was exceeded. An experimental value for the critical electric field for mound formation[11] was obtained by combining the results of two separate measurements; using a W tip and a Pt sample, the value was determined to be $E_c = 0.23$ V/Å. The experimental evidence is consistent with theoretical arguments concerning the sufficiency of the measured value of E_c to remove atoms by electric-field emission[8]. However, in a recent article[12], Tsong has suggested other mechanisms for mound formation than "field evaporation", such as electric-field-gradient-induced surface diffusion initiated by Joule heating due to field emission of electrons. On the other hand, Lang[13] has reported that his theoretical model of electric-field induced transfer gives a reasonable account of atom-transfer experiments in high vacuum[14]. The above results are very interesting because of the challenge to understand the basic physics of the surface modification process and very important because of the application to nanoscale technology. The techniques for nanoscale fabrication using atomic transfer are described in section IIC; characterization of the features so formed are described in Section IV.

II. Scanned Nanoprobes

A. Scanned Probe for Characterization of Large Nanostructures

For the characterization of relatively large nanostructures, atomic resolution is usually not as important as optically-guided initial positioning; thus, probe-design considerations are different from those of the traditonal STM. The large- nanostructure probe (LNP), illustrated in Fig. 1, consists of seven basic parts: 1) a mechanical structure, 2) coarse positioning, 3) an STM tip scanner, 4) an optical guidance system, 5) vibration damping, 6) electronic amplification and control circuitry, and 7) computer-controlled instrumentation. The X-Y coarse translation stage is mounted on a base plate and the fine scanner is mounted in a top piece; these two parts are separated from each other by the vertical coarse translation stage, thus forming a "C"-shaped structure. The sample is mounted on top of an X-Y coarse translation stage. The LNP structure is only 10 cm in height, which provides a low center of gravity for added stability. Coarse X-Y positioning is provided by translation stages driven either by piezoelectric motors or manual differential-screw micrometers. The two motor-driven stages provide computer-controlled horizontal positioning with encoded step sizes as small as 0.1 μm with an overall range of 25 mm; however, they introduce a considerable amount of mechanical and electrical noise to the LNP. The manually-driven stages provide 0.1 μm precision over a range of 25 mm, but contribute very little mechanical noise and no electrical noise. The vertical coarse-translation stage can be adjusted either manually with a micrometer screw or electrically with a piezoelectric micrometer driven by a variable high-voltage source.

Fine positioning of the STM tip and scanning for constant-tunneling-current images were provided by the single-tube, three-dimensional scanner described by Binnig and Smith[15]. The displacement of the tube scanner was determined as a function of applied voltage using the optical-beam-deflection technique described by Wetsel, et al.[16] The uncertainty in tip-sample position (due to system noise) was calculated to be about 1 Å by multiplying the tube-scanner calibration factor determined with the optical-beam-deflection technique by the measured electrical noise in the fine scanner transducer voltage.

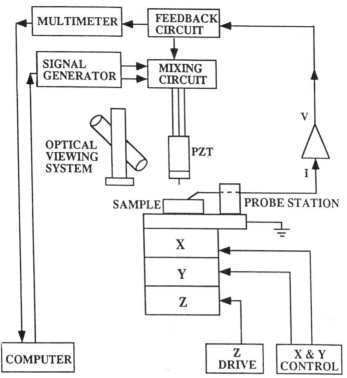

Fig. 1. Large nanostructure probe system block diagram.

The optical viewing system[3,5] consists of an image guiding subsystem, a microscope, and a video (CCD) camera connected to a high-resolution monitor and video recorder. These components are mounted on various translation stages for the lateral, vertical, and angular positioning. The translation system allows scanning of the image guide across the sample surface until the probe tip and reference mark come into view. The tip-to-sample view is also used for vertical coarse approach when tunneling is first established. At an effective magnification of 2000×, the horizontal x or y position of the tip is established to within 1 μm of the desired reference mark on the sample. At lower magnifications, views of the entire sample can be obtained. The STM tube scanner is used for high-resolution imaging of areas as large as 3 μm × 3 μm. Angstrom-unit-resolution imaging can be achieved for the smaller scan ranges of about 10 nm × 10 nm.

The heart of the optical viewing system is the image guiding subsystem, which is composed of a single optical fiber with a quadratic index-of-refraction profile. Self-focusing rod lenses (SRL's) are a common, commercially-available form of quadratic-index fiber, which were originally designed as relay lenses for small-diameter imaging systems. The SRL's are normally used with an imaging lens to form a complete image guide. The imaging lens is an objective lens that gathers light with a wide angle and focus it, thus forming a demagnified image at the back surface of the SRL. For the LNP optical guidance system, the imaging lens was removed from the SRL so that the image guide has a new, narrower, viewing angle of about 9°, and hence much less demagnification. Increased magnification is desirable since the objects to be viewed are micrometer sized.

Inherent with the conventional STM's ability to image very small (Å scale) features is its sensitivity to mechanical vibration and thermal drift. For larger scan ranges, susceptibility to mechanical vibration increases. Therefore, vibration isolation becomes a critical factor in the design of the LNP. Vibration isolation was achieved by various stages of damping using a series of stacked plates with neoprene rubber sandwiched in between each plate, and suspending the assembly and stack of plates by bungee cords from four vibration-damping posts[17]. The posts are mounted on a vibration-isolation table.

The electronic circuitry used in the LNP consists of a precision, instrumentation, transimpedance preamplifier and a feedback loop for control. A computer is used to control the data acquisition and to display the electronic images. For imaging, sweep signals are applied to the tube-scanner quadrants. The topographic information is obtained from the feedback voltage applied to the transducer when the LNP is operating in the constant-current mode.

B. Establishing Nondestructive Electrical Contact in Air

Measurement of electronic as well as topographic characteristics with nanometer spatial resolution is important for the characterization of nanometer-scale surface features and devices. Studies of the transition from tunneling to mechanical and electrical contact between two metals in a vacuum environment have been reported in which a *sharp* transition from tunneling to a contact of atomic dimensions was observed[18]. We have observed a *gradual* transition from tunneling to electrical contact for various samples in air. The gradual transition is probably due to contamination and the associated surface deformation as the tip approaches the surface[19]. Since the z position of an STM-like probe is controlled by sensing tunneling current, the large tunneling resistance of the tip-sample gap is effectively in series with the device if the tip does not touch the sample. If nanostructure device I-V measurements are to be unaffected, the probe tip must make good electrical contact with the surface. A serious difficulty encountered when establishing electrical contact between the tip and sample is device and/or tip damage, particularly when characterization is done in air. For minimized damage, electrical contact must be established by carefully controlling the tip-to-sample approach. We describe below how we have achieved Angstrom-unit-precision tip extension and the transition from tunneling to electrical contact[7].

The two-step technique consists of a coarse tip-to-surface approach (u_c) followed by a fine tip extension (u_f). The majority of tip extension occurs during coarse approach, where the feedback circuit is not interrupted, but is actually used as a servomechanism to bring the tip closer to the surface. We have tried two methods to use feedback regulation in this way: 1) superposition of a ramp voltage (dV_r) on the tunneling-current reference voltage (V_r) in the difference stage of the feedback circuit, and 2) subtraction of a ramp voltage (dV_t) from the quiescent tip voltage (V_t). The second method, which is much less destructive to the surface and tip because of the lower current and voltage levels at small tip-sample spacing (s), consists of decreasing V_t while maintaining constant I. In this case feedback regulation causes the tip to move closer to the surface as V_t is decreased (asin s-V_t spectroscopy). At minimal V_t (corresponding to maximal u_c), feedback is disabled and fine extension is performed. Once fine extension has been completed, V_t is varied while feedback is still being held and electronic measurements are made; u_c-V_t data is recorded during the coarse extension as well. Plots of u_c vs. V_t during coarse approach and I vs. V_t during fine extension reveal when the electrical contact is made with the surface.

If a harmonic component (dV_h) with frequency above the bandwidth of the feedback circuitry is also added to V_t, the resulting harmonic component of tunneling current (dI_h) can provide a very sensitive means of measuring the change in dynamic conductance, g_h (dI/dV_t), of the tunneling gap as the tip is brought closer to the surface. Then, it is possible to monitor the change in g_h during both coarse and fine extension. As V_t is reduced during coarse approach, the amplitude of dI_h is measured using phase-sensitive detection. By simultaneously recording dI/dV_t and u_c as V_t is varied, one can obtain a measure of $(dI/dV_t)_I$ as s

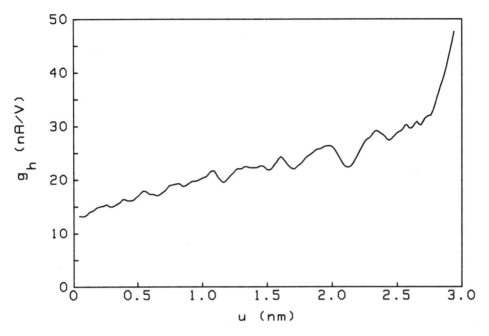

Fig. 2. Gap conductance, g_h, vs. coarse tip extension, u_c, for W tip on Au sample.

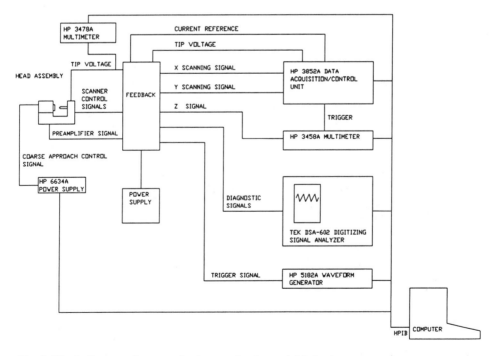

Fig. 3. Block diagram of nanoscale characterization and fabrication system.

changes. A typical experimental curve showing dI/dV_t as a function of u_c is shown for a W tip on a Au surface in Fig. 2. The abrupt change in g_h at about u_c=28 Å indicates that electrical contact has been achieved.

C. Fabrication and Characterization of Nanoscale Features

A flexibly-designed scanned nanoprobe instrument (SNI) has been developed for nanoscale fabrication. Whereas it is capable of atomic resolution in air when used as a constant-current-imaging STM, it is primarily configured for creation of nanoscalefeatures by pulsing the tip-to-sample voltage and recording in-process signals for analysis and control of the feature formation. The block diagram of the SNI is shown in Fig. 3. The controller and data-acquisition system are based on commercial equipment. A Hewlett-Packard (HP) series 300 computer is used as an instrument controller. Data acquisition employs a Tektronix DSA602 digitizing signal analyzer and an HP 3458A multimeter. For signal generation, an HP 3852A data acquisition/control unit with arbitrary waveform generators and an HP 5182A waveform recorder/generator are used. The SNI has the capability to generate any waveform and sample signals at a rate of up to 2 gigasamples per second.

The HP 3852A data acquisition/control unit provides the x and y scanning signals, tip-sample voltage, and current reference. The tip-sample signal could be a DC voltage for imaging, a pulse for surface modification, or a ramp for spectroscopy. The acquisition of the z-displacement transducer voltage (V_z) is accomplished by using an HP 3458A DMM as a fast analog-to-digital converter.

The multimeter allows a sampling rate of 10^5 samples per second and has the capability of storing the data in an internal memory for further analysis. Once all the scans have been completed, the stored data is then transferred to the computer or controller. The Tektronix DSA-602 digitizing signal analyzer is used for diagnostic and data acquisition. In the diagnostic mode the analyzer shows the status of important signals from the SNI. The signals that are usually monitored are the output of the transimpedance amplifier (V_p) proportional to the tunneling current (I), the voltage applied to the transducer (V_z) proportional to the tip displacement (u) and the tip-sample voltage (V_t). In the data-acquisition mode the analyzer is used to acquire u vs. V_t and u vs. I spectroscopic data, and important signals during surface modification.

Topographical images are obtained by measuring V_z for constant I and constant V_t during x-y scans (constant-current mode); typical values of I and V_t are 1 nA and 0.5 V, respectively. Tunneling-spectroscopy measurements are of three types: 1) u as a function of I for constant V_t, 2) u as a function of V_t for constant I, and 3) I as a function of V_t for constant u. For the first measurement, the reference signal (V_t) is ramped at the same time that V_z and V_p are acquired. For the second measurement, V_t is ramped at the same time that V_p and V_z are acquired. For the third measurement, V_t is ramped at the same time that V_p is acquired and V_z is kept constant using a sample-and-hold circuit. Surface modification is accomplished by applying a rectangular pulse (V_p) to the quiescent value (V_q) of tip-sample voltage ($V_t=V_q+V_p$) while the SNI is operating in constant-current mode; the pulse width (τ) is small compared to the response time of the transducer feedback circuit. Another ajustable parameter of the rectangular pulse is the rise time (τ_r). For studies of the threshold for surface modification, V_t is pulsed, the current waveform is noted, and the area is imaged afterwards.

An example of a nanoscale feature formed with the SNI is illustrated by the topographical image shown in Fig. 4. The mound-like feature was formed on an atomically-flat Au surface using a W tip when V_t exceeded a threshold value characteristic of the junction. Also apparent in the image is a monatomic step on the Au surface.

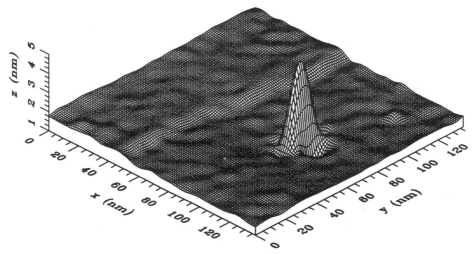

Fig. 4. Topographical image of Au surface using W tip. A monatomic step and a mound formed by pulsing the tip-sample voltage are shown.

III. Quantum-Heterostructure Characterization

A. LNP Images of Identifiable Quantum-Dot Diodes

The LNP was used to obtain high-resolution topographical images of identifiable quantum-dot diodes[6]. The sample to be characterized consisted of pairs of very closely spaced quantum-dot diodes, the fabrication of which had been described by Randall, *et al.*[4]. The nanostructures were fabricated for the purpose of determining the coupling between closely-spaced, vertical, tunneling quantum-dot devices. Understanding of this coupling is important for the creation of an integrated circuit of high functional density based on a cellular-automata architecture. The design spacing between the dots in a pair varied from approximately 20 nm to about 100 nm. The design diameter of each dot was nominally 150 nm. The entire sample was covered by a thick (0.2 μm) layer of silicon nitride except where 10×20 μm rectangular windows had been etched in order to expose the GaAs surface. In each 10 μm × 20 μm rectangle there was a single pair of quantum dots, the site of which within the rectangular field was not precisely known. Once coarse position had been achieved, the tube scanner was used to fine-scan and obtain an STM constant-current image. An image of two closely-spaced dots is shown in Fig. 5. Although it is not readily apparent in the oblique view shown in Fig. 5, line scans and plan views of the image show that each dot is approximatelt 200 nm in height and 150 nm in diameter with several tens of nanometers spacing between adjacent dot edges.[6] This compares favorably with the design parameters used for fabrication. The apparent merging of the two dots near the base shown in Fig. 5 might very well be due to the inability to fabricate two separate devices in close proximity. However, difficulties in scanning the STM tip between two closely-spaced tall features could also account for the lack of resolution between the two dots.

B. Electrical Characterization Using the LNP

A sample whose electrical characteristics are clearly distinguishable from those of the tunneling gap was needed to serve as a reference for verification of the transition from tunneling to electrical contact. A sample containing a series of small GaAs posts with

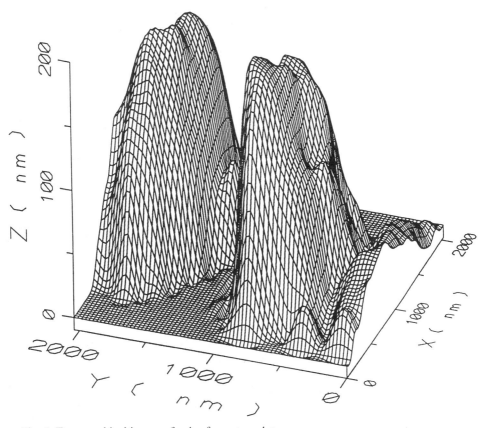

Fig. 5. Topographical image of pair of quantum dots.

Schottky-barrier contacts on the tops was chosen for this purpose. Whereas I varies linearly with V in the tunneling regime for a metal-insulator-metal (MIM) junction[20], I varies exponentially with V for a metal-semiconductor junction (Schottky barrier)[21]. Furthermore, I varies exponentially with T^{-1}, where T is the absolute temperature, for a Schottky barrier, whereas the temperature dependence of the tunneling current is quite weak. The structures containing the Schottky barrier were pedestals of n-type GaAs 700 nm in height ranging in diameter from 0.25 to 1 µm. A Au film with thickness much greater than the electron mean free path covered the top of each pedestal. The tip-air-Au film can thus be considered as an MIM junction while the Au film-GaAs can be considered as a Schottky barrier in series with the MIM junction. As the tip approaches the sample the potential difference across the gap becomes smaller; at electrical contact, essentially all the applied potential difference occurs across the Schottky barrier. Thus, when this contact is achieved, the measured I-V relation should be characteristic of the Schottky barrier.

The measured I-V characteristics of one of the pedestals is shown in Fig. 6 for different values of tip extension, u. For values of u less than about 0.5 nm the tunneling resistance dominates; however, for values of u greater than about 1 nm the exponential dependence of I on V characteristic of a Schottky barrier is evident. Measurement[7] of the temperature dependence of I vs. V confirmed that the data are characteristic of a Schottky barrier and thus that our technique of sensing electrical contact in air is valid.

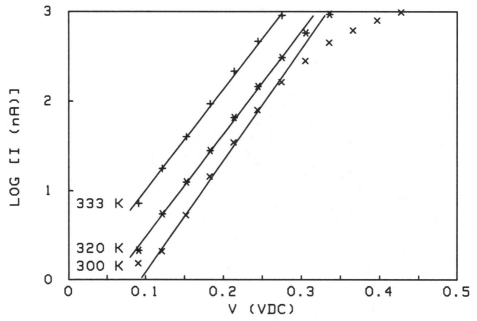

Fig. 6. Current (I) vs. applied potential difference (V) for tip extensions of 0.5, 1.0, and 2.5 nm for a W tip and Au/GaAs pedestal.

C. LNP Electrical Characterization of Identifiable Quantum-Dot Resonant Tunneling Diode

An example of electrical characterization made possible using the LNP is measurement of the I-V characteristics of a single quantum-dot resonant-tunneling diode (QDRTD). The sample, obtained from Texas Instruments, Inc.[22], contained 16 discrete QDRTD's. An isolated QDRTD was known from the design plan to be located somewhere near the center of a 10 μm × 10 μm GaAs field. The optical guidance system allowed the LNP probe tip to be placed within 1 μm of the center of the field. Fine scanning and imaging were then performed to produce the image shown in Fig. 7. The height of the QDRTD was about 0.6 μm and the diameter was about 0.15 μm. The magnification of the LNP fine scanning was increased until the top of the QDRTD entirely filled the image; then a point on the top of the device was chosen to place the probe tip for electrical characterization. After electrical contact was established as described above, the I-V characteristics shown in Fig. 8 were obtained. The region of negative differential conductance is due to the vertical confinement of the quantum well.

IV. Atomic Material Transfer

A. Creation of Nanoscale Features by Pulsing V_t

We have successfully created nanoscale features using W and Au tips on Au, Pt, and C (highly-oriented pyrolytic graphite) samples. The Au and Pt samples used in the fabrication experiments reported here were prepared by feeding 0.5 mm diameter wires into a propane-oxygen flame[23]; small spheres were obtained with very smooth surfaces. The diameter of the spheres varied between 1 mm and 2 mm. Usually, prior to surface-modification experiments,

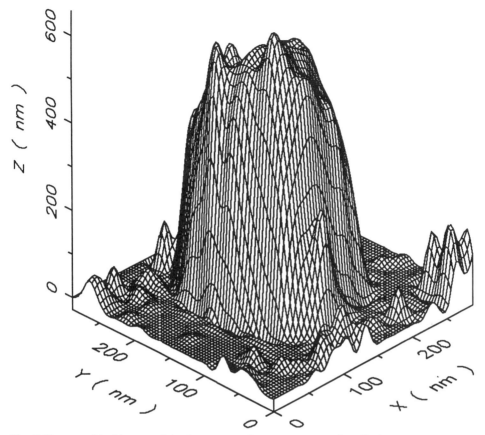

Fig. 7. Topographical image of single quantum-dot resonant tunneling diode.

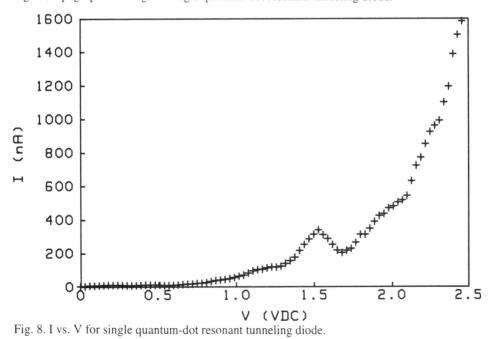

Fig. 8. I vs. V for single quantum-dot resonant tunneling diode.

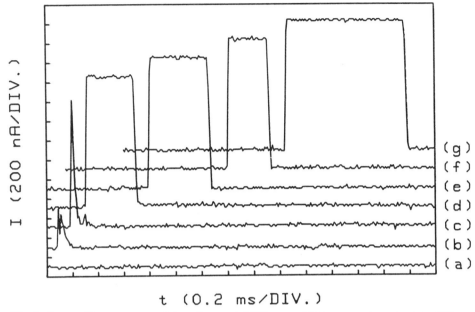

Fig. 9. Current (I) response to pulsing of potential difference (V_t) between (negative) tip (W) and (positive) sample (Au) when: (a) the duration of the V_t=3.0 V pulse was 1.0 ms and the sample was unmodified, (b) the duration of the V_t=3.3 V pulse was 1.0 ms and a crater was formed in the sample, (c) the duration of the V_t=6.7 V pulse was 2.5 ms and a crater was formed in the sample, (d) the duration of the V_t=6.7 V pulse was 2.5 ms and a crater with an adjacent mound was formed on the sample, (e)-(g) the duration of the V_t=6.7 V pulse was 0.25 ms and a mound was formed on the sample.

the prepared sample was imaged in constant-current mode until the tip was positioned over a flat area of the surface. The W and Au tips were electrochemically etched in NaOH and HCl, respectively.

Features were formed in various controlled ways as described above. In the early experiments[2,9], a submicrometer area of the sample was quantitatively characterized before and after pulsing the tip-sample voltage using both tunneling-spectroscopy determination of the apparent mean barrier potential ($\varnothing$) and imaging of the area in constant-tunneling-current mode. The before-and-after images showed that the form of the created features ranged from craters to mounds. The current waveform recorded during a surface modification was indicative of the form of the feature. Our experiments on modification of the surface of Au using a W tip contained four distinctive results: 1) the existence of a threshold of V_t 3-4 V for surface modification (cf. Figs. 9a and 9b), 2) the formation of a crater, revealed by imaging and characterized by a narrow (< 0.1 ms), unsaturated current pulse when the duration of the V_t pulse was 2.5 ms (Fig. 9c), 3) the formation of a crater and an adjacent mound, revealed by imaging and characterized by a broader (0.4 ms), preamplifier-saturating current pulse when the duration of the V_t pulse was 2.5 ms (Fig. 9d), and 4) the formation of a mound, revealed by imaging and characterized by a preamplifier- saturating current pulse with a longer duration (0.3-0.9 ms) than that of a 0.25 ms V_t pulse (Figs. 9e-9g). The presence of a high current after the applied V_t pulse has ended, such as illustrated in Figs. 9e-9g, suggests that tip and sample are close enough to make good electrical contact; the V_z waveforms in these cases show that the tip moves away from the sample during most of the duration of the current pulse. Thus, the formation of a mound evidently involves the transfer of material from

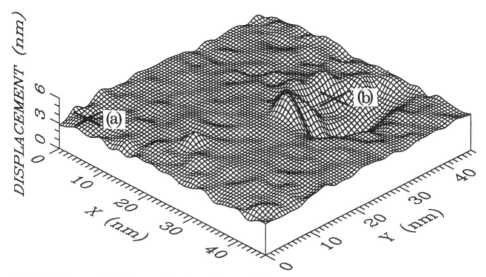

Fig. 10. Topographical image of a Au surface modified with a W tip; area is 46 nm x 46 nm. Crater is 15 nm in diameter and 4.5 nm deep. Spectroscopic measurements were taken on (a) unmodified and (b) modified regions of the surface.

tip to sample during their close proximity. After the modification characterized by Fig. 9g, sequential images of the mound were obtained for a period of 46 minutes. The images revealed that the mound disappeared during this period, leaving a small depression on the surface of the sample.

A topographic image of the neighborhood of surface modification on a Au surface using a W tip to form a crater is shown in Fig. 10. The image corresponds to an area of 46 x 46 nm; the crater formed on the surface has a diameter 15 nm and a depth 4.5 nm. The regions marked (a) and (b) correspond to the places where spectroscopic measurements of I vs. u were taken. The results of I versus u for constant V_t measured at those points on the surface of the sample are shown in Fig. 11. The lower and upper curves correspond to measurements of an unmodified and a modified region, respectively. A least-squares fit of the data to an exponential function is also shown for each curve.

According to Simmons's one-dimensional model of an asymmetrical MIM tunnel junction[24], the theoretical relation between current density (J) and insulator thickness (s) is dominated by an exponential dependence of J on s in the tunneling regime; the coefficient of s in the exponent is related to the mean barrier potential ($\varnothing$) characteristic of models of MIM junctions. From an experimental point of view, an observed exponential dependence of I on s is the signature of the tunneling regime, and values of can be inferred from measurements of I versus tip displacement, u, in that case. The values of $\varnothing$ determined from the fit for the unmodified and modified regions of the Au sample were found to be 0.04 eV and 0.7 eV, respectively. A change in $\varnothing$ of over an order of magnitude was also observed when the measurements were taken before and after surface modification in the same region.

Several investigators have reported that values of determined from measurements of I as a function of u using an STM are less than that calculated from MIM models for samples that have been exposed to the atmosphere[19,25-28]. One explanation is that the diminished values of are due to contamination-mediated deformation of the sample surface so that the relative displacement of tip and sample is less than the transducer displacement[19]. In support of this explanation it was found that as the tip and sample were progressively cleaned in vacuum, the value of $\varnothing$ increased, eventually approaching the value calculated using photoelectric

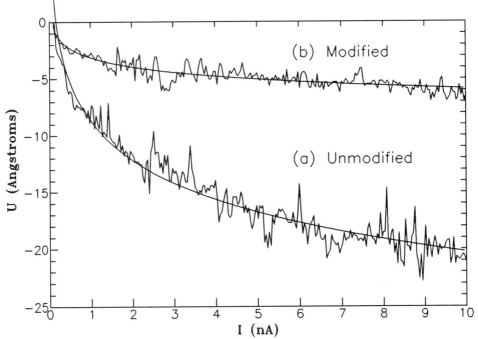

Fig. 11. Displacement (u) vs junction current I for (a) unmodified and (b) modified regions on a gold surface. The dashed curves represent least-squares fits of the data to an exponential dependence of I on u. Inferred values of the apparent mean barrier potentials are 0.04 eV and 0.7 eV for (a) and (b), respectively.

work functions[19]. A low value of ∅ measured in air has thus been associated with the presence of a contamination layer; a value so derived should more properly be called the *apparent* mean barrier potential. The substantial increase in the apparent mean barrier potential shown in Fig. 11 is probably due to the diminution of the contamination layer during surface modification. The magnitude of the increase in ∅ was observed to vary for different regions on the surface; this is consistent with an inhomogeneous contamination layer.

Using a procedure similar to that described by Mamin, *et al.*[8], we have been able to form reproducible mounds and patterns on Au samples using Au tips. Only mounds were formed when V_t=-3.8 V, τ=150 s. Formations of both craters and mounds were observed when V_t=-4.0 V, τ=250 μs.

Mounds have been successfully formed with a Au tip and a Pt sample when V_t=±3.8 V, τ=150 μs. During these experiments it was noticed that there is interference in the formation of new mounds with previously-formed mounds when they are too close together. The interference is observed as the formation of nonuniform mounds. Sometimes only one image of a mound is observed when an attempt is made to form a second mound in the proximity of the first, even though the current waveform indicates that a second mound was formed; apparently in such cases the second modification occurs at the location of the previously-formed mound.

B. Investigation of the Threshold Conditions

The effects of varying V_p and τ on the formation of mounds on Pt samples using W tips have been investigated. First, pulses with τ =150 μs and V_t decreasing from -3.0 V in small

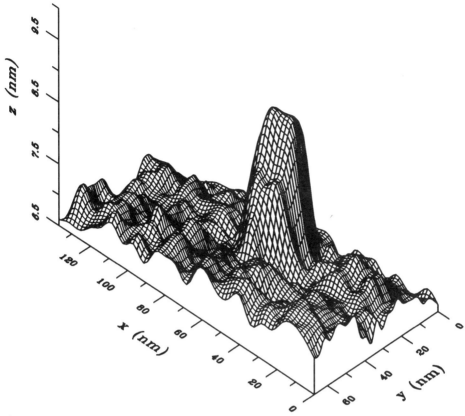

Fig. 12. STM image of mound formed with W tip and Pt sample.

steps were applied and the corresponding waveforms were stored. No surface modification was observed for $V_t > -3.5$ V. When V_t reached -3.6 V, the current waveform characteristic of a mound was observed and the image of a mound was obtained, an example of which is shown in Fig. 12. The mound is about 2.5 nm in height and 24 nm in diameter. Surface modification was achieved for $V_t < -3.6$ V; values of V_t down to -6 V were applied. Second, τ was varied between 150 μs and 2.5 ms; no effect of the duration of the applied pulse was observed. By analyzing the current waveforms we determined that a mound was formed in every case that the duration of the saturated current pulse exceeded 200 μs.

In an experiment similar to one reported for Au tip and Au sample by Mamin, et al.[8], pulses were applied 10 times for each value of V_t between -3 and -4 V in -0.1 V decrements with I=1 nA and various values of V_q. The number of times a mound was formed at each value of voltage was recorded. The results for $V_q = 0.5$ V, shown in Fig. 13, display a sharp transition in the probability of mound formation when $-V_t$ is equal to a critical value (V_c) of approximately 3.6 V; for $V_t = -6$ V the probability of forming a mound is unity.

Values of the critical voltage, V_c, for mound formation have been observed to depend on I and V_q for Au tip and Au sample[1] and for W tip and Pt sample[10]. If the nature of the surface modification reported here is an electric-field effect, then V_c should depend on the tip-sample spacing (s); thus, measurement of V_c as a function of s is desirable. According to models of metal-insulator-metal (MIM) junctions such as those discussed by Simmons[20,24], I is linearly proportional to V_t and exponentially dependent on s in the tunneling regime, so that lnR is linearly proportional to s, where R is the tunneling resistance, $R \equiv V_q/I$. The proportionality factor is $\kappa = 2(2m)^{1/2} \varnothing^{1/2}/\eta$, where m is the electronic mass and $\varnothing$ is the mean barrier potential.

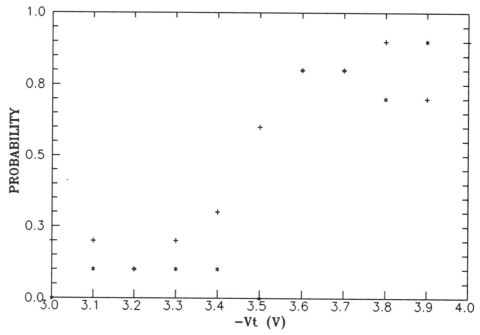

Fig. 13. Probability of mound formation vs V_t for W tip and Pt sample.

We have measured V_c as a function of R for values of R from 10^7 to 8×10^9 ohms. Measurements of V_c for various combinations of I and V_q were made for 7 different sites involving different W tips and different Pt samples. An example of one of the data sets is shown as V_c vs logR in Fig. 14 along with a least-squares fit of the data to a straight line. All such data have been fit to straight lines; since V_c lnR, then V_c s. This is direct experimental evidence that a critical electric field, $E_c=dV_c/ds$, is characteristic of the mound formation using W tip and Pt sample. A similar result was reported by Mamin, et al.[1] for Au tip and Au sample.

Determination of E_c from V_c requires knowledge of κ and hence of ∅. Values of ∅ can in principle be inferred from the slope of a plot of measured values of lnI as a function of u. This has been successfully accomplished for measurements in ultrahigh vacuum on carefully prepared samples[19,26,27,29]. However, as explained above, the value of so determined in air[19,25,28] is substantially less than that expected from models of MIM junctions.

We have found that a more realistic value of ∅ can be inferred from measurements of u as a function of V_t for constant I when V_t is varied through the value (V_{tf}) for the transition from electron tunneling to electron field emission. The barrier height obtained from lnI versus u measurements is very sensitive to the inequality of u and Δs, since ∅ is inferred form the slope of the plot of lnI versus u. Evidently, the value of V_{tf} is not particularly affected by the inequality of u and Δs; thus, the fit of the u versus V_t data to theory is guided principally by the value of V_{tf}. Values of the barrier potentials, $\emptyset_1$ and $\emptyset_2$, at electrodes 1 and 2 were obtained by fitting the experimental data to Simmons's model for an asymmetric tunneling junction with rectangular potential barriers[24]; the best fit was obtained using the model without image potential. The value of 10^7 A/m² used for the current density was based on I and the size of the image of the mound (Fig. 12). The experimental data and the theoretical curve are shown in Fig. 15; the fit yielded $\emptyset_1=1.7$ eV and $\emptyset_2=2.5$ eV, so that ∅=2.1 eV. Thus, using the corresponding value of κ, the mean value of E_c for the 7 sets of data on W tip and Pt sample was determined to be 0.23 V/Å with a standard deviation of 0.08.

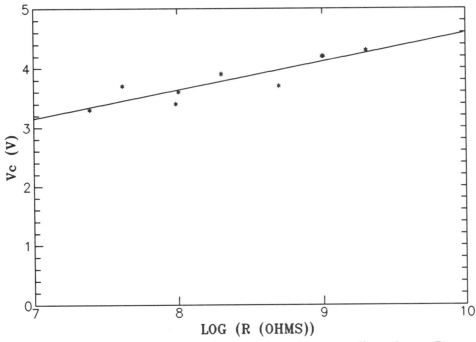

Fig. 14. Threshold tip-sample voltage (V_c) for mound formation vs. tunneling resistance (R) for W tip and Pt sample, V_q=0.5 V.

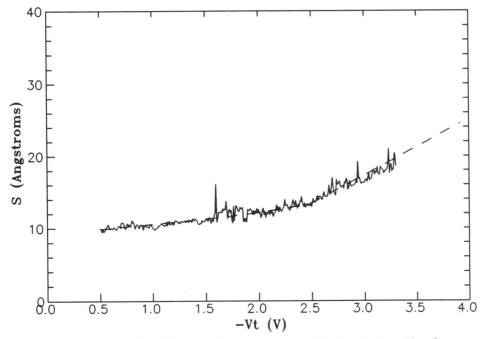

Fig. 15. Tip-sample spacing (s) versus tip-sample voltage (V_t) showing transition from tunneling to field emission for W tip and Pt sample. The solid curve is experimental tip displacement (u) translated by s_0=1 nm; the dashed curve represents a theoretical model of a metal-insulator-metal tunnel junction with rectangular barriers: $\varnothing_1$=1.7 eV, $\varnothing_2$=2.5 eV.

We have observed during measurements of u vs V_t for W tip and Pt sample, which were obtained during the relatively low frequency scans (allowing the feedback to regulate), that craters were formed in the sample surface when $V_t<-4.2$ V. Transient records of V_z show that the tip moves away from the sample during the interval $-0.5>V_t>-4.2$ V. When $V_t=-4.2$ V, the tip starts moving towards the sample for a short time and comes back away from the sample when $V_t<-4.2$ V. An image of the area revealed that a crater was formed in the surface. Thus, analysis of the data of Fig. 15 to obtain $\varnothing_1$ and $\varnothing_2$ includes only the data below the threshold for crater formation.

We have attempted to address the question of what material is actually transferred during the mound formation process. One possibility is that sample material or contamination adheres to the tip during imaging prior to the V_t pulse and is deposited back on the sample during the pulse. A fresh Pt sample and W tip were prepared using the method described above. Tunneling was established with $V_t=-0.5$ V, I=1 nA; with no prior imaging, a single pulse was applied with $V_t=-3.8$ V, $\tau=150$ μs. The current signature during the pulse showed that a mound was formed (current pulse duration ≥ 200 μs). The formation of a mound was also verified by imaging. This experiment was repeated five times with freshly prepared W tips and in a different place on the sample each time. It was observed that 4 out of 5 times a mound was formed. In one of the instances, no mound was formed as shown by the current signature. (An image of the area showed that the surface was very rough in this case.) The results of this experiment suggest that the features are probably not formed by material that was picked up by the tip during previous surface contact.

Mounds have been formed on a graphite surface using both W and Au tips with a pulse, $V_t=6$ V and duration $\tau = 150$ μs. For a Au tip, the formed mounds were ≈ 30 nm in diameter and ≈ 6 nm in height. After the mound was formed, I-V measurements were performed at different places on the surface. The I-V curve measured on the top of a mound differed from that measured away from the mound; this result also indicates that the surface had been modified.

C. Control of Form of Nanoscale Features

The experiments described above demonstrate that surfaces in air can be modified on a nanometer scale by pulsing V_t, and that the modification process probably involves the transfer of material between tip and sample. The features appear to be stable for periods of at least hours. For purposes of nanoscale patterning and device fabrication, the process of material transfer must be reliably controlled and should be understood. Our present understanding is incomplete; however, it appears that a threshold electric field is involved in each type of feature formation. As part of our continuing effort to obtain a quantitative understanding of the physics of material transfer, we have recently investigated the effects of varying the rise time (τ_r) as well as the amplitude of the V_t pulse and the initial tip-sample separation on the creation process. The initial results have not been as conclusive with regard to understanding the physics as planned, but progress toward predictable and reliable formation of a desired feature has been better than expected.

The initial separation (s_0) of the tip and sample in air is not known precisely. Since, as discussed above, the actual mean barrier potential cannot be determined when the surface deforms as the tip approaches the sample, the initial value of s_0 cannot be calculated from the tunneling equation. Furthermore, even if some procedure for sensing contact—such as that described in Section IIB—is used, the surface relaxes as the tip is withdrawn. We have used a pulsed laser beam to heat the surface of the sample prior to pulsing V_t. The photothermal heating[30] of the sample causes elastic expansion toward the tip. Since the thermal expansion varies with the laser power, the displacement of the sample and hence s_0 can be controlled

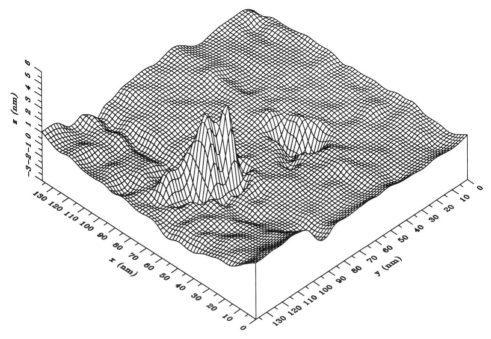

Fig. 16. Topographical image of a Au sample using a W tip showing a mound and a crater formed by V_t pulses with $\tau_r = 0.134$ µs and 69.2 µs, respectively.

prior to pulsing V_t. For sufficient laser power, the sample and tip form a good electrical contact, which is indicated by saturation (at about 12 µA) of the tunneling-current transimpedance amplifier. The contact occurs because the response of the feedback circuit that controls the tip position is slow compared to the surface thermal expansion.

Experiments were performed to create mounds using a Au tip and a Au sample with and without the laser pulse; the quiescent tunneling parameters were $V_q = -200$ mV, I=1 nA. The digitizing signal analyzer recorded V_t, I, V_z, and the laser pulse during the attempts to form features. With no laser pulse, mounds were readily formed by pulsing V_t as described above. The objective of the first experiment was to determine if mounds could be formed (without pulsing V_t) by causing good electrical contact using only the laser pulse. Many attempts to form mounds in this way at several different places on the sample were unsuccessful; imaging revealed that no features of any kind were formed. In a second experiment, a V_t pulse of 150 µs width and a wider, overlapping laser pulse were applied simultaneously. Although the V_t pulse was sufficient to form mounds in the absence of a laser pulse, no mounds as described above were formed even though the transimpedance amplifier saturated. Thus, formation of mounds is quenched by good electrical contact during pulsing of V_t. This result is consistent with the description of the mound-formation process as electric-field-induced atomic transfer.

We have recently discovered how to exercise a degree of control of the nature of the nanoscale feature formed when V_t is pulsed.[31] Features from craters to mounds on a Au sample have been predictably formed with a W tip by controlling the pulse parameters and initial conditions. A topographical image of a mound and a crater so formed is shown in Fig. 16. Experiments indicate a correlation between the shape of the formed feature and τ_r. The experiment was repeated many times; a pulse with $\tau_r = 0.134$ µs always formed a mound and a pulse with $\tau_r = 69.2$ µs always formed a crater. Furthermore, it was also found that there is a correlation between the size of the mounds and the rise time of the pulse.

V. Projected Research

The feature-size scale between about 1 nm and 50-100 nm appears to be of great importance in nanoelectronic technology. Future requirements in this size range will more likely be met by new developments than by advances in present lithographic techniques. We believe that the controlled transfer of material (of the order of hundreds to thousands of atoms) between a precisely-positioned tip and surface, and the during-process or post-process chracterization of the transfer will be an important part of those new developments. In addition to creation of device patterns, atomic transfer can be used for such tasks as modification of existing patterns, deposition of nanoscale conducting paths for alternative configuration of nanoelectronic circuits, repair of devices or circuits, localized doping, and completion of devices the initial fabrication of which is more traditional. Our present fabrication goals include deposition of high-Z X-ray absorbers such as W and Au on X-ray transparent membranes to form X-ray-lithography masks, and completion of the fabrication of quantum-dot resonant-tunneling diodes with lateral confinement of several nanometers using tip-deposited Au features.[32] The feasibility of laser interactions in the tip-sample region as a means of process modification or control is also being investigated.

We are continuing to conduct a thorough, quantitative investigation of the physical process or processes involved in the formation of nanoscale features by pulsing V_t. While it is presently believed that high electric fields are important in the material transfer, other causes may be involved; furthermore, one process may dominate for crater formation while another process may dominate for mound formation. The primary goal is to establish a scientific understanding of the physical processes important to this type of nanostructure fabrication.

The identification of the features formed by electric-field effects as due to the removal of sample material (craters) or the deposition of tip material (mounds) is based on reasonable but circumstantial evidence. We are also investigating the possibility of nanoscale material characterization (chemical species) of the formed features.[33] The lateral resolution required for the task will be provided by using the probe tip as exciter or as detector. Physical effects being considered for the identification of the feature material are characteristic photon emission during transfer atomic transfer and optical spectroscopy using a near-field optical-microscope probe[34].

This material is based in part upon work supported by the Texas Higher Education Coordinating Board Advanced Technology Program.

VI. References

1. R.T. Bate, "The quantum-effect device: tomorrow's transistor?", *Sci. Am.* **258**, 96 (1988).

2. S.E. McBride, "Application of scanning tunneling microscope technology to characterization and lithography", Ph.D. dissertation, The University of Texas at Dallas, August, 1991.

3. H.M. Marchman, "Scanned probe characterization of nanostructures", Ph.D. dissertation, The University of Texas at Dallas, August, 1992.

4. J.N. Randall, M.A. Reed, and Y.-C. Kao, "Fabrication of closely spaced quantum dot diodes", *J. Vac. Sci. Technol.* **B8**, 1348 (1990).

5. H.M. Marchman and G.C. Wetsel, Jr., "Optically-guided scanned probe for characterization of large nanostructures", *Rev. Sci. Instr.*, May, 1993.

6. H.M. Marchman, G.C. Wetsel, Jr., M.A. Reed, J.N. Randall and Y.-C. Kao, "Scanning tunneling micro-scope images of identifiable quantum dot diodes", *Superlattices and Microstructures* **11**, 333 (1992).

7. H.M. Marchman and G.C. Wetsel, Jr., "A technique for establishing nondestructive electrical point contact in air using a scanned nanoprobe", submitted for publication in *J. Appl. Phys.* (1992).

8. H.J. Mamin, P.H. Guenther, and D. Rugar, "Atomic emission from a gold scanning-tinneling-microscope tip", *Phys. Rev. Lett.* **65**, 2418 (1990).

9. S.E. McBride and G.C. Wetsel, Jr.,"Quantitative characterization of physical processes during nanometer surface modification", *Appl. Phys. Lett.* **57**, 2782 (1990).

10. G.C. Wetsel, Jr. and S.E. McBride, "Stable nanometer scale patterns produced by high electric fields", *Scanned Probe Microscopies: STM and Beyond*, pp. 467-479, K. Wickramasinghe, Ed., American Institute of Physics, N.Y. (1991).

11. S.E. McBride and G.C. Wetsel, Jr., "Nanometer-scale features produced by electric-field emission", *Appl. Phys. Lett.* **59** 3056 (1991).

12. T.T. Tsong, "Effects of an electric field in atomic manipulations", *Phys. Rev.* **B 44**, 13703 (1991).

13. N.D. Lang, "Field-induced transfer of an atom between two closely-spaced electrodes", *Phys. Rev.* **B 45**, 13599 (1992).

14. I.-W. Lyo and P. Avouris, *"Field-induced nanometer-to-atomic-scale manipulation of silicon surfaces with the STM"*, *Science* **253**, 173 (1991).

15. G. Binnig and D.P.E. Smith, "Single-tube three-dimensional scanner for scanning tunneling microscopy", *Rev.Sci.Instrum.* **57**, 1688 (1986).

16. G.C. Wetsel, Jr., S.E. McBride, R.J. Warmack, and B. Van de Sande, "Calibration of scanning-tunneling-microscope transducers using optical beam deflection", *Appl. Phys. Lett.* **55**, 528 (1989).

17. Newport Corporation, P.O. Box 8020, 18235 Mt. Baldy Cir., Fountain Valley, CA 92728.

18. J.K. Gimzewski and R. Möller, "Transition from the tunneling regime to point contact studied using scanning tunneling microscopy", *Phys. Rev. B*, **36**, 1284 (1987).

19. H.J. Mamin, E. Ganz, D.W. Abraham, R.E. Thomson, and J. Clarke, "Contamination-mediated deformation of graphite by the scanning tunneling microscope", *Phys. Rev. B* **34**, 9015 (1986).

20. J.G. Simmons, "Generalized formula for the electric tunnel effect between similar electrodes separated by a thin insulating film", *J. Appl. Phys.*, **34**, 1793 (1963).

21. S.M. Sze, *Physics of Semiconductor Devices*, John Wiley & Sons, New York, (1981).

22. Dr. J.N. Randall, Texas Insruments, Inc., Dallas, TX, 75265.

23. J. Scheir, R. Sonnenfeld, O. Marti, P.K. Hansma, J.E. Demuth, and R.J. Hamers, "Tunneling micrography, lithography, and surface diffusion on an easily-prepared, atomically flat gold surface", *J. Appl. Phys.* **63**, 717 (1988).

24. J.G. Simmons, "Electric tunnel effect between dissimilar electrodes separated by a thin insulating film", *J. Appl. Phys.* **34**, 2581 (1963).

25. G.C. Wetsel, Jr., Z.M. Liu, S.E. McBride, T.L. Weng, and W.M. Gosney, "Scanning-Tunneling-Spectroscopy Determination of Barrier Potentials in Air or Moderate Vacuum", in *Review of Progress in Quantitative Nondestructive Evaluation*, **9B**, pp. 1177-1183, D.O. Thompson and D.E. Chimenti, Eds., Plenum, N.Y. (1990); Z.M. Liu, G.C. Wetsel, Jr., T.L. Weng, W.M. Gosney, and R.J. Warmack, Bull. Amer. Phys. Soc. **34**, 671 (1989); S.E. McBride and G.C. Wetsel, Jr., Bull. Amer. Phys. Soc. **35**, 209 (1990).

26. G. Binnig, H. Rohrer, Ch. Gerber, and E. Weibel, "Tunneling through a controllable vacuum gap", *Appl. Phys. Lett.* **40**, 178 (1982).

27. G. Binnig, N. Garcia, H. Rohrer, J.M. Soler, and F. Flores, "Electron-metal-surface interaction potential with vacuum tunneling: observation of the image force", *Phys. Rev.* **B30**, 4816 (1984).

28. G.C. Wetsel, Jr., Z.M. Liu, T.L. Weng, W.M. Gosney, and R.J. Warmack, "Effects of Environmental Gases on Tunneling Spectroscopy of Gold Films", *High Resolution Microscopy of Materials*, Vol. **139**, pp. 303-307, W. Krakow, F.A. Ponce, and B.J. Smith, Eds., Materials Research Society, Pittsburgh (1989).

29. J. Gómez-Herrero, J.M. Gómez-Rodriguez, R. García, and A.M. Baró, "High values in scanning tunneling microscope: field emission and tunnel regimes", *J. Vac. Sci. Technol.* **A 8**, 445 (1990).

30. F.A. McDonald and G.C. Wetsel, Jr.,"Theory of photoacoustic and photothermal effects in condensed matter", pp. 167-277, *Physical Acoustics*, Vol. **18**, W.P. Mason and R.N. Thurston, Eds., Academic Press, San Diego, CA (1988).

31. G.C. Wetsel, Jr., S.E. McBride, and K.J. Strozewski, "Effects of Pulse Parameters and Initial Conditions on Electric-Field-Initiated Nanoscale Material Transfer", presented at the Engineering Foundation Conference on Atomic and Nanoscale Modification of Materials, Fundamentals, and Applications, Ventura, CA, Aug. 16-21, 1992; S.E.McBride, G.C. Wetsel, Jr., and K.J. Strozewski, "Controlled Formation of Nanoscale Features Using a Scanned Nanoprobe Instrument", *Bull. Amer. Phys. Soc.* **38**, 184 (1993).

32. G.C. Wetsel, Jr., S.E. McBride, M.D. Taylor, H.M. Marchman, A.C. Seabaugh, L.A. Files, Y.-C. Kao, J.N. Randall, and G.A. Frazier, "Observation of Periodic Conductance Oscillations in the dI/dV Characteristics of Near-Surface Resonant-Tunneling Nanostructures Using a Scanned Nanoprobe Instrument", *Bull. Amer. Phys. Soc.* **38**, 813 (1993).

33. K.J. Strozewski, S.E. McBride, and G.C. Wetsel, Jr., "Characterization of Nanoscale Features Produced by Electric-Field Emission", *Bull. Amer. Phys. Soc.* **37**, 618 (1992); K.J. Strozewski and G.C. Wetsel, Jr., "Optical Characterization of Nanoscale Features", presented at the 18'th International Quantum Electronics Conference, Vienna, Austria, June 17, 1992.

34. E. Betzig, P.L. Finn, and J. S. Weiner, "Combined shear force and near-field scanning optical microscopy", *Appl. Phys. Lett.* **60**, 2484 (1992).

Ballistic Electron Emission Microscopy: From Electron Transport Physics to Nanoscale Materials Science

H. D. HALLEN
Physics Division
AT&T Bell Laboratories
600 Mountain Avenue
Murray Hill, NJ 07974
USA

1. Introduction

The production of very small structures is now performed with a variety of approaches. The works described in this volume are a tribute to the growth of the field. But as the field advances it becomes increasingly important to optimize the creation of these structures, and to analyze the properties of the structures. The latter reason has not received as much attention as the former, but becomes critical when applications of the structures are considered. Small structures imply a high surface to volume ratio. In use, they are likely to be exposed to large current densities. Furthermore, the operation of small devices will probably depend strongly on the chemistry and physics of these structures, hence their characteristics must be measured and the relevant parameters identified for control.

It is not the aim of this chapter to solve all of the problems just mentioned. Rather we will discuss a new method with which one can do materials science at nanoscale resolution. Some examples will be given to illustrate the power of the method. The work focuses mainly on hot electron interactions with gold. It is shown that one can image structures on a buried interface and thus quantitatively measure hot electron stimulated atomic motion. The study of such properties at a buried interface is important from several considerations. The most obvious is that tip-sample or probe-sample interactions are absent. The high electric fields produced near an STM tip are also shielded. Thus the hot electron effects can be studied independently on the technologically relevant passivated interface. Vacancies produced in the hot electron scattering process build up in the film. Their presence can be detected. This opens the possibility of studying vacancy diffusion at very high resolution and at a known location -- be it well within a single grain or near a grain boundary. The question of how the structure will react to high current densities can also be considered on a local scale. Our method is based on a recent extension of scanning tunneling microscope (STM) to observe subsurface properties, and is known as ballistic electron emission microscopy (BEEM)[1]. Although initially used to study Schottky barrier heights[1-5], the technique has also been demonstrated to be sensitive to interfacial transport[6,7], scattering in the film[6,8,9], diffraction at the interface[10], and interfacial structures[10-14]. One can think of the method as a near field, few eV electron transmission microscope. The source of electrons is an STM tip, and the detector a Schottky barrier, as described in Section 2.3. Hot electrons injected by the STM tip can produce modifications of the gold film or its surfaces. Lower energy electrons are used to image the structures without further intervention. This paper will focus on modification of a gold film deposited on silicon. A carbon-based passivation layer lies between the film and substrate. The system will be described in more detail below.

The types of hot electron induced effects can be grouped into two categories: those that occur at a surface and those in the bulk of the film. Examples of each type are seen. By surface I refer to the outside of a grain: the top surface is the side of the grain scanned by the STM; the inner surface is opposite, i.e. at the interface of the gold and silicon; the grain boundaries are lateral surfaces which can be important during grain growth. At surfaces, the hot electrons induce the formation of adatom-vacancy pairs. The adatom is a gold atom which has moved out onto the surface. It may combine with other adatoms to form a terrace or diffuse to a sink such as a step edge, vacancy or grain boundary. The vacancy diffuses into the film. We have observed[10,11,13] terrace growth on the inner surface, grain growth on the lateral surfaces, and mound growth on the top surface of the films. The stability of the structures are found to depend strongly on the properties of the gold film.

The hot electrons can scatter from vacancies once the vacancies are in the bulk of the film. This can result in an enhanced, non-thermal motion of these vacancies. We have observed[10,11,13] the creation of large areas of defect-filled film. Such areas strongly scatter even lower energy electrons. The reaction of the system to this modification can be studied in real time. The system will begin to react to the induced changes even while still under hot electron bombardment, and continue after the stress is removed. The reaction of the system can be a measure of the stability of the structures which were created, but also can reflect further development of the structures. One example of the latter occurs when several layers of terraces have been grown on the inner surface of the gold film, bringing the gold into contact with the silicon. As has been well documented[15], silicon will diffuse into a gold film evaporated into intimate contact with it, provided that the terrace is large enough. The resulting gold-silicon alloy scatters all electrons strongly. The resultant structure is very stable, much more so than the layers of terraces produced directly by the hot electron mechanism. A benefit of the BEEM technique is that the stability of bulk and subsurface interfacial structured can be observed in addition to the changes of the film topography. Through correlation between the observations one can gain an understanding of the mechanism by which mounds on metal surfaces are formed, and what the important parameters are in their decay. We believe that the BEEM technique will also be of aid in understanding other systems.

2. Experimental Technique

We have alluded to the fact that the BEEM technique can be used to detect subsurface structures. Section 2.1 will give a heuristic discussion of the essential requirements for such measurements, and will illustrate that such measurements are possible with some experimental data. This is intended to give the reader a picture of the process without all the details, so the reader can better understand the need for the detailed sample preparation. The rest of Section 2 will expand on this heuristic picture. First the preparation of the real samples will be discussed, noting the reasons for the choices of materials. Next the details of the BEEM process will be noted in Section 2.3. Much of this discussion is not necessary for understanding the measurements here, but illustrates other uses of BEEM and provides a foundation for the simple electron source/detector picture. Section 2.4 goes beyond the subsurface imaging discussion preceding it, and gives the details of the experimental method for subsurface modification.

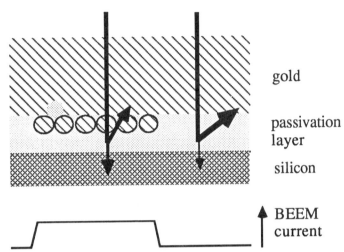

Figure 1. A schematic drawing which illustrates how an inner surface terrace can increase the BEEM current due to a high scattering rate in the passivation layer material which it displaces. The BEEM electrons pass through a thinner layer of passivation material where a terrace exists. The electrons which are scattered back, in addition to many which cross the passivation layer but are not able to surmount the Schottky barrier, are eventually collected to maintain the tunnel current. [From Ref. 14.]

2.1 Detection of a Subsurface Terrace

In order to study the subsurface structures produced by the inelastic scattering of the hot electrons, it is necessary to be able to detect, with good sensitivity, buried structures which are only a single monolayer in height. The method we use is shown in Fig. 1. The experimental situation will be described below, here we concentrate only on the contrast mechanism. A narrow beam of electrons is directed through the thin metal film. Those electrons which reach the interface must pass through a passivation layer which strongly scatters electrons. The thickness of the passivation layer will typically be a few monolayers and will scatter most of the electrons (i.e. experimentally we find that most of the electrons are scattered). The electrons which do not scatter are collected from the semiconductor. To a good approximation, the passivation layer scatters electrons independent of the electron's energy[10]. Thus a variation in the thickness of this passivation layer, caused by the interposition of a terrace layer, induces a large change in

Figure 2. 1. Gray scale 1.4 V constant 1 nA current STM (left) and corresponding BEEM (right) images illustrate enhancement type modifications of a type 1 sample. The images are 800 Å square. (a) shows the STM topograph which did not visibly change as a result of the modifications. (b) is the BEEM image where all the whitish areas were individually created by stressing with the STM current. The BEEM image before any stressing was uniformly gray. Clockwise from the two largest (just touching, in the lower left) the modifications were created with 2.5 V for 3.8 sec, a voltage sweep 0.4->2.88 V in 5.7 sec, 2.1 V for 6.7 sec, 2.0 V for 6.5 sec, 2.25 V for 3.3 sec, and the modification at the bottom center: 2.25 V for 7.0 sec. [From Ref. 14.]

the magnitude of the transmitted current. Therefore, a measure of this current acts as a sensitive detector of single atom high terraces on the inner metal surface. Note that some amount of transmitted current is present in all parts of the trace. An image produced by measuring this current as the incident electron beam is rastered will provide a map of the interfacial or inner surface of the metal film, as is schematically indicated at the bottom of Fig. 1.

To provide some experimental data to motivate this picture, consider Fig. 2. Fig. 2(a) shows a standard STM topograph of a sample described below consisting of a metal/passivation layer/semiconductor. Imaged at the same time is the transmitted current, measured from the semiconductor, shown in Fig. 2(b). Each of the lighter regions in Fig. 2(b) corresponds to a single atom high terrace on the inner surface of the metal. In this case, the terraces were produced by stressing the interface with hot electrons from the STM tip, and the electron beam for imaging is also supplied (at lower electron energy) by the same tip. The transmitted current is approximately doubles when the tip is situated such that current will pass through a terrace region.

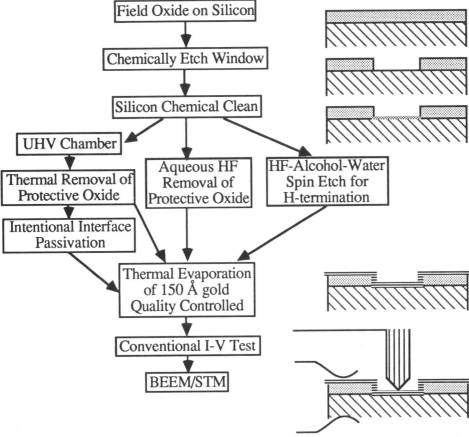

Figure 3. A flow chart illustrates the various sample preparation steps. At the right is a schematic drawing of the sample at each stage.

2.2 Sample Preparation

The general requirements of the sample have been outlined above: a thin metal layer, the system under investigation, lies on top of a few monolayer thick passivation layer with a semiconductor substrate. In this section, the material choices for each component will be discussed, in addition to other experimental considerations.

The metal is the system under study. We chose to study gold because of its technical importance and the fact that it is easy to study under ambient conditions -- most of this work was done on an STM operating in air. The thickness of the gold was chosen to be 150 Å, which insured continuity while remaining on the order of one mean free path (~128 Å for this energy electrons[4]) thick. The choice of semiconductor is determined by availability, quality, and the Schottky barrier height. Silicon was used in this study, giving a Schottky Barrier height of ~0.8 eV. As can be expected, the passivation layer has a dramatic effect on the interfacial modification processes. If the passivation layer consists of 1-2 monolayers of SiO_2, no evidence of terrace growth is observed[11]. We attribute this to the structure of the oxide -- it is not easily deformed to accommodate the terrace. Conversely, a monolayer of hydrogen[16] (or no passivation layer[11]) is not stable, and the entire sample degrades by intermixing of the gold and silicon to the point where no BEEM current is observed within a couple of days (or faster). We have grown samples in more elaborate structures than the metal/passivation layer/silicon described here to verify that the contrast in Fig. 2(b) is due to inner surface terraces[13], but will only describe the simpler case here. The work described in this chapter utilized carbon-based passivation layers. These were formed by allowing the sample to sit in a dirty vacuum (liquid nitrogen trapped diffusion pumped to a pressure $\sim 10^{-6}$ Torr) for a few hours, and were analyzed with XPS[10], to find their composition and thickness (1-2 monolayers). Samples produced with evaporated carbon passivation layers[13] (same thickness range) showed similar properties and magnitudes of transmitted current. An analysis of the magnitude of the BEEM current[10] indicates that the carbon passivation layers (presumably amorphous) behave as thin layers of a material with a short scattering length and not like a tunneling barrier. This type of passivation layer was chosen since it was very reproducible[10], seemingly easy to deform to allow terrace formation, and the samples lasted a long time[10,14].

A flow chart illustrating the sample preparation process is shown in Fig. 3. As will be shown in Section 2.3, a small area device is necessary for experimental reasons. The small device is defined, the silicon rigorously cleaned[17], passivation layer added[10], and 150 Å thick gold film evaporated. The Schottky diode was then tested in a conventional probe station, where the current-voltage characteristic of the entire device was measured to obtain the average Schottky barrier height and the ideality factor (which was always close to one for these diodes). The sample was loaded into an STM/BEEM microscope described elsewhere[10], which was operated in air. The microscope used a mechanical lever-arm reduction mechanism for the coarse approach. Mechanical coarse motion was also provided for the lateral (XY) directions, and was viewed with a long focal length microscope. Two current preamplifiers, one for tunnel current and one for BEEM current, were mounted directly on the microscope stage. Two qualitatively different types of gold films were used in this study. The first (type 1) was evaporated at 1 Å/sec in a high vacuum system. The grain size was typically one or a few hundred angstroms. Type 2 gold films were evaporated in UHV at a much slower rate 0.1 Å/sec to allow the film to anneal while growing. The grains were much larger (a few thousand angstroms), and presumable were cleaner and contained fewer point and line defects. It will be shown below that the stability of the structures depends strongly on the film characteristics, illustrating the relationship between bulk and surface in small structures and that this method can thus be a probe of bulk film characteristics.

2.3 BEEM

The BEEM process, including models which predict the shape of the BEEM current as a function of sample-tip bias, has been discussed previously[1,6,7]. Therefore a short and rather heuristic treatment will be given here. The sample configuration for BEEM is the same as that used here, although the passivation layer is not required for all types of BEEM samples. A schematic drawing of the BEEM configuration is shown in Fig. 4(a). If one concentrates on just the tip and metal film, one finds that it is simply an STM operated in the constant tunnel current mode. The additional requirements for BEEM include the semiconductor below and another current preamplifier to detect the current that

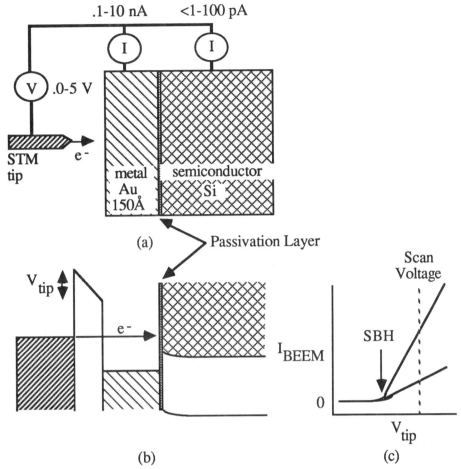

Figure 4. The BEEM measurement technique is illustrated by (a) a schematic drawing of the sample and amplifier configurations. The STM tunnels into a thin metal film. The BEEM current is collected from the semiconductor substrate. Typical parameter values are shown. (b) A band structure drawing illustrates the energetic considerations of BEEM. It can be used to estimate the structure of a BEEM current vs. sample-tip voltage spectra as is shown in (c). The BEEM spectra shown in (c) have a threshold at the Schottky barrier height, and then increase nearly linearly. The two curves illustrate a change in the scaling of the BEEM current with position as is observed in this work. The dotted line shows the origin of contrast in the BEEM image, taken at constant voltage and tunnel current.

enters it. Thus, in BEEM, one can always acquire a constant tunnel current STM image of the surface of the film simultaneously with the BEEM current image, where the BEEM current is that current collected from the semiconductor. To determine the behavior of the BEEM current as a function of sample-tip voltage, consider Fig. 4(b). The tunnel bias controls the relative positions of the Fermi levels of the tip and metal film under investigation. The Schottky barrier is kept at zero bias, so one expects no BEEM current until the tip bias exceeds the Schottky barrier height (SBH). This is the case as indicated in the figure. The threshold current follows a $(eV-SBH)^2$ form at threshold then becomes linear in a simple effective mass model[1]. Consideration of quantum mechanical reflection at the interface[6,7] corrects the threshold form to (eV-SBH)$^{5/2}$. The images for this work were usually taken at 1.4 V and 1 nA, which is in the linear region of the BEEM spectra. Modifications were done at a higher bias where the simple model, which neglects resonant scattering in the metal, is questionable, as will be seen below.

From the above arguments, it is obvious that the BEEM technique is sensitive to the SBH of an interface. Other effects of the Schottky barrier are not as obvious, and vary for different types of samples. The Schottky interface provides a barrier to filter the electron distribution so that only those electrons within the proper (interface specific) transverse wave vector and energy range pass into the silicon. The BEEM current is that current which is collected in the silicon after having passed the Schottky barrier, so reflects the properties of the (sample dependent) filter. If one uses the BEEM technique on epitaxial sample for which the symmetry, size and orientation of the lattices on either side of the Schottky barrier is the same, symmetry arguments imply that the transverse wave vector of the electrons must be conserved as the electrons cross the interface. For systems with less symmetry, such as the Au/Si interface, one expects some relaxation of the momentum constraints, although some constraints remain[10]. These transverse momentum constraints have an effect on both the shape of the BEEM spectra and the magnitude of the BEEM current[1,8]. Thus, BEEM is an ideal tool to study the spatial variations of interface quality (or epitaxy) on a metal/semiconductor system. The shape of the BEEM spectra can be analyzed[6] to yield energy dependent transport properties of the interface. This chapter will not focus on the features of BEEM as a tool for Schottky barrier transport studies. Rather, it focuses on the effects of scattering within the metal film, which has effects on the scaling and shape of the BEEM spectra. This paragraph serves as a reminder of other effects to be considered during the measurement analysis.

The measurement criteria for BEEM require a quiet tunnel current and a high gain current preamplifier[10]. The tunnel current is controlled by the microscope design and the condition of the sample surface and tip. The sample has been described. We used electrochemically etched Pt tips. These performed well in air on the rough gold grains and was robust enough to perform reliably while operating at high bias for extended periods during sample modification. The noise in the measurement of the BEEM current is dominated by the Johnson current noise produced by the (zero bias) Schottky diode[10]. To minimize this noise one must increase the resistance which can be done by lowering the temperature[4] or by decreasing the physical size of the diode[10]. We chose the latter method which is the reason for the small metal contact discussed above.

2.4 Hot Electron Modification

One of the advantages of this technique is that the system can be modified by turning up the tunneling bias but imaged non-invasively when lower energy electrons are used. The damage induced by the hot electrons can be the creation of terraces on any surface of a grain, or the creation or stimulation of motion of defects within the film. These processes will be described in later sections. Here the method used to stress the system with hot electrons will be outlined.

The STM is assumed to be tunneling to the sample with the sample to tip bias at the level used for imaging (typically ~1.4 V) which was chosen to be below any thresholds for modifying the system. The tip is then scanned to the point where stressing is to be done. The sample to tip bias is swept to the stressing level at a rate slow enough that the feedback loop is able to maintain the constant current state. The voltage is held at this higher level for a specified amount of time -- always with the feedback maintaining constant current. The tunnel bias is then reduced to the imaging value at the same rate as it was increased. An area scan measuring both the topography and a BEEM image is then taken, within a few minutes of the stressing. Subsequent images of the same region are often to taken to observe the stability of the structures. As will be discussed below, the created structures are usually stable, but with important exceptions to be discussed below. The BEEM current is monitored during the entire time that the voltage is above the imaging value to provide real-time modification data which will be discussed below. The tunnel current is also monitored during the stressing and imaging to insure that it is being held constant by the feedback loop.

One can also modify a sample by sweeping a BEEM spectra to a very high bias. Some data of this sort will be shown below, but the interpretation of the real-time data is complicated by the usual variation of the BEEM current with voltage. One is also stressing with electrons of a range of energies.

An important insight into the nature of the process is found from calculating the number of electrons which are incident on the sample before it has time to relax. The tunnel currents used for modification were between 0.1 and 10 nA, typically 1 nA. For a 1 nA tunnel current, the average time between electron arrivals is ~10^{-10} sec for a region the size of the BEEM resolution. This time is much longer than electron relaxation times and phonon periods. Thus one would expect the electron system to have settled into an equilibrium or metastable state before the next electron arrives. The process results from the scattering of a single electron. As is expected for a single electron event, the relevant parameter is the dose of electrons, i.e. the product of the current and the time. This was verified experimentally[10].

3. Inner Surface Adatom Production

One of the most intriguing aspects of the present work is the fact that electrons injected into the gold when the bias is as low as ~1.8 V can stimulate the motion of gold atoms onto the inner surface. These adatoms then coagulate into terraces. The dynamics of the nucleation and initial growth periods can be observed in real-time by measuring the BEEM current while the stressing bias is applied. The growth of the terrace after it has become larger can directly yield a bound on the production rate of these adatoms. All these measurements can be made as a function of incident electron energy. As discussed in Section 2.2, we assume that the passivation layer has little effect on the process (as was discussed, this is true for this C-based but not all passivation materials). The passivation layer must be compressed or incorporated into the terrace to make room for the terrace as it grows.

The process will be described in steps. First in Section 3.1, the data acquired from a small region under the tip as the terrace growth process takes place will be discussed. These data give us information about the initial stages of terrace formation, and lead naturally into Section 3.2 which discusses the larger scale structures for which a more steady state growth is observed. A model for the detailed physics of how a hot electron can stimulate the formation of an adatom-vacancy pair is then given. The control with which these structures can be created is striking to the experimenter. We try to convey some of this feeling in Section 3.3, where some examples are given. Section 3 ends with an extension of the single atomic layer terrace notion, and due to the nature of the sample, connects to Section 4.

The same adatom production process occurs at all surfaces of the film. Evidence is found in lateral grain growth and mound formation on the outer or free surface. The advantage of studying the buried surface is that the effects of the tip and electric fields are shielded. Combining this with the comment at the end of Section 2.4: the phenomena is stimulated by a single hot electron at a time which scatters inelastically in a field free region.

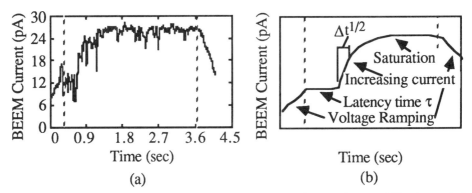

Figure 5. The BEEM current is monitored in real-time during modifications of the sample. The sample-tip bias was increased from the imaging value of 1.4 V linearly to the stressing value, at which it was held constant for a period of time, then decreased linearly back to the imaging value. The dotted lines indicate the times at which the bias reached and left the highest stressing value. (a) Experimental data for which the stressing voltage was 2.25 V and tunnel current 1 nA. (b) A schematic drawing of the real-time BEEM current plot with labels on the various stages.

3.1 Real-Time Observations of Terrace Nucleation

The modification which occurs due to stressing with hot electrons as described in section 2.4 can be followed in real time by monitoring the BEEM current. Experimental data are shown in Fig. 5(a). The results of this stressing was a net increase in the BEEM current in a region surrounding the tip location. Subsequent area scans showed terrace formation. Note that there are several distinct phases which occur while the voltage is held constant. These are labeled in a schematic drawing Fig. 5(b). Before describing the properties of

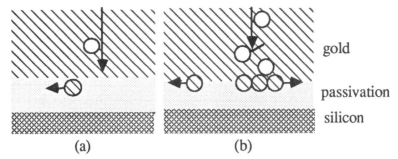

Figure 6. Schematic drawings of the state of the interface at various stages of the terrace growth process. (a) During the latency period, the adatoms which are created diffuse away. (b) Eventually some adatoms coagulate to form a small terrace, but the terrace is not thermodynamically stable since the curvature is very large. The evaporation rate of atoms is high so the terrace grows slowly. The small terrace allows the BEEM current to increase somewhat. This is the situation as the BEEM current rises in the plots of Fig. 5. [From Ref. 14.]

each phase in detail, it is important to recall that the BEEM current is only sensitive to what is occurring in a region the size of the BEEM resolution (~10-20 Å in diameter) at the interface. We have seen that the modification process often creates structures much larger than this. The real time BEEM current plot will reflect only a fraction of this region, usually the center of the region unless the tip is nearly over a grain boundary. Following along Fig. 5(b), one finds that a latency period during which the BEEM current does not change follows the ramp to the stressing bias, then the current rises and saturates.

The latency period is best described by the picture in Fig. 6(a). Hot electrons are stimulating the production of adatoms at the inner surface of the gold. A vacancy must also be produced and is shown in the figure, its fate will be discussed in Section 4.1. When no terrace exists, the adatom will tend to diffuse away -- inhibiting the formation of a terrace. It is found that the latency time for a given dose depends on voltage in a roughly exponential manner as $\exp(-6.6 \pm 1.7 \times V)$. This is shown in Fig. 7(a). One expects that the voltage dependence reflects the adatom production rate in some form, as will be verified below, but the latency time can be considered to reflect the time until a sufficiently large cluster of adatoms is built up to prevent a new one from freely diffusing away. When a new terrace is produced within 10-30 Å of a previous modification, a zero latency period is observed. In these cases, the nearby terrace presumably prevents the adatoms from freely diffusing.

When the BEEM current begins to rise, some adatoms must be localized at the inner surface underneath the tip. Fig. 6(b) depicts the situation. A small terrace has formed. The fact that the BEEM current is increasing implies that the terrace grows but is still smaller than the area to which the BEEM current is sensitive. The evaporation rate from such a small terrace is expected to be very high due to its thermodynamic instability[18]. Thus, the initial growth rate of the terrace will be much slower than the growth rate when the terrace is larger. This is simply checked by comparing the time it takes for the real-time BEEM plot to saturate compared to the time it would take to grow a region the size

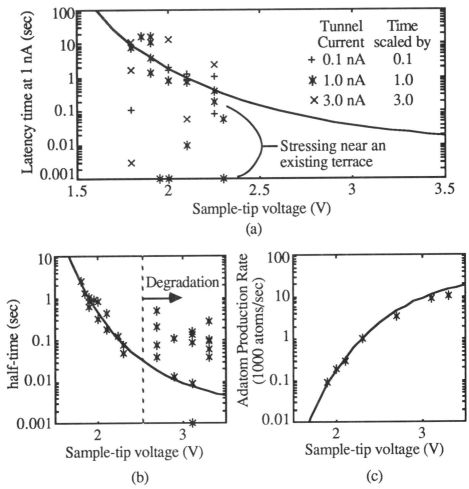

Figure 7. (a) The latency time is plotted as a function of voltage. Different symbols correspond to measurements at different currents, i.e. +, *, and x correspond to 0.1, 1.0, and 3.0 nA tunnel current, respectively. The latency times have been scaled (using the relation that the latency time is inversely proportional to the tunnel current) to their equivalent values at 1 nA tunnel current. The solid line is from a model calculation described in the text. (b) The rate of rise of the BEEM current is shown as a function of bias voltage by plotting the time it takes for the BEEM current to get half way to its saturation value. The solid line is the model calculation. (c) A lower bound on the adatom production rate at 1 nA calculated from the areal growth rate of the terraces is shown as a function of sample-tip voltage. The solid line is from the model calculation. The initial slope is ~5.92. Note the change in voltage dependence as the center of the threshold energy distribution is reached.

of the BEEM resolution using the rates measured for larger terraces. One finds that the initial growth is ~10 times slower. A measure of the growth rate during nucleation is given by the time it takes for the BEEM current to reach half its saturated value, a quantity which behaves $\sim\exp(-6.8 \pm 0.4 \times V)$ up to ~2.5 V stressing bias, as is seen in

Fig. 7(b). The functional form of the current rise is approximately proportional to the square root of time.

After the real-time BEEM current plot saturates, the terrace is larger than the resolution of the BEEM technique. This provides a convenient way to measure the BEEM resolution. The stressing is stopped as soon as the BEEM current is found to saturate. A BEEM image shows the size of the structure, which fills a region the size of the BEEM resolution. Subsequent images reveal if the terrace is evaporating or if it retains its size, i.e. if a size correction must be applied. This is the origin of the ~10-20 Å BEEM resolution quoted above.

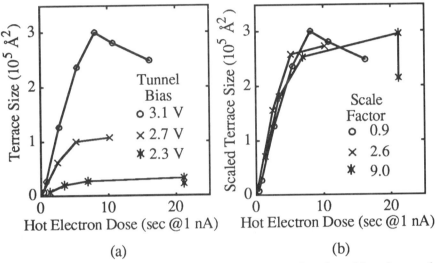

Figure 8. The increase in terrace area is shown as a function of hot electron dose is shown for several voltages. Note the initial increase is linear. (a) shows the raw data. (b) shows the same data scaled to emphasize the similarity of the functional forms at different stressing biases.

3.2 Adatom Production

The growth of the terrace is measured in a BEEM image such as Fig. 2 as a function of electron dose and bias voltage. The initial increase of the area of the terrace is linear with the dose as is expected if the growth is limited by the adatom production rate. This can be seen in Fig. 8. The size of the terrace eventually saturates, probably due to annihilation of adatoms with vacancies from within the grain or loss of adatoms to nearby grain boundaries. The initial terrace growth rates can be converted into a lower limit on the adatom production rate. It is a limit since some adatoms are lost through annihilation with vacancies. The adatom production rate depends upon voltage as $\exp(5.9 \times V)$ at lower voltages, and begins to saturate ~2.5 V as is seen in Fig. 7(c). It is not surprising that the voltage dependence of the latency time, the rate of BEEM current rise and the adatom production rate all share the same voltage dependence. They all depend on the rate at which adatoms are produced.

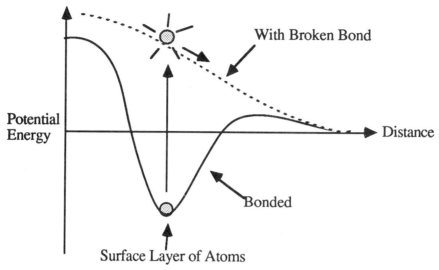

Figure 9. When a hot electron inelastically scatters from a gold atom, it may break a bond causing the atom to become unstable in its present location. A schematic potential energy vs. position is shown indicating how an atom can be accelerated onto a surface by a potential energy gradient if the bond remains broken long enough. [From Ref. 14.]

One might wonder why diffusion does not play a role. The adatoms are created in the center of the terrace and must diffuse to the edge to increase the size of the terrace. The diffusion of gold atoms on gold is very fast for 'dirty' gold surfaces[18], i.e. outside of UHV. The presence of the passivation layer puts the inner surface in that class. Another, experimental, reason to believe that diffusion is fast on the inner surface is the length of the latency times (they would be very short if diffusion was slow) and the fact that it takes a long time for a second terrace to nucleate (more on this below). Note that the heuristic picture of adatoms being created at the center of the terrace, diffusing to the edge, and then falling off (or the equivalent vacancy diffusion from the edge to the center) may not be correct. Recent molecular dynamics calculations[19] suggest that a better picture for the decay of an upper gold terrace is that it falls straight down into the lower terrace, accompanied by a concerted motion of atoms in the lower terrace to make room for it. Thus the adatoms end up in the center of the terrace as opposed to the edges. If this picture is true, then very little diffusion is required, but the terrace atoms must make room for the adatom as it joins them. It is important to realize that the adatom must be an adatom (alone at the top of the terrace) at some point. Otherwise there would be no nucleation of the first and (as we will see below) further terrace layers in contradiction to experimental observations.

So far we have attributed the adatom-vacancy production to 'inelastic scattering of hot electrons.' A model has been proposed for this process[13]. It is similar in nature to electron stimulated desorption (ESD), which has recently been reviewed[20]. The difference from ESD is that the electron energies are much lower in the present investigation, and the ions are not ejected from the surface. The similarities are that an injected electron breaks a bond of an atom in the sample. This makes the atom unstable in the lattice -- it undergoes a Franck-Condon transition. A picture is given in Fig. 9. The atom was

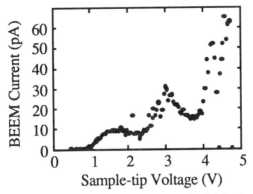

Figure 10. A BEEM I-V spectra taken to a very high bias is shown. Note the dips in the BEEM current at higher voltages. These are attributed to electron-hole pair formation in the gold by inelastic scattering of the injected hot electrons. The scattering reduces the number of injected electrons which reach the interface so causes a reduction in the BEEM current near resonance.

originally sitting in a potential well formed by the bonding to its neighbors. When a bond is broken, it shifts to the upper potential energy curve which is shown for a surface atom. If the bond remains broken for a long enough period of time for the atom to be accelerated by the potential energy gradient and move out onto the surface, an adatom-vacancy pair will have been created. Typically a bond will not remain broken for such a long time unless it has been stabilized by a lattice distortion. Instead, the hole (broken bond) will jump from one atomic site to another through the lattice until either it is localized or decays. Atoms near surfaces or defects in the bulk (e.g. vacancies) do not have as high a coordination number so generally have softer bonds than other atoms. These bonds can therefore deform more easily to localize the hole than others. Thus a hole created in the bulk of the film can move to a defect or surface and get localized. If it is too far away, however, it is likely to decay on the way. We hope to be able to measure such decay lengths and localization probabilities in the future using this BEEM technique with an appropriate range of samples. Some discussion of hole localization is given in Ref. 20. One remark is in order before leaving the subject of localization. In a metal, one normally thinks of electrons as being very delocalized, so the idea of a localized hole is counter-intuitive. What we are proposing is a hole in the d-band of the gold. This band has a narrow width (it is more like a core electron) so the electrons are in fact more localized than the conduction electrons.

A model to predict the voltage dependence of the adatom production rate was given in Ref. 13. The model used an approximate calculation of the hole formation rate at finite temperatures for a system with a gaussian broadened threshold energy. A broadened threshold for hole creation is not unrealistic: the bonds involved are near defects so are strained and in a variety of environments. The model predicts a nearly exponential voltage dependence below threshold, with the exponential factor depending on temperature and the width of the threshold distribution. A good fit to the latency time, half-time for BEEM current rise and the areal growth rate of the terraces, all of which depend on the adatom production rate, is found for a broadened threshold centered at 2.5 eV with a full width of 0.4 eV. This fit is shown as a solid line through the data in Fig. 7. Theoretical band structure calculations for gold[21] are consistent with a ~2.5 eV energy between the d-bands and the Fermi level.

There is another measurement which can be done to give an independent estimate of the threshold energy. This is to sweep the sample-tip bias to very high values and look for a decrease in the BEEM current which ought to occur when the incident electrons resonantly scatter to produce electron-hole pairs. If a significant fraction of the secondary electrons do not cross the Schottky barrier to become part of the BEEM current (Section 2.3 gives good evidence why they shouldn't in this energy regime), then an obvious dip should be seen. This type of measurement is not easy to make since the resonant scattering can easily change the sample so that the BEEM current drops or goes away (more on this in Section 4). Thus the spectra are not always repeatable and can have a variety of shapes[10]. Nevertheless, one particular type of BEEM spectra is seen much more frequently than the others. It has been seen on several samples with several tips so can be considered the typical case. It is shown in Fig. 10. The BEEM current does indeed decrease in the neighborhood of 2.5 V in agreement with the above threshold value. There is also another decrease in the BEEM current at ~3.8 V. This is close to the value of another vertical transition between two high density of states points in the band structure[21], giving further evidence that these BEEM spectra features arise from electron-hole pair formation (electron-electron scattering) in the gold.

Figure 11. Gray scale 1.4 V constant 1 nA current STM (left) and corresponding BEEM (right) images illustrate various modifications of a type 1 sample. The images are 1800 Å square. (a) shows the STM topograph which exhibited some grain growth as a result of the intermixing modifications. (b) is the BEEM image which was uniformly gray before stressing with the STM current. The top, left and center intermixing modifications were created with 3.1 V and 3 nA for 5.4 sec, 0.1 nA for 5.4 sec, and 3 nA for 2.7 sec, respectively. The creation of the lines near the bottom and right is described in the text. [From Ref. 14.]

3.3 Control and Reproducibility

At the bottom and right hand sides of Fig. 11 are some terraces which appear as lines. These are literally the first attempts at making any structures besides dots with the technique. First the line near the bottom was made by positioning the tip at the left end, raising the tip bias to 2.3 V@1 nA tunnel current, and moving 700 Å to the right at 230 Å/sec and back to the starting position at the same rate. When no increase in BEEM current under the tip was observed, the tip was swept again to the right at 47 Å/sec, leaving the line in the figure. The lines at the right of Fig. 11 were formed by moving to the right hand end of the upper line, where the tip bias was increased to 2.3 V@1 nA tunnel current. The line segments were then swept in order -- the last one is vertical in the image right at the edge of the figure. The horizontal lines are 350 Å long and the vertical lines 400 Å long, all swept at ~30 Å/sec. This demonstrates the extreme control and reproducibility of the phenomenon.

In principal, arbitrarily shaped and sized structures can be created. The limitation is thermodynamic stability of the structures, i.e. they must be large enough that surface energy does not cause them to evaporate, and diffusion rates of both the terrace atoms on the surface and vacancies within the film. Stability issues will be discussed in Section 5.

3.4 A Second Layer

Since a single atomic layer high terrace can be grown so easily, it is natural to expect that a several atomic layer high terrace could be grown. Since the adatom sink into the first terrace is much faster than adatom loss to grain boundaries, one would expect that a significantly higher adatom production rate would be required to nucleate a second layer. This is true experimentally. One might also expect that the BEEM current under the second layer would be much higher than even that where a single atomic layer high terrace exists, i.e. one would expect a tiered structure from where there are no, one, and two atomic layer high terraces. This is not found experimentally. Rather, a region of very small or no BEEM current is found where the two atomic layer high terrace exists. This can be understood in terms of gold-silicon interdiffusion[15], which occurs when enough gold comes into intimate contact with silicon. Images of such structures are seen in Fig. 11. The behavior is due to diffusion of silicon into the film, so will be discussed in Section 4.2.

4. Bulk Effects

In the previous sections, we have concentrated on structures at the inner surface -- their creation and observation. In this section we will illustrate how BEEM can be used to study effects in the 'bulk' of the film. There are two main types of effects in the system studied here. Section 4.1 discusses vacancies within the bulk, but formed in conjunction with the adatoms discussed above. A different effect, interdiffusion across the buried interface as hinted at in Section 3.4, will be the topic of Section 4.2. The following section describes the methods which can be used to experimentally distinguish between the two types of bulk effects.

Scattering in the bulk of the film reduces the number of electrons which reach the interface to attempt to cross it, so therefore reduces the BEEM current (although this seemingly obvious statement is true in the gold-silicon system, momentum conservation at the interface and the incident electron distribution can conspire to reverse the result in some systems[8]). A schematic drawing of the situation is shown in Fig. 12. Two types of bulk scattering processes are shown. One results from scattering throughout the film, and one from scattering in a near-interfacial layer. Both types are observed in the Au/passivation layer/Si system.

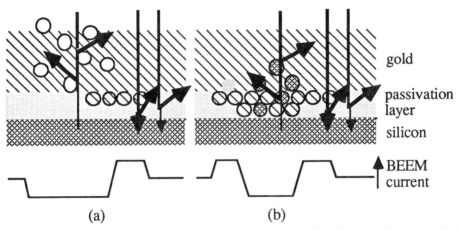

Figure 12. Sketches of mechanisms which can result in the local decrease of the BEEM current from surrounding values. (a) A high density of defects such as vacancies within the bulk of the film can strongly scatter injected electrons leading to a decrease in the BEEM current. (b) An interfacial layer can also scatter electrons to reduce the BEEM current. [From Ref. 14.]

4.1 Vacancies

A vacancy production always accompanies the production of an adatom if the terrace made from the adatoms is to grow. This vacancy diffuses into the 'bulk' of the film. A hot electron can scatter from a vacancy similarly to the way it scatters at the inner surface, so one would expect that whenever a vacancy gets into the (resolution-sized) region in the path of the incident electrons, it would decrease the BEEM current. One would expect to observe this phenomena during and after the modification process. Bursts of such decreases in the BEEM current are seen during all parts of the real-time BEEM plot. Examples are shown in Fig. 13. Some are found near the end of the latency period in Fig. 13(a), and some in the saturated region of Fig. 13(b). Note that the vacancy scattering stops after a short period. This means that the vacancy has become trapped or has moved out of the path of the incident electrons.

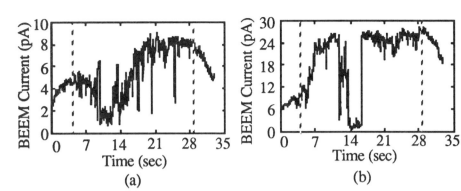

Figure 13. Plots of the BEEM current taken in real-time during modifications of the sample. The sample-tip bias was increased from the imaging value of 1.4 V linearly to the stressing value, at which it was held constant for a period of time, then decreased linearly back to the imaging value. The dotted lines indicate the times at which the bias reached and left the highest stressing value. The stressing voltages and tunnel current values for the plots are: (a) 2.1 V, 0.3 nA, and (b) 1.9 V, 1 nA. Although the various stages shown in Fig. 5(b) are observed, periods of time when the BEEM current is strongly decreased from its expected value are found. We expect that vacancies produced in conjunction with the adatoms are within the path of and scattering the injected electrons during these periods. These are blatant examples of a major source of noise in the real-time measurements -- the interface through which the BEEM electrons pass is being modified.

It is not surprising that the effects of vacancies are observed during the modification, when high energy electrons are injected. The electrons can scatter nearly resonantly from the vacancies so even a small vacancy density should be observable. It is also possible to see the effects of a high density of vacancies with the (~1.4 V bias) electrons used for imaging. This requires a high density of vacancies. The mechanism is strong elastic scattering which decreases the number of electrons which can cross the Schottky interface. Such a high density of defects is unstable, and will anneal. Experimentally, the annealing time (observed by changes in the BEEM image) is of order minutes, and is accompanied by changes in the topography (observed in the constant current STM image). This has been observed on both type 1 (high vacuum deposited at a high rate) and type 2 (clean, annealed) gold films.

An example of such a high density of vacancies in a type 2 film is seen in Fig. 14. The STM tip was centered over the region shown in the image of Fig. 14(a-b) while the bias voltage was held at 3.5 V for 3 sec while a constant current of 1 nA was maintained. Comparing the topographic images Fig. 14(a,c), one finds that the grain has been puffed up. In reality it has been filled with vacancies. The corresponding BEEM images Fig. 14(b,d) show how the lower energy (1.4 V bias) electrons are scattered by the vacancy structure. In subsequent images, the region of decreased BEEM current is seen to decrease until it is eventually gone. The lateral size (diameter) of this region decreases linearly in time[13] from the edges inward at a rate of ~32 Å/min. The grain in the STM image which was puffed up by the hot electron stressing decreases in size until it no

Figure 14. A sequence of gray scale 1.4 V constant current STM (top) and corresponding BEEM current (below) images show a region 1800 Å square of a type 2 sample (a-b) before, (c-d) just after, and (e-f) just under one half hour after the tip was held over the grain near the center of the image while the tunnel bias was increased to 3.5 V for 3 sec at a 1 nA tunnel current. (a,c,e) are the STM topographs with gray range 95 Å. The gold grains are up to 60 Å tall. (b,d,f) are the corresponding BEEM images shown with a 12 pA range. The enhanced features (~11-15 pA) appear whitish, background features (5-8 pA) gray, and degraded features (0 pA) black. Note that the changes in the BEEM images are correlated to the changes in the STM images. The enhanced region in (f) later disappeared. [From Ref. 13.]

longer appears as a lump by the time the BEEM image has healed. This observation underlines the reversibility of the vacancy motion process.

In type 1 films, the situation is less dramatic. Usually a second layer terrace has grown and intermixing occurred before such a high density of defects has been created (Section 5). Some additional decrease in the BEEM current is seen due to vacancies, which heals at about the same rate as the type 2 structures. The entire grain does not expand. Instead, a mound forms over part of the grain, or several nearby grains. This mound changes slightly during the annealing process, but does not disappear completely.

4.2 Highly Scattering Interfacial Layers

The formation of an interfacial layer which scattered electrons strongly was discussed in Section 3.4. The second layer terrace brings the gold into intimate contact with the silicon over an area large enough that the silicon bonding is disrupted and diffuses into the gold. The resulting high density of impurities in the gold greatly increases the elastic scattering rate resulting in a decreased BEEM current. A schematic drawing is shown in Fig. 12(b). Examples of BEEM images are shown in Fig. 11.

Interfacial reactivity such as that observed here with very high lateral resolution is common. BEEM is a powerful technique to probe such systems, as the transmission of electrons through an interface will usually be strongly effected by any interfacial reaction. In contrast to the unstable vacancy structures, this type of modification is irreversible at room temperature by the second law of thermodynamics.

Information about the thickness and growth of the interfacial layer may be obtainable from the real-time BEEM current as such a modification is produced. Such a real-time BEEM current plot is given in Fig. 15(a). The BEEM current decreases to a saturation level. This decrease is exponential in time as is expected if a layer of material with a short mean free path was growing in thickness. In this case, the material is silicon intermixed with gold. The method is equally suited to measure other materials.

4.3 Identifying the Type of Modification

Two distinct mechanisms which produced local degraded regions in the BEEM image have just been described. We saw that one type (high defect density within the film) is reversible whereas the other is not. This provides one method of determining the type of modification. The highly scattering interfacial layer is produced by terrace layer growth so is found in connection with an enhanced region. This also serves as an indication of which type of modification is present. But one would like to have other methods to distinguish between the two. It is probable that both types of modification are present in some cases. If one observes a stressed region for several minutes after stressing, one often finds a slight change in the BEEM image accompanied by a change in topography[10]. This is almost certainly due to the healing of vacancy structures within the gold film as described above in Section 4.1. The remaining structure (none in the case of Fig. 14) could be due to intermixing at the interface or the stabilization of a vacancy structure by some other defect (such as a line defect).

There is another way to distinguish the two processes. The BEEM current can be monitored during the production of the degradation structure. Well-defined signatures allow one to predict whether the mechanism will be intermixing at the interface or vacancy structures within the gold. In general, the interfacial mixing process has a less dramatic effect on the BEEM current than does dense vacancy formation. This is not surprising since the strongly intermixed layer remains near the interface (or there would be no reason for there not to be enough lateral growth to cover and obscure the enhanced region where only one terrace layer exists) whereas the defects can scatter the injected electrons for the entire thickness of the film. This can be seen in Fig. 15. Real-time BEEM current plots during stressing, such as those discussed in Section 3, are shown in

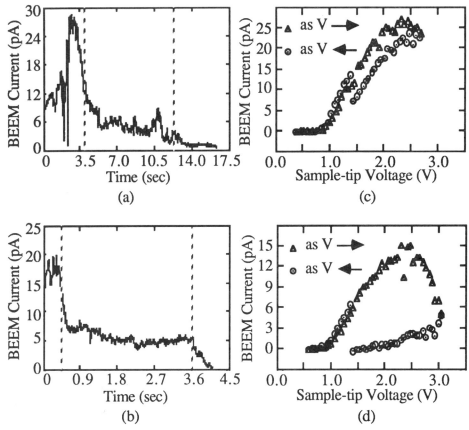

Figure 15. Plots of the BEEM current taken in real-time during modifications of the sample help to determine the type of modification which is produced. (a,b) show the BEEM current as the interface is modified at a constant voltage. The dotted lines indicate when the voltage arrived at and departed that constant value. The starting and ending voltage values were the 1.4 V imaging bias. The current during stressing was 1 nA and the biases (a) 3.1 V and (b) 3.5 V. Note the different time axes. (c,d) show BEEM I-V spectra swept to high voltage to produce modifications. As above, the BEEM current is recorded for the entire time the bias voltage is away from the 1.4 V imaging value. Different symbols indicate different directions as the voltage is varied from 1.4 V to 0.4 V (circles), up to the maximum voltage (triangles), and back down to 1.4 V. The discontinuity in the circle data at 1.4 V represents the change in the BEEM current caused by the modification. Both processes degraded the local BEEM current. (c) produced a region with intermixing at the interface, later images resemble those in Fig. 11. (d) produced a large gold mound in addition to a BEEM current decrease.

Fig. 15(a-b). A much faster and sharper drop is observed in Fig. 15(b), which left a mound and presumably many defects within the film, compared to Fig. 15(a), which left an intermixed interfacial layer type modification. Differences can also be seen when the modification is created by a voltage sweep. BEEM I-V spectra are shown in Fig. 15(c-d). The data are shown for the entire time the sample-tip voltage was different from the imaging bias of 1.4 V. Different symbols are used to differentiate the different directions of voltage sweep. Comparing the starting and finishing values of the BEEM current in Fig. 15(c-d), both taken at the imaging bias, one finds that the BEEM current has decreased in the region under the STM tip. The result of the stress shown in Fig. 15(c) was an intermixed region. A highly defected film was obtained, as evidenced by the STM and BEEM images taken after the sweep in Fig. 15(d). Note the sharp drop in the BEEM current as the vacancy structure has grown before the bias is decreased. A subsequent BEEM I-V extending to a lower maximum bias exhibits a different spectral shape near the Schottky barrier height threshold, further evidencing a large change in the film.

Finally, we comment that in the case of a type 1 gold metal layer under study, it appears that a second terrace layer (and hence a highly intermixed interfacial region) can be nucleated at a lower bias than is needed to create a substantial defect structure within the film. Note the very high bias (3.5 V) required to fill the grain (of type 2 = clean gold) in Fig. 14 with vacancies compared to the value (~2.5 V) at which the second layer formation is first observed in the type 1 samples. Type 1 samples stressed at very high biases show a large amount of mound formation and some subsequent healing similar to that discussed here for type 2 samples. These observations illustrate that one has some degree of control of which type of structure is produced by managing both the sample and stressing characteristics.

5. Stability and Materials Dependence

We have discussed interface phenomena in Section 3 and bulk phenomena in Section 4. The issues addressed in this section tie the two types of modification together. Also, the stability of the structures created within the film or on any of the surfaces is strongly tied to the nature of the film itself. This is why the title of this section couples the two together. Two types of gold films were used in this study as discussed in Section 2.2. They are the type 1 and the cleaner and less defected type 2. The effects of the interface chemistry and structure on both the stability of the BEEM interface and the terrace growth phenomena have been discussed in Section 2.2 and elsewhere[10,11]. Thus the discussion here will be limited to the effects of the gold film.

The inner surface structures are quite stable on a type 1 film. The terraces of Fig. 2 and the lines and other modifications in Fig. 11 stay close to their original form for at least hours (as long as we cared to observe them). In contrast, inner surface terraces on type 2 samples are not stable but decay very quickly[13]. The rate of decrease of the diameter can be as fast as 1 Å/sec. Afterwards, the interface is much as it was before production. We interpret the difference in stability to the difference in vacancy diffusion rates. Vacancies can diffuse much more easily in the clean type 2 films. Once a vacancy reaches the inner surface, it can recombine with a terrace atom to decrease the size of the terrace.

Vacancy structures within the films anneal at least somewhat in both types of samples. The time constant of change seems to be about the same, a few minutes, in both cases. An example of this behavior in a type 2 film was shown in Fig. 14. In the type 2 films, where interfacial intermixing is not observed, the BEEM image returns to its initial (before stressing) form. Thus one can rule out the existence of a stabilization of remaining defects. Such a stabilized structure cannot be ruled out in the type 1 films, as a degradation from an interfacial intermixed layer remains.

Mound formation on the top surface of the film is also seen to differ. Whole grains are effected uniformly in the type 2 films as seen in Fig. 14. Type 1 films show mounding within a grain, Fig. 11. The mounds on a type 1 sample do not anneal away completely (just slightly) on the minutes time scale, but instead last at least a few hours. The differences in mound formation between the two types of gold probably reflect differences in the quality of the film. They may partially be due to length scales, however. It may be possible to create a mound within a grain of a type 2 film if the grains were large enough, e.g. if the sample was epitaxial.

The question of why no interfacial intermixing modification was observed in type 2 samples deserves further comment. If a second layer terrace grows on the inner surface bringing the gold into contact with the silicon, we expect intermixing. Thus we can reword the question to : Why doesn't a second layer nucleate on the inner surface of type 2 films? The answer is related to the stability of the terraces as discussed above. Terraces are rapidly being annihilated on the type 2 films, even as the terraces grow. Thus a much higher adatom production rate is required to produce the same final result on a type 2 inner surface than a type 1 inner surface. The problem is that vacancy structures form within the bulk at these higher biases. The vacancies scatter hot electrons even before they reach the inner surface, so effectively decrease the rate of adatom production. Thus one is not able to maintain the rate of adatom production to nucleate a second layer in the type 2 samples. The terrace annihilation rate is very small for the type 1 samples, so it does not need to be compensated for and terraces can be nucleated.

6. Electrons or Holes

The discussion of a mechanism for the hot electrons to stimulate atomic motion in Section 3.2 emphasized the creation of a broken bond or hole. It is interesting to consider the experimental situation in which hot holes are injected into the sample rather than hot electrons. A technique related to BEEM, but using the reverse tunnel bias from that which one would normally expect to get a BEEM current, has recently been demonstrated[9]. In our samples, this corresponds to injecting holes into the gold from the tip. We use the term Auger BEEM to describe the measurement since the BEEM current arises from secondaries (Auger electrons) which are formed as holes scatter and are filled by electrons from near the Fermi level. Thus one obtains a current with the same sign as that observed with normal-BEEM, but the tip bias is reversed. The detailed threshold shape has been predicted[9], and images are collected in much the same way as in normal-BEEM.

It is interesting to compare the modification behavior of the sample under the injection of hot holes to that observed under hot electron injection. Real-time BEEM plots during Auger BEEM stressing do not look like those in Figs. 5,13. The BEEM current rise is lacking. Instead, the BEEM current drops after a latency period[11]. As above, the latency time and rate of BEEM current decrease depend strongly on voltage. Subsequent Auger-BEEM images show a degraded region, but no change is observed in before and after normal-BEEM images[10]. This suggests that the imaging mechanism differs between normal and Auger BEEM. The contrast in normal BEEM images of Au/Passivation layer/Si samples is dominated by the interface, although very high defect densities can effect the images. In contrast, we believe that the formation of the Auger electrons dominates the contrast in the Auger-BEEM images. The absence of terrace growth (as imaged by normal BEEM) by Auger-BEEM or hole-injection suggests that the holes scatter much closer to the top surface than the inner surface. The fact that holes scatter more strongly is reasonable considering the ESD model where the electrons create holes that stimulate the atomic motion. It also suggests a source of the contrast in the Auger-BEEM images. The atomic motion process would compete with or substantially alter the formation of Auger electrons. Thus even a small density of defects could strongly decrease the Auger-BEEM signal while not greatly effecting the imaging energy electrons used to form the normal-BEEM image. This accounts for the paradox of why normal-BEEM does not detect a hole injection modification detected by Auger-BEEM, and suggests the physical nature of the modification produced by hole injection.

7. Mound Formation

The creation of mounds on gold surfaces with STMs has received much attention in the past few years, a few examples are given in Ref. 22-23. In particular, by varying the parameters to make a 'recipe', extremely reproducible results are obtained[23]. It is interesting to compare the other mound formation work to that presented here. Mound formation has been observed in this work, as has motion of gold at the inner surface of the samples. The motion of gold at the inner surface is the result of inelastic scattering of the hot injected electrons. There are no tip-sample or tip electric field effects at the inner surface. We suggest that the motion of gold to produce the mounds is caused by the same mechanism, as is strongly suggested by this work. This represents a new model for mound formation. A previous paper[23] suggested that the cause was field induced evaporation of gold from the tip to the sample. Few previous studies controlled the tunnel current as the bias was raised. The tunnel currents during mound formation were very large. Voltage was controlled, however, and mound formation was observed above a threshold near 3.2-3.5 V. The field evaporation model[23] was not able to explain such a low threshold voltage. The threshold value is roughly what one would expect for massive gold movement within the model presented here (Fig. 7,14), which suggests that the same mechanism may explain all the mound formation work. We have observed a bias sign dependence on the threshold at which mound formation is clearly evident. A bias of ~2.5 V positive sample-tip bias or ~2.1 V for negative sample-tip bias. This is consistent with the idea of holes scattering closer to the top surface than electrons, as discussed in the last section. The nature of the gold must also be considered. We observed a strong dependence of the modification and mound formation properties as a function of gold quality(Section 5). We do not know the cleanliness of the gold in the other works. Note that the fact that mound formation can be observed on very thick or bulk gold samples does not preclude the inelastic scattering and vacancy motion model described here. We expect adatom-vacancy pairs to be generated at all gold surfaces, including the top one which is present in all the studies.

8. Summary

The BEEM technique has been shown to be a powerful technique for studying hot electron scattering in a thin metal layer in addition to providing high lateral resolution measurements of diffusion of defects within the film and interdiffusion of species at a buried solid-solid interface. We have shown that hot electrons injected by an STM tip scatter and modify a gold film throughout its volume and at all its surfaces. The BEEM measurement technique is found to be a powerful method for probing such buried structures. With an appropriate choice of passivation layer between a gold film and silicon substrate, small single atom high terraces on the inner surface of the gold can be studied. This is important as it allows observation of hot electron effects at high spatial resolution without the effects of a high electric field from an STM tip or other tip-surface interactions. It is also quite general, so other systems can be studied in the same way. The nucleation and initial growth of buried-surface terrace structures is observed in real-time. One can measure the effects of the thermodynamic instability of the initially very small terraces by comparing the initial growth rate to the growth rate of larger terraces. The growth rate is related to the rate of adatom production, which can be determined quantitatively as a function of voltage from such measurements. Vacancy formation accompanies adatom formation. The structures formed by the vacancies can effect the BEEM images, which then allows a glimpse into their evolution. The BEEM image is correlated with an STM image to relate vacancy structure changes and topographic changes. The mechanisms of mound formation and stability can be addressed. The stability of the created structures is shown to be related to the physical properties of the gold film.

9. Future Outlook

The work discussed in this chapter shows many possible applications of the BEEM technique for studying transport and materials science at nanoscale resolution. The healing of defect structures made of a high density of defects and the annihilation of terraces by vacancy diffusion suggest a new way to study vacancy diffusion within a single grain. One can see where the grain boundaries are in the STM image, so their effect can be studied as well, but independently. This will require careful analysis of the BEEM current image change as a function of time after modification. Diffusion of gold on the inner surface could be investigated, as can grain growth.

Electron transport issues can be addressed. BEEM can measure the Schottky barrier heights[1]. Quantum effects in interface transport can be studied[6]. By choosing an appropriate system, e.g. a lattice matched interface and a particular substrate orientation, transverse momentum constraints can become crucial for electron transport. This can emphasize elastic scattering in the film[8], or can be used to probe interface quality -- in regions where the interface is rough, transport constraints will change so one can expect a change in the BEEM current magnitude and spectral shape. Interfacial reactions may also be evident as a change in BEEM current as discussed in this chapter. Questions such as how hot electrons scatter and what is the effect on the film microstructure can be answered. One can even hope to gain some better understanding of the electromigration process using BEEM as a tool.

For the purposes of this book, we should comment that the structures produced by proximal probe techniques are very small but need to be characterized. BEEM related techniques may be able to do that, and further may be able to pinpoint mechanisms for growth, such as the mound formation discussed herein. The high surface to volume ratio of the structures we are producing may change the materials characteristics from bulk-like to surface dominated. This is a new regime for materials study, and tools need to be developed to provide physical insight and data for theoretical testing. The small structures will be asked to carry high current densities. We need to characterize the effects of such electron transport to identify which processing methods yield materials suitable for use, and what are the expected lifetimes of the structures. Note that the BEEM technique described here does not require very specialized samples. Metal on top of silicon occurs frequently in devices; an area of silicon could be doped and such devices fabricated on-chip if need be to address film quality issues.

Finally, one might consider direct uses of the technology presented here, for example the use of the BEEM contrast as an electronic storage device. The technique has promise as is evidenced by the control and reproducibility with which the inner surface terrace structures can be produced. The density of the storage could be made very large, limited by the stability considerations (Section 5), and would require careful choice and preparation of materials. Other considerations include tip stability. We found the Pt tips to easily last several days of experiments if they were properly made, but several years of rapid mound making is a stronger constraint, and would require further investigation. The final consideration is speed. With any proximal probe, a problem is to get to different bit locations quickly, but without destroying or wearing the probe. This method is superior to a topographic method in this respect since the probe does not have to be continuously scanned. Even the simple sampled discussed here showed remarkably consistent BEEM currents. If all the information one needs is which of several BEEM current levels exists at a particular place on the sample, one needs only to touch down long enough to read the current (or modify the interface if writing). Motion between bits can be accomplished with the tip retracted a safe distance, so is limited by the servo mechanism. The dwell time per bit would be limited by the magnitude of the current to be read, or writing time, both of which are reduced when the tunnel current is increased.

10. Acknowledgments

The experimental aspects of this study were completed while the author was at Cornell University, in collaboration with Tony Huang, Andres Fernandez and R. A. Buhrman. We wish to acknowledge constructive discussions with David Peale, Wilson Ho, John Silcox, and Dan Ralph. We thank Bill Kaiser, Carl Kukkonen and the Jet Propulsion Lab for helping us to get started in STM. During part of the time this work was done, the author was supported by an IBM Graduate Fellowship. Research support was provided by the Office of Naval Research. Additional support was provided by the National Science Foundation through use of the National Nanofabrication Facility and the facilities and equipment of the Cornell Materials Science Center.

11. References

* Current Address: Physics Department, North Carolina State University, Raleigh, NC 27695-8202.

1. W. J. Kaiser and L. D. Bell, 'Direct investigation of subsurface interface electronic structure by ballistic-electron-emission microscopy,' Phys. Rev. Lett. **60**, 1406-1409 (1988); and L. D. Bell and W. J. Kaiser, 'Observation of interface band structure by ballistic-electron-emission microscopy,' Phys. Rev. Lett. **61**, 2368-2371 (1988).

2. M. H. Hecht, L. D. Bell, W. J. Kaiser, and F. J. Grunthaner, 'Ballistic-electron-emission microscopy investigation of Schottky barrier interface formation,' Appl. Phys. Lett. **55**, 780-2 (1989); M. H. Hecht, L. D. Bell, W. J. Kaiser, and L. C. Davis, 'Ballistic hole spectroscopy of interfaces,' Phys. Rev. **B42**, 7663 (1990); and .T. H. Shen et al., 'Ballistic electron emission microscopy, current transport, and p-type δ doping control of n-isotype InAs-GaAs heterojunctions,' J. Vac. Sci. Technol. **B9**, 2219-24 (1991).

3. Y. Hasegawa et al., 'Ballistic electron emission in silicide-silicon interfaces,' J. Vac. Sci. Technol. **B9**, 578-80 (1991); A. Fernandez et al., 'Ballistic electron emission microscopy studies of the $NiSi_2$/Si(111) interface,' J. Vac. Sci. Technol. **B9**, 590-93 (1991); and Philipp Niedermann et al., 'Ballistic electron emission microscopy study of PtSi-n-Si(100) Schottky diodes,' J. Vac. Sci. Technol. **B10**, 580-5 (1992).

4. L. D. Bell, W. J. Kaiser, M. H. Hecht, and L. C. Davis, 'New electron and hole spectroscopies based on ballistic electron emission microscopy,' J. Vac. Sci. Technol. **B9**, 594-600 (1991).

5. A. E. Fowell et al. Semicond. Sci. Technol. **5**, 348 (1990); A. E. Fowell et al., 'Ballistic electron emission microscopy studies of Au-CdTe and Au-GaAs interfaces and band structure,' J. Vac. Sci. Technol. **B9**, 581-84 (1991); and M. Prietsch, A. Samsavar and R. Ludeke, 'Structural and electronic properties of the Bi/GaP(110) interface,' Phys. Rev. **B43**, 11850-6 (1991).

6. H. D. Hallen, et al., 'Scattering and spectral shape in ballistic electron emission microscopy of $NiSi_2$-Si(111) and Au-Si samples,' Phys. Rev. **B46**, 7256-59 (1992).

7. M. Prietsch and R. Ludeke, 'Ballistic-electron-emission microscopy and spectroscopy of GaP(110)-metal interfaces,' Phys. Rev. Lett. **66**, 2511-14 (1991); L. J. Schowalter and E. Y. Lee, 'Role of elastic scattering in ballistic-electron-emission microscopy of Au/Si(001) and Au/Si(111) interfaces,' Phys. Rev. **B43**, 9308-11 (1991); M.D. Stiles and D. R. Hamann, 'Kinematic theory of ballistic electron emission spectroscopy of silicon-silicide interfaces,' J. Vac. Sci. Technol. **B9**, 2394-8 (1991); and E. Y. Lee and L. J. Schowalter, 'Electron-hole pair creation and metal/semiconductor interface scattering observed by ballistic-electron-emission microscopy,' Phys. Rev. **B45**, 6325-28 (1992).

8. A. Fernandez et al., 'Elastic scattering in ballistic electron emission microscopy studies of the epitaxial $NiSi_2$/Si(111) interface,' Phys. Rev. **B44**, 3428-31 (1991).

9. L. D. Bell, M. H. Hecht, W. J. Kaiser, and L. C. Davis, 'Direct spectroscopy of electron and hole scattering,' Phys. Rev. Lett. **64**, 2679-82 (1990).

10. H. D. Hallen, Ph.D. thesis, 'Ballistic electron emission microscopy studies of gold-silicon interfaces,' Cornell University, January 1991.

11. H. D. Hallen, et al., 'Gold-silicon interface modification studies,' *Proceedings of the Fifth International Conference on Scanning Tunneling Microscopy/Spectroscopy*, (Baltimore, July 23-27, 1990) in J. Vac. Sci. Technol. **B9**, 585-589.(1991); and 'Ballistic electron emission microscopy of metal-semiconductor interfaces,' *Proceedings of the Ballistic Electron Emission Microscopy Workshop 1990*, Jet Propulsion Laboratory, California Institute of Technology, Pasadena, CA, March 9, 1990.

12. M. H. Hecht et al., in *Proceedings of the Ballistic Electron Emission Microscopy Workshop 1992*, Death Valley, CA, January 27, 1992.

13. H. D. Hallen, A. Fernandez, T. Huang, R.A. Buhrman, and J. Silcox, "Hot electron interactions at the passivated gold-silicon interface," Phys. Rev. Lett **69**, 2931 (1992); H. D. Hallen and R. A. Buhrman, 'Hot electron induced atomic motion and structural change at the passivated gold-silicon interface,' (submitted to Phys. Rev. B).

14. H. D. Hallen and R. A. Buhrman, 'BEEM: A probe of nanoscale modifications,' to appear in the *NATO Advanced Studies Institutes* series, edited by Ph. Avouris (Kluwer, Dordrecht).

15. For example L. J. Brillson, A. D. Katnani, M. Kelly, and G. Margaritondo, J. Vac. Sci. Technol. A **2**, 551 (1984).

16. W. J. Kaiser and L. J. Schowalter, private communications.

17. A. Ishizaka and Y. Shiraki, J. Electrochem. Soc. **133**, 666 (1986).

18. D. Peale and B. H. Cooper, (private communication); and D. Peale, Ph.D. thesis, 'Diffusion and mass flow dynamics on the gold (111) surface observed by scanning tunneling microscopy,' Cornell University, January 1992.

19. U. Landman, 'Atomic-scale dynamics and stability of interfacial systems and nanostructures,' to appear in the *NATO Advanced Studies Institutes* series, edited by Ph. Avouris (Kluwer, Dordrecht).

20. R. D. Ramsier and J. T. Yates, Jr. 'Electron-stimulated desorption: principals and applications,' Surface Science Reports **12**, 243-378 (1991).

21. N. Egede Christensen and B. O. Seraphin, 'Relativistic band calculation and optical properties of gold,' Phys. Rev. **B4**, 3321-44 (1971).

22. David W. Abraham, et al., 'Surface modification with the scanning tunneling microscope,' IBM J. Res. Develop. **30**, 492 (1986); Y. Z. Li et al. 'Writing nanometer-scale symbols in gold using the scanning tunneling microscope,' Appl. Phys. Lett. **54**, 1424 (1989); and J. Schneir et al., 'Tunneling microscopy, lithography, and surface diffusion in an easily prepared, atomically flat gold surface,' J. Appl. Phys. **63**, 717 (1988).

23. H. J. Mamin, P. H. Guethner, and D. Rugar, 'Atomic emission from a gold scanning-tunneling-microscope tip,' Phys. Rev. Lett. **65**, 2148-21 (1990).

Part III
Metrology

METROLOGY

Fabrication of lithographically patterned devices requires tight control of the size, the shape and the position of the pieces that comprise the desired objects. Control requires measurement, so metrology is an essential element in proximal probe lithography. A widely accepted rule of thumb is that the measurements must be good to within 10% of the dimensions of the patterned objects. Traditional optical and electron beam metrology tools often do not provide precision this high for the tiny sizes considered in this volume. Fortunately, probe microscopes can themselves be used as metrology tools. Application of a probe microscope to metrology requires many refinements in the microscope. Errors in an apparently simple number, such as the width of a line or the roughness of a substrate, can arise from many sources, all of which must be kept under control. The papers in this section describe how accurate numbers can be obtained.

Clayton Teague (NIST) discusses the global aspects of the problem. He shows how to establish a coordinate reference frame and how to realize a metric within the frame. An important source of error, Abbe offset error, is discussed in detail. Joe Griffith (AT&T) and his co-workers cover issues associated with the scan head. Because of hysteresis in the piezoceramic material used to move the tip, an independent monitor of the tip position is necessary. In addition, the geometry of the probe-sample interaction is analyzed to reveal sources of measurement error. They describe surprisingly subtle effects appearing in surface roughness measurement. Finally, they describe a way to measure the probe shape with the probe microscope itself. Leigh Ann Files-Sesler, John Randall and Francis Celii from Texas Instruments describe some typical applications of probe metrology in lithography today. Images of quantum dots, films grown from molecular beam epitaxy and lateral resonant tunneling transistors are shown. These authors describe applications of surface roughness measurement, and they discuss the application of fractal analysis to surface roughness.

The ultimate performance of scanning probe metrology tools depends strongly on the topography to be measured. With careful attention to the techniques described in these volume, nondestructive measurements on an extraordinarily wide range of materials can be obtained. In many cases, the measurements can be obtained with no other tool.

Generating and Measuring Displacements Up To 0.1 m To An Accuracy of 0.1 nm: Is It Possible?

E. Clayton Teague
National Institute of Standards and Technology
Gaithersburg, Maryland 20899

Abstract

Four major tasks of accurate positioning are examined: generating highly repeatable motion in a workspace, constructing a metrology frame from highly-stable, accurate lines and planes, realizing a metric in the workspace, and linking the workpiece to the coordinate system with a probe. Optical scales, x-ray interferometry, optical heterodyne interferometry, Fabry-Perot etalons, and impedance-based transducers are evaluated as possible means to realize a metric. The major error sources, practical limitations, and guidelines for establishing a coordinate reference frame are described. The repeatability of sliding bearings and flexure bearings is examined and this combined with cosine and Abbé offset errors is used to estimate uncertainties in generating and measuring linear motion. Finally, the effects of probe-substrate forces and energies of interactions on positioning accuracy are discussed. Estimates of the overall uncertainty for positioning a probe over an area of 100 mm × 100 mm indicate that achieving the goal given in the title will be very difficult; uncertainties of 1 - 10 nm being more likely. The paper concludes with a discussion of how the availability of large-area, atomically-flat surfaces could improve our abilities to perform the four key tasks of accurate positioning with significantly greater accuracies.

Introduction

An essential requirement for any type of machine designed to shape or structure matter is the ability to move a forming tool or probe in an accurate, prescribed path throughout a workspace. Historically, we have evolved in our capabilities to perform this function from hand-controlled carving tools, to machine-controlled cutting edges, to cutting with lasers and ions, to recent demonstrations of controllably positioning individual atoms on a suitable substrate with scanning tunneling microscope probes. This volume concentrates on the technology of proximal probe lithography, the use of a scanning tunneling microscope (STM) tip, an atomic force microscope (AFM) tip, an array of such tips, or a low-voltage electron beam produced by a proximal probe to form features in a mask for subsequent structuring of a substrate surface or to directly induce such features on a substrate by spatially structuring the physical and/or chemical environment. Any of these approaches to proximal probe lithography requires that the probe or probes be moved and located in the workspace with a positional accuracy significantly smaller than the dimensions of the smallest feature being fabricated, and that the positioning be accomplished within a time scale to make the fabrication process practical.

All accurate positioning -- regardless of scale -- requires proper execution of four tasks during design and operation of the fabrication machine:

* Generating highly repeatable motion in the workspace,
* Constructing a metrology frame -- a coordinate reference frame -- from highly-stable, accurate lines and planes against which the motion can be referenced,
* Realizing a metric -- a measure of displacement -- in the workspace, and
* Linking the workpiece to the coordinate system by a probe.

For proximal probe lithography to become a practical manufacturing process, these tasks must be executed with the utmost accuracy and must be integrated successfully with the probe-substrate processes used for structuring. This paper reviews current capabilities for accomplishing these tasks, examines some of the factors that set bounds on the accuracies that can be achieved, and finally, estimates the overall minimum uncertainties that should be achievable for positioning and measuring over a 100 mm × 100 mm area. A final section discusses how the availability of large-area, atomically-flat surfaces could improve our abilities to perform these key tasks with significantly greater accuracies.

About five years ago, we started a high-risk project at NIST to explore the question posed in the paper's title. The Molecular Measuring Machine[1] project incorporates a scanning tunneling microscope and high-accuracy heterodyne interferometry that will, if all goes reasonably well, be capable of positioning and measuring to accuracies of 1 nm over an area of 50 mm × 50 mm. This paper is based in large measure on what we have learned as we've attempted to bring this machine into operation. ("We" being used here to mean the author and his colleagues listed in the acknowledgements.)

Do We Need the Capability to Generate and Measure Displacements Up to 0.1 m With Accuracies of 0.1 nm?

Let's attempt to answer this question by looking at the demands posed by current advanced lithographies. Most projections[2] about minimum critical dimensions as a function of DRAM density give a minimum critical dimension of 180 nm as being required for a 1 gigabit DRAM with a chip size of 500 - 600 mm². The accuracies in generating and measuring displacements needed to support the fabrication and measurement of masks used for manufacturing these chips are estimated to be 6 nm for the industrial process control metrology and 1.5 nm for the national laboratory as a reference for the industrial metrology systems over first the chip dimensions and then finally the dimensions of the full wafer.

These values are obtained by the analysis shown in figure 1. The minimum critical dimensions shown there for a specified DRAM density are typical of the values commonly used in the literature.[2] The first step in arriving at the estimated metrology and positioning requirements for the advanced

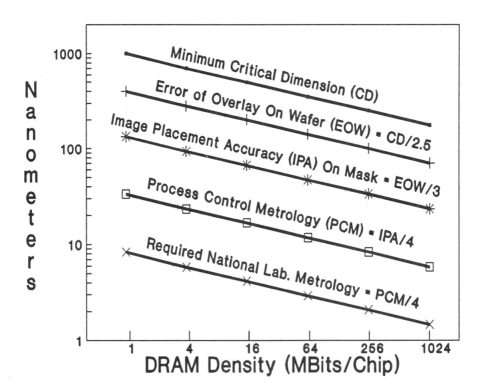

Figure 1: An Analysis of Metrology Requirements for Advanced Lithographies.

lithographies was to assume that the error of overlay on a wafer between two lithography steps, as for example between the via hole through a dielectric and the underlying conductor, may be taken, conservatively as 1/2.5 of the minimum critical dimension being used for the circuit being fabricated. The error of overlay on a wafer between two lithography steps is determined by: (a) the accuracy of image placement on the two masks used for the steps, i.e., the accuracy of mask points relative to an absolute grid; (b) the precision of alignment of the second mask with respect to the exposed pattern of the first mask; and (c) the dimensional changes in the wafer resulting from processing during and between the two exposures. Rules for combining these three components to obtain the overall overlay error budget vary depending on the type of devices being formed, the particular fabrication processes being employed, and the decisions of the manufacturer. If one lets each factor have equal weight, then the accuracy of placing images on a mask or a reticle must be less than 1/3 of the error of overlay on the wafer. Taking the Military-Standard factor-of-four relationship of process-control measurement accuracy to design tolerances yields the 6 nm and 1.5 nm values needed for the metrology systems.

Advanced lithographies such as some of the processes described in this volume for proximal probe lithographies may not utilize the sequence of processing steps just outlined. However, the equivalent concepts of overlay, the need to precisely match the patterns generated by two sequential processing steps and the need to account for material variations during and between processing steps will remain. If we project the requirements for minimum critical dimension, error of overlay, image placement accuracy, and process-control metrology out to densities of 1 terabit, assuming the chip size doesn't increase, image placement accuracy demands are 4 nm and process-control metrology is 1 nm. Such capabilities are beyond current capabilities of any laboratory in the world and pose serious challenges for metrologists who seek to meet the needs of these prospective technologies. The analyses of this paper and the potential for improved accuracies offered by the proximal probes access to the metric and geometry of single crystal surfaces are possible steps toward addressing these needs.

Why is accuracy of image placement on a mask relative to an absolute grid important? Without taking a major digression from the subject of the paper, four reasons can be given as justification for why image placement accuracy relative to national or international standards of length and angle is required as opposed to measuring the mask with a highly repeatable measurement system. First, an accurate, absolute grid (or an artifact which has been characterized relative to an absolute grid) is required to determine the distortions produced by mounting of the mask in the metrology systems being used as a reference system for a particular production line, hereafter called the process control metrology system. Second, an absolute grid is required to monitor the temporal stability of the process control metrology system. Third, the absolute grid is required to determine the temporal stabilities and accuracies of multiple process-control metrology systems that may be being used for a production line, i.e., both a Leitz LMS-2000[3] and a Nikon 3i[3] or two Leitz LMS-2000s[3] could be being used to qualify the masks. Finally, the absolute grid is required to determine the accuracy of NX versus 1X masks that will be used in the next generations of mix and match optical, x-ray lithography, and proximal probe technologies.

It is particularly important to note that temperature accuracy and stability of the metrology systems, steppers, and writers is critical to achieving the dimensional tolerances previously stated. For a silicon substrate or a silicon membrane, a change in temperature of 0.01 K will produce a change in spacing of 1 nm for two points spaced apart by 50 mm. Temperature stabilities of this level are at the current state-of-the-art for mask metrology systems; establishing a temperature reference at an accuracy of 1 mK requires significant care, but can be achieved by using a platinum resistance thermometer calibrated against the gallium melting point.[4]

Realizing A Metric

Since 1983, path length has been defined in terms of the speed of light;[5] that is: 1 meter is the length of the path traveled by light in vacuum in $1/c$ seconds. "c" has a defined value of 299,792,458 m/s. This definition combined with a very accurate clock allows us, at least conceptually, to measure path length as accurately as we can measure time. Today, with cesium-beam clocks the accuracy of time measurements[6] is about 1 part in 10^{14}. Current realizations of the definition are, however, based upon accurately determining stable laser frequencies in terms of the cesium-beam clock frequency and then realizing metrics by using the wavelength of this laser radiation as an intermediate standard against which displacements can be compared by interferometry.[7] The most stable laser available commercially is an iodine-stabilized helium-neon laser. The frequency of this laser[8] has been determined to an uncertainty of about 1 part in 10^{10}. The long frequency synthesis chain needed to relate the frequency of the cesium-beam clock and the helium-neon laser is described by Evenson.[8] The paper shows how the complexity of this synthesis now causes four orders of magnitude in accuracy to be lost.

In the context of this paper, **realizing a metric means providing a measure of displacement with subnanometer accuracy.** Developments in this field over the last ten years have produced several methods which have sub-nanometer resolution. Accuracy is determined by how well the displacement measurements are linked to the definition of the meter, via the de facto intermediate standard of the iodine-stabilized helium-neon laser wavelength. To describe how this linking produces limitations on displacement measurements, some background on the basic principles of operation of the different methods is necessary. In this section, displacement measurements by optical scales, x-ray interferometry, optical heterodyne interferometry, frequency-tracking Fabry-Perot etalons, and impedance-bridge transducers will be briefly reviewed.

Optical Scales and X-Ray Interferometry: The diffraction grating interferometer configuration (Fig. 2) is often used for both optical scale measurements and x-ray interferometry. These interferometers use three identical and equally spaced diffraction gratings. Some optical forms use a biprism for one of the elements. A splitter grating, S, breaks the incoming radiation into two coherent parts traveling over different spatial paths; a mirror grating, M, rediffracts the two beams back into coincidence at the plane of an analyzer grating, A. In operation, the analyzer grating is displaced relative to the other two along the direction indicated by the arrow. Describing the interferometer's operation in terms of a simple model, the two recombining beams create a set of interference fringes fixed in space at the analyzer. Then as the analyzer is translated, interaction between this set of fringes and the analyzer produces intensity changes in both the A and B beams which vary sinusoidally with the periodicity of the diffraction gratings.

Note that the periodicity is dependent only on the diffraction gratings period and not on the wave-length of the incoming radiation. For x-ray interferometry this naive model would need to be modified to include finite thickness gratings and x-ray interaction effects with solids as described in the dynamical theory of x-ray diffraction.[9]

With optical scales, the minimum grating spacing is about 1 to 5 μm. Thus if the analyzer is displaced 1 to 5 μm, one period of intensity change will occur at a detector in beam A or B. Using multiple-phase measurement techniques,[10] one can interpolate to about 1/1000 of a period to achieve resolutions of between 1 and 10 nm. Accuracy will depend on how well the grating period has been determined and how uniform and similar are the gratings.

X-ray interferometry was first proposed and demonstrated by Bonse and Hart.[11] To achieve the necessary grating uniformity and atomic-scale alignments, they formed the three gratings by cutting two wide grooves in a silicon crystal block. By mounting the crystal on a mechanism which could very precisely move one of the blades, Hart[12] realized an "angstrom ruler." Deslattes and Henins[13] completed the first measurements to determine the silicon (220) lattice plane spacing in terms of optical wavelength standards. Similar

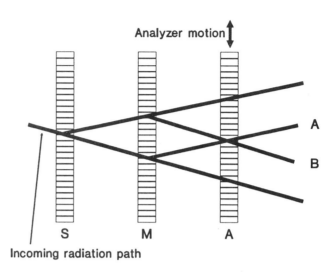

Figure 2. Ray paths in a typical diffraction grating interferometer.

measurements have now been realized by other major national laboratories: PTB in Germany,[14] NPL in the U.K.,[15] and NRLM in Japan.[16]

The silicon (220) lattice plane spacing is about 0.2 nm which means that in an x-ray interferometer using gratings made of the silicon lattice, a displacement of this amount will produce one period of intensity change at beams A and B. Deslattes and Kessler[17] used a Fabry-Perot etalon to achieve a very high level of interpolation of the 0.2 nm period. More typical and less sophisticated schemes based on quadrature detection of beams A and B are capable of dividing the 0.2 nm into 50-100 parts.[18] Resolutions of approximately one pm are then achieved. As with the optical gratings, accuracy is limited by the knowledge of the lattice plane spacing in terms of international standards. The most recent measurements have achieved uncertainties, $\Delta d/d$, of between 10^{-6} and 10^{-7}. (As a personal aside, during the mid 1970's I had the opportunity to visit Deslattes' laboratory. On the wall was displayed a 20 foot long strip chart recording containing one optical fringe with about 2000 x-ray fringes superimposed on it. That chart recording represented a major accomplishment in precision measurement whose significance is now being realized in many applications in nanometrology.)

The intensity changes in beams A and B can be observed only if the planes of the gratings S, M, and A remain closely parallel to each other during the displacement of the analyzer A and if the vibration amplitudes of the gratings are less than a few tenths of the grating period. Mendlowitz and Simpson[19] give first order expressions for the effects of angular misorientations and vibrations on the performance of diffraction grating interferometers.

The most critical angular rotation of the three gratings is about an axis normal to the grating surfaces. If the analyzer grating is rotated by an angle $\Delta\theta$ relative to the other two, Moiré type fringes are produced in a plane parallel to the gratings at A and B. For a diffraction grating period, d, these fringes have a spacing, $x_s = d/\Delta\theta$. To obtain high contrast at one detection point in the output beams, x_s must be comparable to or larger than the detector aperture. Assuming an aperture of 3 cm and a grating spacing of 1 μm, typical of optical scales, $\Delta\theta$ must be less than about 30 microradians. With a grating spacing of 0.2 nm for the silicon lattice, and the same detector aperture, $\Delta\theta$ must be less than 5 nanoradians. For optical scales, these constraints do not pose serious problems for the motion generation systems. However, the constraint for x-ray interferometry is very demanding and thus far has only been maintained for translations of the gratings of less than about 200 μm.[20]

Optical Interferometers: Optical interferometers for displacement metrology may be divided into two main classes.[21,22]

(1) Two beam interference, such as the Michelson interferometer and,
(2) Multiple beam interference, of which the Fabry-Perot etalon is a representative form.

Assuming flat and parallel reflecting surfaces, both configurations produce ring-shaped interference patterns in the image plane of an extended source. For radiation of wavelength λ_o in vacuum, output maxima occur when:

$$m\lambda_o = 2\ \mu L\ \cos\theta \qquad (1)$$

where m, considered a real variable, is the order of interference, L is the effective separation between the test and reference reflecting surfaces, μ is the refractive index of the medium in the optical paths, and θ is the angle between the reflecting surface normals and the direction of incoming radiation.

Non-flatness and non-parallelism of the reflecting surfaces and non-planar wavefronts of the source radiation beam will alter the ring pattern at the interferometer output. However, as long as the reflecting surfaces are sufficiently parallel and flat that multiple fringes are not generated across the effective detector aperture, displacement measurements can be deduced reliably from intensity measurements in the image plane. In practice, mean departures from flatness and parallelism are about $\lambda/10$ to $\lambda/20$ for Michelson interferometer reflecting surfaces and $\lambda/100$ to $\lambda/200$ for Fabry-Perot etalon surfaces.

A change, $\Delta m \lambda_o$, in the optical separation between the reflecting surfaces produces a displacement of the fringe pattern -- in space and, at one location, in time -- through Δm orders. Displacement of the test surface relative to the reference surface may then be determined from Eqn. 1:

$$\Delta L = L_1 - L_2 = \Delta m \frac{\lambda_o}{2\mu \cos\theta} . \qquad (2)$$

Thus, uncertainties in measuring displacements by optical interferometry result from:

(1) Uncertainty in determining the change in order of interference, $\epsilon(\Delta m)$,
(2) Uncertainty in the vacuum wavelength of the radiation, $\epsilon(\lambda_o)$,
(3) Uncertainty in knowledge of the index of refraction of medium along the optical paths, $\epsilon(\mu)$,
(4) Uncertainty in angle between the direction of incident light and normals to the reflecting surfaces, $\epsilon(\theta)$.

As stated earlier in this section, $\epsilon(\lambda_o)/\lambda_o$ is $\geq 10^{-10}$. Recent investigations[23,24,25] have shown that careful measurements of the properties of air along the interferometry paths can reduce $\epsilon(\mu_{air})/\mu_{air}$ to as low as 10^{-7} to 10^{-8} when operating in air.

The angle θ in Eqn. 2 should not be confused with the angle more conventionally associated with a cosine error. Figure 3 is an attempt to illustrate the distinction. Note that in case (1), corresponding to Eqn. 2, uncertainty in θ produces an uncertainty in realizing a metric by an interferometer. If one assumes $\theta = 0$ and infers a displacement $D_i = \Delta m \lambda_o / 2\mu$, the true D will always be greater, i.e., $D = D_i / \cos\theta$. Here, with plane mirrors, the returning beam will shift laterally as the test mirror is moved.

In case (2) the metric is correctly realized but the axis of the interferometer is not parallel to the motion coordinate system, discussed in the third section. Here if one assumes $\beta = 0$ and infers a displacement $D_i = \Delta m \lambda_o / 2\mu$, the true D will, as in case (1), always be greater and be given by

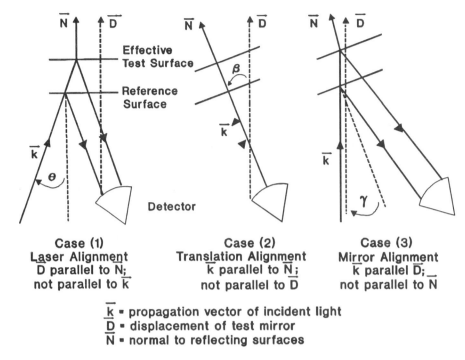

Figure 3. Three forms of misalignment which produce errors in deducing a mirror or reflector displacement from intensity changes at a detector of an optical interferometer.

$D = D_i/\cos\beta$. The returning beam will however not shift as the test mirror is moved. If the test reflector is a retroreflector, the behavior of the two cases is reversed. The returning beam does not shift in case (1) but does shift for case (2).

In case (3), tilting of the reflecting surfaces (parallel to each other) produces a second source of uncertainty in realizing the metric by an interferometer. Here, if one assumes $\gamma = 0$ and infers a displacement $D_i = \Delta m \lambda_o/2\mu$, the true D will always be greater and be given by $D = D_i/\cos^2\gamma$. The returning beam shifts similarly to that of case (1).

The uncertainty, $\epsilon(\theta)$ is generally tractable because of the high sensitivity of both types of interferometry to this angle. But for displacement measurements aimed at achieving subnanometer accuracies, θ must be carefully reduced to as near zero as is feasible. For small angles, Eqn. 2 can be written as,

$$\Delta L = \Delta m \frac{\lambda_0}{2\mu} (1+\theta^2/2) . \qquad (3)$$

To achieve uncertainties in ΔL, $\epsilon(\Delta L)/\Delta L$, of 10^{-9}, then requires that $\theta \leq 5 \times 10^{-5}$. With plane mirror Michelson interferometers, Michelson interferometers based on retroreflectors, and Fabry-Perot interferometers, a nonzero θ is manifested as a shift in the output beam as a function of the displacement, $D = \Delta L$, of the test reflecting surface. The output beam shift, s, is for small angles θ, given by,

$$s = 2 kD\theta \qquad (4)$$

where $k = 1$ for Michelson interferometers and is typically 100 or more for Fabry-Perot interferometers.[26] The effective angle θ for a retroreflector results from its internal angular imperfections; generally it is less than 10^{-5} rad. With plane mirror reflectors, one must use great care in monitoring s to minimize θ during setup and alignment. For example, if $D = 50$ mm, s must be ≤ 5 μm to insure that $\theta < 5 \times 10^{-5}$ rad. Note that there is a comparable error for case (2) when retroreflectors are used.

Beam shifts in Fabry-Perot interferometers have a higher sensitivity to θ. If one could use Fabry-Perots over D's of 50 mm, then the beam shifts should be 100 or more times greater, for a given θ, than would be present with Michelson's. Because of the difficulties in maintaining parallelism of the two mirrors, D is normally less than 200-400 μm in Fabry-Perots. Thus to achieve a 0.1 nm uncertainty component due to a nonzero θ, $\epsilon(\Delta L) = 10^{-10}$ m/4×10^{-4} m $= 2.5 \times 10^{-6}$, requires $\theta \leq 2 \times 10^{-3}$ rad. The beam shift produced by such an angle, for a displacement of 400 μm is by Eqn. 4, only 160 μm. Thus, because D is usually small, alignment with Fabry-Perots is also relatively critical to achieve low uncertainties. There are also other techniques which can be used with Fabry-Perots which bypass the problems discussed here.[26]

Beam shifts may be monitored by projecting cross-hairs or graticules and adjusting for no parallax, or by using quadrature detectors and adjusting to minimize beam shift. Autocollimators may also be used to monitor changes in θ as a function of reflector position. Typical estimates for minimum shifts detectable with cross-hairs and graticules are 0.1 mm; with quadrature detectors, 5-10 μm. Thus, precise set-up with one of these approaches should enable one to reduce s to be within the bounds to keep displacement uncertainties arising from this factor $\leq 10^{-10}$ m.

Heterodyne Interferometry: The principal focus of most interferometric techniques for displacement measurements is to obtain a high degree of resolution in determining the order, m. Of the many techniques which have been explored for measuring the order of interference,[27] heterodyne interferometry and frequency-tracking, Fabry-Perot etalons will be discussed here. By the use of heterodyne interferometry,[28,29] intensity measurements are replaced by measurements of electrical phase, which can be made more accurately and with higher speed. The heterodyne configuration also provides a high degree of immunity to slow (< 10 kHz) variations in laser power in the test or reference arms.

A typical heterodyne interferometer configuration is shown in Fig. 4. Light from a laser source is split into two frequencies, f_1 and f_2, and two polarizations either using Zeeman splitting in the laser cavity or externally with acousto-optical modulators. The time-varying phase or frequency of the signal obtained from the interfering beams from the interferometer is compared with a reference phase obtained from a mixing of the two beams before passing through the interferometer. Difference frequencies in common use range from 200 kHz up to 20 MHz. Circular polarization states from the Zeeman splitting are converted to linear orthogonal polarizations by $\lambda/4$ plates assumed integral to the laser mixer.

In the ideal implementation of the heterodyne interferometer, the two output beams of the laser mixer would be in pure polarization states. Any subsequent mixing of the two beams would then only occur at the detectors resulting from the linear polarizer placed at 45° to the two pure states. For this instance, a signal will be produced at the measurement arm

$$V_m(t) = \cos[2\pi \, \delta f \, t + \phi(t)] \tag{5}$$

detector, where $\delta f = f_1 - f_2$ and $\phi(t)$ is the time varying phase shift resulting from motion of the measurement arm retroreflector relative to the fixed reference arm retroreflector. The detector signal obtained from the beams not passing through the interferometer, provides a reference signal:

$$V_R(t) = \cos[2\pi \, \delta f \, t] . \tag{6}$$

The electrical phase difference between $V_m(t)$ and $V_R(t)$ is a direct measure of the length $L(t)$ of the measurement arm relative to that of the reference arm. $\phi(t)$ and $L(t)$ are simply related by the equation

$$\frac{\phi(t)}{2\pi} = \frac{2\mu L(t)}{\lambda_o} . \tag{7}$$

Another common, and equivalent, way of expressing Eqn. 7 is in terms of Doppler shifting of the frequency f_2 in the measurement arm[30]. If $v = dL(t)/dt$ is the mirror or retro speed, then the angular argument in Eqn. 5 can be expressed as $[2\pi \, \delta f \, t + 2\pi \, f_D t]$, where f_D is the Doppler shift:

$$f_D = 2 \frac{v}{c/\mu} f_2 = \frac{2\mu v}{\lambda_o} . \tag{8}$$

$\phi(t)$ is then the time integral of $f_D(t)$,

$$\phi(t) = 2\pi \int f_D(t) \, dt . \tag{9}$$

Depending on the maximum speed required, commercial interferometer systems can measure the relative phase of two signals in the 1 MHz range to resolutions of about 1 to 2 degrees. One degree resolution in ϕ by Eqn. 7 corresponds to a displacement resolution ΔL of $(1/720) \lambda$, (μ assumed equal 1) or about 1 nm for He-Ne laser wavelengths.

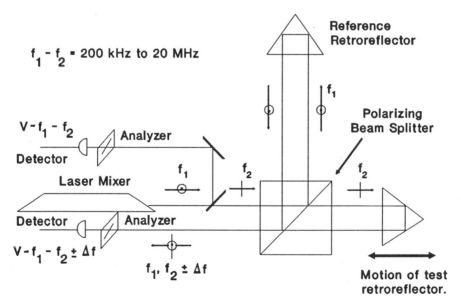

Figure 4. A typical heterodyne interferometer configuration. Polarization states perpendicular to the plane of the page (·) and parallel to the plane of the page (¦) are indicated.

Stray light reflections from various surfaces in the interferometer optical components,[31] undesirable mixing of the two polarization states,[30] and diffraction corrections,[32] ultimately limit the achievement of accuracies comparable to resolutions of less than 1 nm unless great care is taken to minimize or correct for these effects. The optical configurations of Tanaka et al.,[33] and Reinboth and Wilkening,[34] have been demonstrated to achieve accuracies of $\lambda/1000$ or better at He-Ne laser wavelength. Tanaka et al. use the electronic interpolation technique described above with a new optical configuration to bypass polarization mixing. Reinboth and Wilkening use a two-frequency laser interferometer but a measuring device for the combined beams returning from the interferometer determines the phase shifts with mechanically rotating polarizers.

Frequency-Tracking Fabry-Perot Etalons: The final technique to be described for realizing a metric is frequency-tracking, Fabry-Perot Etalons (FPE). This method has been used for some time to determine the long-term dimensional stability of materials.[35] Recently, it has been applied to displacement measurement.[14] A schematic of instrumentation, based on Patterson's system,[35] for a frequency-tracking FPE is shown in Fig. 5. As shown, light from a tuneable laser is transmitted through a FPE. The frequency of the laser is then locked on one of the transmission peaks given by Eqn. 1. The frequency at which lock occurs is then mixed with the output of a reference laser such as an iodine-stabilized laser. Counting of the frequency difference then enables one to directly relate change in cavity length to frequency.

The change in frequency, δf, of a given transmission peak due to a length change δL is given by:[36]

$$\frac{\delta f}{\text{FSR}} = - \frac{\delta L}{\lambda_o /2\mu}, \qquad (10)$$

where FSR is the free spectral range of the Fabry-Perot cavity. FSR is the frequency difference between adjacent modes of the cavity. For a parallel plane-mirror FPE, FSR = $c/(2\mu L)$, where L is the optical length of the cavity. A 1 cm long FPE has a FSR of about 15 GHz which corresponds to a δL of 316 nm with He-Ne laser light. By Eqn. 10, the change in frequency δf per nanometer change in cavity length, for a 1 cm cavity is 47 MHz. Thus, if frequency changes of one or less Hertz could be resolved, a one nanometer displacement could be subdivided into 47 million parts or more. However, the ability to track a peak, amplitude noise in the laser sources, and detector noise all contribute to limiting the interpolation to more like several thousand parts.[35] Resolutions in the picometer range are certainly achievable.

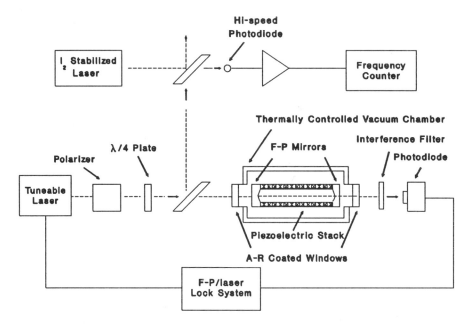

Figure 5. Schematic of apparatus for measuring displacement by tracking the frequency of a transmission peak of a Fabry-Perot etalon with a tuneable laser. The frequency shift is determined by comparison to a stabilized laser.

Range of displacement is a primary problem area to be overcome for more general application of the frequency-tracking FPE wide technique. The gain curve of a typical He-Ne laser is about 800 MHz wide, thus allowing a range of only 16 nm for a 1 cm FPE. By increasing the cavity length, the FSR may be decreased to this value or less. One might then be able to tune across the cavity FSR and measure displacements of many optical wavelengths. The other problem in measuring longer displacements with a FPE is that the angular orientation demands of the translating mechanism are almost as exacting as those for the x-ray interferometer. An out-of-parallelism of $\lambda/200$ with mirror diameters of 25 mm is equivalent to only 10^{-7} radians angular misalignment.

Summary of Means to Realize a Metric: A comparison of the different techniques described in this section is presented in Fig. 6. The structure of the graph follows that by H. Kunzmann.[37] Absolute measurement uncertainty is plotted on the ordinate axis as a function of the magnitude of displacement. Relative measurement uncertainty subdivision of the graph enables convenient entry of the bounds imposed by wavelength uncertainty, knowledge of the silicon lattice constant, etc. A rough estimate of the potential of scales based on STM readout of crystal

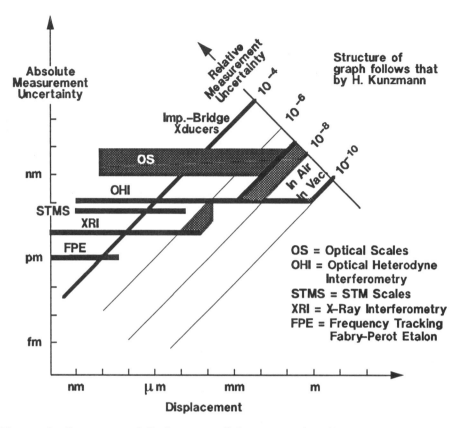

Figure 6. Summary of limits to realizing a metric with the various means shown. Note that all entries shown on the graph are rough estimates based on the arguments presented in the text. Bar widths indicate the spreads in best estimated values.

surfaces in a manner similar to Kawakatsu et al.[38] is shown for comparison with the more conventional methods described earlier. Also shown is an estimate of the performance of impedance-bridge based transducers including capacitance gages, linear-variable-differential transducers, and induction sensors. All have relative measurement uncertainties of about 10^{-4} of their range which for capacitance gages and induction sensors is comparable to the standoff distances. All the impedance-bridge based transducers share the need for calibration; i.e., it is very difficult to predict displacement characteristics accurately from device properties alone. Typically, gain is also strongly dependent on range of operation. However, one should note that the relative measurement uncertainty extends over a very wide range of displacements.

The range of each technique is extended from 10 times its absolute measurement uncertainty out to its bound of relative measurement uncertainty. Optical heterodyne interferometry (OHI) and optical scales are limited first by the uncertainties in the index of refraction of air, and finally for OHI in vacuum by knowledge of the He-Ne wavelength. X-ray interferometry, at displacements of only 10-100 μm, is limited in maintaining its achievable absolute uncertainties by lack of knowledge of the silicon lattice constant. The upper limit on x-ray interferometry was set ≈ 200 μm since this is the maximum range reported for such an instrument. Frequency-tracking FPEs were limited by tuning ranges of available lasers. STM scales were limited in range to about 30 μm since this was the maximum size of atomically-flat crystal surfaces likely to be available.

Table I gives a more quantitative comparison of the characteristics of the different techniques and emphasizes their differing abilities to measure at different rates of change of displacement. The last column estimates the maximum rate of change currently obtainable while still achieving the resolutions drawn in Fig. 6. X-ray interferometry requires the slowest speeds, about 10^{-3} nm/s or less. The main rate limiting aspect of this technique is the relatively slow count rate available with typical x-ray sources used with x-ray interferometers.

Establishing A Coordinate Reference Frame

Once a metric has been realized within the measurement space, it must be related to a coordinate system with a fixed or known origin and fixed or known geometry. This task is: **to physically construct points, lines, and planes which embody, to the highest degree of perfection justifiable, the ideal geometry to provide a reference for the coordinate axes and the motion axes of the metrology system**. In most measuring machines and fabricating machines not designed for achieving the highest accuracies, the reference points, lines, and planes for coordinate axes and motion axes are implicit and integral parts of the machine structure. As an example, in a stacked slideway machine for generating motion of a probe or test object in a plane, one slideway's path of travel would define the x-axis; the path of the slideway stacked on an orthogonal to the first would define the y-axis. Obvious flaws with this approach are: (1) that the straightness of the two axes are determined totally by the quality of the slideways and by the bearings used for the carriages, (2) the two axes are not coincident at one point which can serve as an origin -- conceptual or otherwise, and (3) the paths of motion are highly susceptible to self-induced structural deformations produced by movement of the carriages. The locations of

Table I
Properties of Metric Systems for Nanometrology

	Resolution (nm)	Accuracy (nm)	Range (nm)	Range / Resolution	Max. Rate of Change nm/s
Optical Heterodyne Interferometry	$0.1^{(1)}$	0.1	$5 \times 10^{7(2)}$	5×10^8	$2.5 \times 10^{3(3)}$
X-Ray Interferometry	$5 \times 10^{-3(4)}$	10^{-2}	$2 \times 10^{5(5)}$	4×10^7	$3 \times 10^{-3(6)}$
Optical Scales	$1.0^{(7)}$	5.0	5×10^7	5×10^7	$10^{6(7)}$
Inductive Xducers	$0.25^{(8)}$	---	$10^{4(8)}$	$2.5 \times 10^{5(11)}$	10^4
LVDTs	$0.1^{(9)}$	---	2.5×10^2	$2.5 \times 10^{3(11)}$	$\sim 10^{4(9)}$
Capacitive Xducers	$10^{-3(10)}$	---	$25^{(12)}$	$2.5 \times 10^{4(11)}$	10
STM Scales	$0.05^{(13)}$	0.05	$10^3 - 10^{4(14)}$	$2 \times 10^4 - 2 \times 10^5$	~ 10
Fabry-Perot Etalon (Freq. Tracking)	$10^{-3(15)}$	$10^{-3(15)}$	5	5×10^3	5-10

(1) Ref. 33. (2) The range of optical heterodyne interferometry can be up to at least 10 m without serious degradation in performance except for effects from air turbulence. (3) Entry assumes a phase interpolation rate of only 25 kHz since the degree of interpolation is substantially greater than in most commercially available systems. (4) Ref. 18. (5) Ref. 20. (There is no theoretical limit to the range of x-ray interferometry. The range specified here is based on the maximum values reported in the literature and was determined by the flexure bearing properties. By servo-control of the motion of the moving element one should be able to use x-ray interferometry up to displacements determined only by the size of silicon crystals with sufficiently low lattice distortions - possibly as large as centimeters. Private communication with R.D. Deslattes.) (6) Ref. 18; 20 points per fringe at 5 s per point implies a rate of ~0.3 nm/100 s = 3×10^{-3} nm/s. This speed does not assume use of their proposed three-phase method. (7) Ref. 10. (8) For example, see literature from Kaman Instrumentation Corp. (9) D.J. Whitehouse, D.K. Bowen, D.K. Chetwynd, and S.T. Davies; J. Phys. E: Sci. Instrum. 21, 46-51 (1988). (10) R.V. Jones and J.C.S. Richards, J. Phys. E.: Sci. Instrum. 6, 589 (1973). (11) All three impedance-bridge based transducers have a range/resolution of 10^4-10^5 which can be made to extend over larger ranges than indicated but with a resultant loss in resolution. (12) T.R. Hicks, N.K. Reay, and P.D. Atherton, J. Phys. E.; Sci. Instrum. 17, 49-55 (1984). (13) Ref. 37. (14) Limited by the size of atomically flat and smooth, crystallographically perfect surfaces. (15) Ref. 35.

any reference points integral to the slideways or to their supporting structure relative to a fixed and stable reference frame are also subject to thermally induced deformations of the structure due to internal or external heat sources.

To overcome these problems, there is a long tradition[39] of using a reference frame -- often called a metrology frame -- which is a physical object (or objects) independent of the basic machine structure. Figure 7 illustrates, for a one-dimensional machine, the effects of slideway distortion

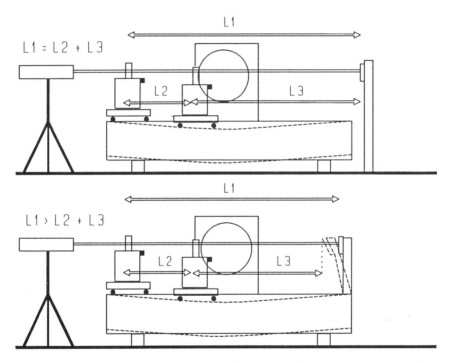

Figure 7. Machine deformations causing errors in measurement. Lower diagram; deformations cause the measured carriage motion, L1 - L3, to be greater than the true carriage motion, L2. Isolating the reference point, top, prevents this interaction.

and the utility of an independent reference point.

One of the clearest illustrations of the reference frame concept is that of the Zeiss M400[3] high-resolution coordinate measuring machine[40] shown schematically in Fig. 8. The machine has a working volume of 400 mm x 600 mm x 200 mm. Three large blocks of Zerodur[41] polished to a high degree of flatness and finish serve as the reference planes for the coordinate reference frame. The design gives measurement results which are not related to the quality of the machine motion axes that move the probe head but to the quality of the reference mirrors. Linear and angular motions of the probe head are measured relative to these planes using heterodyne interferometry. With this design, 10 nm resolution measurements over the working volume have been obtained; slope measurements of surface profiles of high quality mirrors have shown repeatabilities of better than 0.5 microradians rms.

Generally reference points, lines, and planes are provided for only the most critical degrees of freedom of motion elements. Only rarely are all six degrees of freedom of critical moving elements referenced to a coordinate

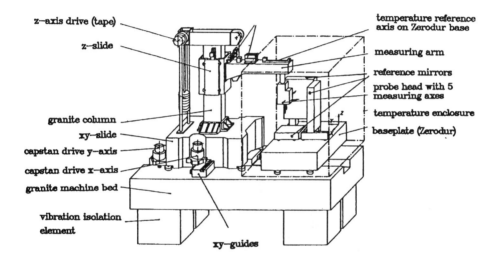

Figure 8. Schematic drawing of Zeiss M400 High-Resolution Coordinate Measuring Machine. Drawing provided by Klaus Beckstette and Kurt Becker and used by permission.

reference system. An example in which this approach was adopted is that of Shimokohbe et al.[42] who designed and built a single linear axis system with all degrees of freedom referenced to a metrology frame system using an array of autocollimators, straightness interferometers, etc.

Guidelines for establishing a coordinate reference frame (CRF) are then:

(1) Realize the CRF in an independent physical object or objects,
(2) Make the CRF as geometrically perfect as possible,
(3) Decouple paths through elements of metrology system as much as possible from force or structural paths of machine, and

(4) Make reference frame from materials of high stability and low thermal expansion coefficient.

Figure 9 summarizes, using the same graphical form as Fig. 6, limits encountered in trying to generate a coordinate reference frame. The graph, strictly interpreted, only applies to Cartesian coordinate frames.

However, similar limitations will apply for other coordinate systems. For a Cartesian coordinate reference frame, one must generate three straight lines, establish three origins or reference points for the metric along each line, and make the lines orthogonal. Generally, the three origins/reference points cannot be made coincident, so one must be concerned with the thermal and temporal stability of the structural material connecting them.

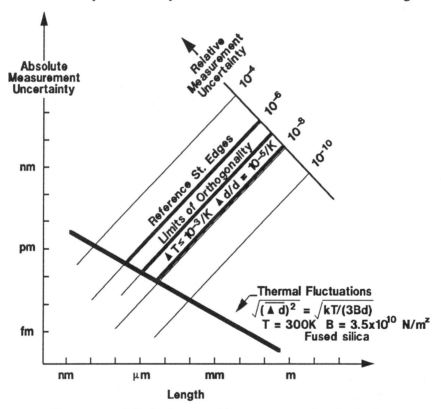

Figure 9. Summary of limits imposed by present technological capabilities and fundamental properties of matter on generating a coordinate reference frame.

In addition to affecting the coincidence of the origins for each axis, thermal changes also affect length measurements through shifts in the origin relative to points on the object to be measured -- even though the reference frame and metric are temperature independent. To measure the spacing between points P_1 and P_2 in Fig. 10, one would measure the displacement $M_1(t_1)$ at time t_1 and the displacement $M_2(t_2)$ at time t_2 in the system metric, relative to the origin 0. But, as will generally be true, if the temperature of the object, machine base, or reference frame changes between t_1 and t_2, the position of P_1 will shift relative to the origin. The effect of the origin shift on measurements of the spacing between P_2 and P_1 is then determined by

$$\begin{aligned} M_2(t_2) - M_2(t_1) &= [L_{RM}(t_2) + L_o(t_2)] - L_{RM}(t_1) \\ &= L_{RM}(t_2) - L_{RM}(t_1) + L_o(t_2) \\ &= CL_{RM}[T(t_2) - T(t_1)] + L_o(t_2) \ , \end{aligned} \quad (11)$$

where L_{RM} is the effective material length between the origin on the reference frame and point P_1 on the object, C is the effective thermal expansion coefficient of this material, and T is the temperature of this part of the structure at the designated times. Thus, to account for an arbitrary location of P_1, effects of thermal expansion of the material to be measured cannot ultimately be eliminated unless $T(t_2) = T(t_1)$. Recall that M_1 and M_2 were assumed independent of temperature.

The practical limit of realizing 20°C in the laboratory-like workspace of a specialized machine, achievable with high performance equipment such as an 8 1/2-digit voltmeter and best laboratory practice, is about 1 mK.[43] With a typical expansion coefficient of 10 ppm/K, one is then limited to an origin stability of about 1 part in 10^8.

Reference lines are best established by straight edges;[23,44,45,46] values given here indicate that, at least up to lengths of 200 mm, mean deviations from straightness of about 1 part in 10^6 or 1 μm/m can be achieved with conventional practice. Deviations from straightness with spatial wavelengths greater than about 0.2 m and less than about 100 μm are generally more difficult to eliminate. For the most demanding applications, i.e., a Fabry-Perot etalon flat, mean deviations can be reduced to as little as 1.5 parts in 10^8 or 0.015 μm/m. Orthogonalities of 0.1 microradian can be achieved by manufacture or by extreme care in adjustment.[46]

Thermal length fluctuations, the dimensional equivalent of Johnson noise, are indicated on the lower part of the graph. These fluctuations are shown because they will be the ultimate limitations in locating origins of the different axes relative to each other and to the test object. They are also given as an indication of the limitations on how well length of an object can be defined. The equation for the length fluctuations is

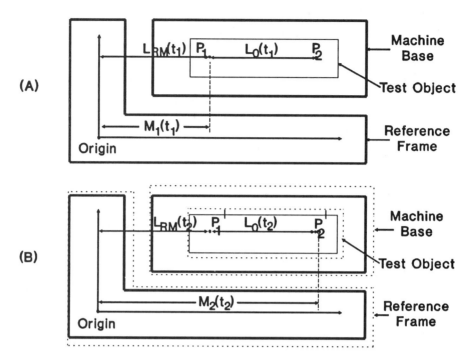

Figure 10. Effects of time varying thermal expansion on measuring the distance between two points on a testpiece. The dotted lines in the lower drawing indicate schematically thermal expansions and shifts in the respective objects.

$$\overline{(\Delta d)^2} = kT/(3Bd) , \qquad (12)$$

where Δd is the fluctuations in the length d of an object, T is the absolute temperature of the object, and B is the bulk modulus of the object material. Equation 12 is derived from the more intuitive equation[47]

$$\overline{(\Delta V)^2} = kT/B \; V \qquad (13)$$

where $(\Delta V)^2$ is the volume fluctuations in an object of volume V. Eqn. 12 is obtained from Eqn. 13 by assuming all three dimensions of a cube fluctuate equally.

Generating Repeatable Motion

Generating motion of a probe relative to a test object is an essential task in dimensional metrology. The distance between two points on an object must ultimately be measured by sensing one point with a probe and then moving the probe or the object such that the probe is positioned over the second point while simultaneously determining the displacement of the moving member. Comparator systems are the only alternative. Imaging microscopes of all forms are comparators since these instruments perform an implicit or explicit comparison of the test object to a calibrating artifact. But, the dimensions of the calibrated artifact will at some point have to be determined by the canonical measuring procedure just described.

At this point in our efforts to conduct dimensional measurements, we have developed a means to create a scale along linear axes with units and subdivisions established in terms of the defined unit of length and constructed physical objects with a high degree of geometrical perfection, straightness, orthogonality, etc. against which these axes can be referenced. Along with these two Cartesian coordinate systems we must now create a third one -- the motion coordinate system. As an illustration of realizing these three coordinate systems, those for the two major axes of the NIST Molecular Measuring Machine are shown in Fig. 11.

Ideally, all three coordinate systems would have coincident origins, perfectly straight and stable axes, all corresponding axes parallel, and orthogonal axes orthogonal. Certainly for the regimes of concern in this paper, the metric axis defined by a light path is straight. The previous section discussed the problem of constructing references for the coordinate axes. Generating straight line motion is a problem one step more complex than making reference axes. In most instances one must construct straight bearing surfaces and mount and drive a carriage on these surfaces such that it moves in a straight line or more generally with one degree of freedom.

To the extent that the moving elements may be considered as rigid bodies, they each have six degrees of freedom. Thus, during motion all six will be manifest unless undesired degrees of freedom are constrained and constrained in a very proper manner. (Many readers will find J.C. Maxwell's discussion[48] of kinematic constraints of interest.) However, in even the best of mechanical designs there will be deviations from the ideal straight line motion with no changes in the other five degrees of freedom. In recognition of this limitation, current practice[39,49,50,51] in precision instrument design is to **design for maximum repeatability of motion of a metrology system's moving elements.** Deviations arising from non-perfect motion in the desired degree-of-freedom are then measured to the level of their repeatability. Accuracy in the overall metrology system is achieved by correcting for the propagation of these motion errors into the system using look-up tables calculated from known machine geometry.[52,53]

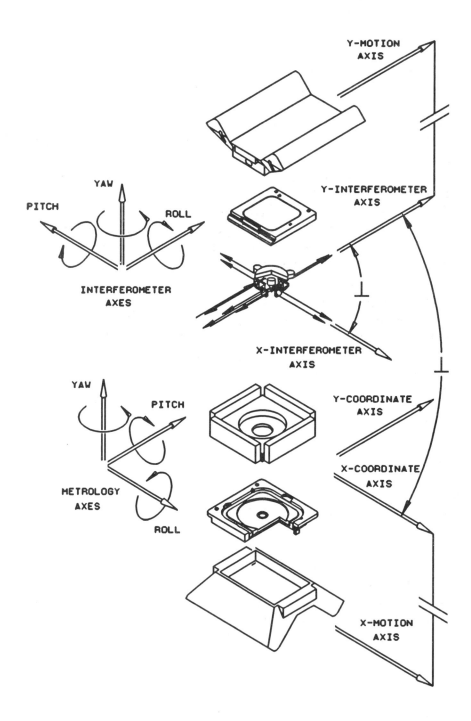

Figure 11. Schematic of the three coordinate systems of the metrology system for the NIST Molecular Measuring Machine's two major axes.

Conceptually, the only uncertainty produced in carrying out this task should be that due to lack of repeatability of the constrained motion. However, lack of repeatability gives rise to two associated uncertainties, those from cosine errors and Abbé offset errors, which depend on machine adjustment and machine design. As discussed in the second section on realizing a metric, (case (2) in Fig. 3) any angle between the metric axis (which for an optical interferometer is the direction of light propagation and normals to reflectors) and the direction of motion of the test reflector produces a cosine error in the measurement. While not essential, the test element of all metric systems with which the author is familiar is always attached to the moving carriage. Thus motion of the metric test element follows that experienced by the carriage. (Alternatives such as an additional motion system for the metric test element appear at first thought to only compound the problem.) Any imperfection in the motion axis or in the error-correcting scheme, or any lack of repeatability in the motion, generates a cosine uncertainty in knowledge of the displacement of the metric test element. As shown by Eqn. 3 and the discussion just preceding it, uncertainty due to this effect is

$$\epsilon(D) = D - D_i = D \overline{(\delta\beta)^2} / 2 , \qquad (14)$$

where D is the magnitude of the test element displacement, D_i is the value of D inferred, assuming no cosine uncertainty, and $\delta\beta$ is the mean angular uncertainty in β. A mean angle, β, between k and D, in the terminology of Fig. 3, case (2) is assumed to be accounted for in the error correction process.

The second source of uncertainty produced by an imperfect motion axis or lack of repeatability is due to any offset between the metric axis and the functional point whose displacement is to measured. If one is measuring the motion of the testpiece, the functional point is the feature of concern on the testpiece. If motion of the probe is to be measured, the functional point is the sensing point on the probe. Abbé first discussed the advantages of having the path of the functional point coincident with the metric axis. Thus this design concept is usually referred to as the Abbé principle.[54]

The importance of the principle is readily seen as follows. Let A be the offset distance between the metric axis and the path of the functional point (Fig. 12) and α the **change** in angle of the carriage being used to translate the functional point between two positions in the plane defined by the two paths (those of the metric test element and the functional point). An error between the true displacement, D_f, of the functional point and that measured, D_m, then results,

$$D_f - D_m = A \tan \alpha \quad . \tag{15}$$

Arguments similar to those presented for cosine error can be made that if the motion axis of the relevant carriage is not free of angular motion then angular motion data can be used to correct for the error given by Eqn. 15. Lack of repeatability in the angular motion data would then result in an uncertainty in α and a resultant uncertainty in D_m.

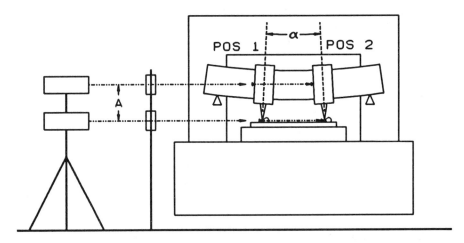

Figure 12. Causing measurement errors with an Abbe offset, A. With metric axis in upper position, rotation of the probe carriage by an angle, α, causes error in Eq. 15. Putting the metric axis in the lower position minimizes such errors.

The task of generating repeatable motion is tightly coupled with the task of establishing a coordinate reference frame. All, or at least the critical, degrees of freedom of each moving element will be measured relative to the coordinate reference frame. Thus the motion repeatability or our knowledge of it will be dependent on the perfection of the reference frame. In the most sophisticated of machines, the critical degrees of freedom will be servo-controlled from transducers using the reference frame as a platform from which angular motion errors are measured and controlled.[39,40]

Linear deviations from the ideal motion axes are termed straightness errors. Such motion errors are treated in a fashion similar to that just described for the angular errors. In the rigid body descriptions of these motions, straightness motion errors would only occur if the center of mass of the moving element departed from a straight line. Practically, the measurement of center of mass motion is generally difficult if not impossible. Usual conventions for straightness either measure the deviations of a fixed probe against a straightedge mounted to the moving carriage or vice versa. Such a measurement makes the results dependent on where the measurement path is located on the moving carriage. These measurements, however, do have the merit that they are indicative of the path errors produced as a carriage moves a testpiece or a probe relative to each other.

Repeatability of bearing systems for constraining carriage motion to one desired degree of freedom at nanometer levels of motion noise has received very little careful study. Flexures, sliding bearings, air or liquid bearings, magnetic bearings, and rolling element bearings are bearing systems that seem most appropriate for the tasks of nanometrology. Flexures and sliding bearings are the ones to be briefly described here, both because they appear the most promising and because they are the ones for which reliable performance data were available.

Configuration of the constraints is a critical issue for repeatable motion generation. However, it is a subject that would require a major digression. Teague and Evans[55] give a review of this subject; others[56,57,58,59] also discuss the importance of kinematic constraints and the contrasting philosophy of elastic averaging.

For sliding bearings, the design of the U.K. National Physical Laboratory Nanosurf instrument[60] has produced the most repeatable motion reported. Two successive traces, in the mode described for straightness measurements, were taken with a high-resolution stylus instrument traversing across a highly polished glass surface. For traverse lengths up to 40 mm, differences between successive traces were shown to be less than 1.5 nm rms. Subsequent improvements in decoupling the drive system further from influencing carriage motion have reduced differences to about 0.3 nm. The Nanosurf consists of a simple Vee slideway with the carriage kinematically mounted on the slideway with five polymer pads. The slideway and carriage are made of Zerodur. Slideway surfaces are polished to a high quality optical finish; rms roughness was less than 1 nm.

Flexure hinges as bearings were pioneered by R.V. Jones.[56] Since his early work they have been widely used in a variety of ultra-high precision instruments, including x-ray interferometers (see references in section on Realizing a Metric) and Fabry-Perot interferometers. The minimum repeatability achieved by flexures may be deduced from the angular requirements for these interferometers. Throughout the motion of the moving blade of an x-ray interferometer angular orientation must be maintained to better than a nanoradian. Repeatability of motion must therefore be at this level or better. For a displacement of 200 μm this corresponds to a level of repeatability of 10 pm or less.

Figure 13 summarizes the effects of angular and linear motion repeatability on our ability to generate and measure translation. The graph stresses the extreme importance of Abbé offset errors. With only a 1 mm offset and angular changes between measurement points of 1 microradian, we have

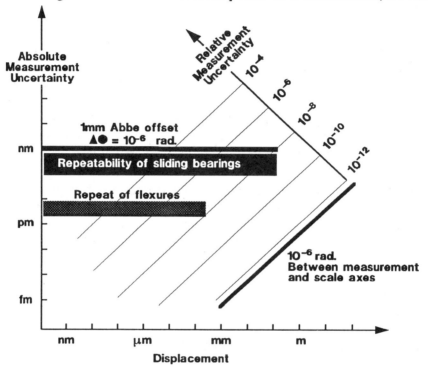

Figure 13. Summary of limits to generating and measuring translation resulting from current technology, repeatability of motion constraint systems, or resulting from improper design and adjustment.

reached an uncertainty of 1 nanometer being produced by what is typically thought of as an obscure component of an overall measurement system. Cosine errors, resulting from a similar magnitude angular error are insignificant. However, for small displacements, reducing the cosine errors to this level is extremely difficult. More typical for centimeter displacements would be angles of 10^{-4} - 10^{-3}. Here cosine errors could be comparable to Abbé offset errors.

Linking The Testpiece To The Coordinate Reference Frame/Metric By A Probe

A probe is any means by which features, edges, and surfaces are located or created. Examples would be any of the scanning probe tips discussed in this volume as well as the on-axis rays from a conventional confocal optical scanning microscope, and the electron beam in a scanning electron microscope. The task of dimensional metrology to be discussed in this section is **to locate in space a point on a testpiece and to link this point to the coordinate-reference frame/metric**. Primary concerns in performing this linking of the coordinate-reference frame and the testpiece are: (1) the dimensions of the interaction between the probe and the testpiece, (2) the physical size and geometry of the probe, (3) the forces and energies of the probe interaction, and (4) the offset between the sense or functional point and the metric axis of the metrology system. Griffith et al.[61] give a thorough analysis of these concerns as they apply to the measurement of feature dimensions with scanning probe microscopes. This section examines the impact of these concerns on positioning errors and gives an elementary comparison of scanning probe microscopes with optical and electron microscopes.

There are three basic types of length measurements of artifacts (Fig. 14). The first and simplest is called a line spacing measurement. Here, one attempts to measure the distance between corresponding points of two similar features. To the extent that the features are similar, the probe response function should be equal at the two points and the measurement reduces to determining the corresponding value of the probe response and the displacement of the probe between the two points. The properties of the probe-specimen interaction are relatively unimportant for this type of measurement. Probe-specimen interaction properties only enter in determining the sensitivity of the probe to different object dimensions. Even if there are back forces of the interaction on the probe resulting in bending and distortion of the probe, symmetry will null any errors, assuming the interactions are the same at both features.

Line width is a more troublesome quantity to measure. Here the full role of the probe-testpiece feature interaction comes into play. If modeling of the interaction has introduced some unknown shift, a, between the defined sensing point of the probe and the true edge of the testpiece feature, then an error of 2a in determining the feature width will be produced. If there are significant back forces on the probe, they will be asymmetric and thus contribute to the resultant measurement uncertainty. For the complete spectrum of probe types, from mechanical probes through electron microscopes to STMs and AFMs, these asymmetries and strongly model-dependent errors must be carefully evaluated. Representative studies of these effects for different methods are Song and Vorburger for high resolution stylus techniques,[62] Nyyssonnen[63] and Rosenfield[64] for optical methods, Postek and Joy[65] for electron microscopy, and den Boef for AFM.[66] Teague et al.[67] and Stedman[68] give comparisons of many of the methods in terms of a height sensitivity-wavelength resolution space. For measurements of more complex shapes than those illustrated in Fig. 14 such as a tall thin feature, the overall geometry of the probe will influence the measurements. (See discussion by Griffith et al.[58])

Extension measurements determine the distance between two points on opposing faces of an object or, in a more restricted sense, the distance between opposing surfaces of an object. Such measurements have an additional source of uncertainty since lack of knowledge of the effective radius of the probe must be added to the others discussed for line spacing and line width. For the calipering configuration shown in Fig. 14, the probe diameter uncertainty is eliminated. However, an uncertainty in the zero of measurement is set by a probe-probe interaction distance, d_{PP}, and by how this distance differs from that of the interaction distance between the probe and testpiece surface, d_{PT}.

Explicitly, if δd_{PP} and δd_{PT} are the uncertainties in these distances, then the net uncertainty from these probe effects is

$$\delta E_P = \delta d_{PP} + 2\delta d_{PT} . \quad (16)$$

At the accuracies being considered in this paper, there are major problems with realizing the ideal measurement just described. Probe axes may not be aligned, the probe-probe contact points may not be the same probe points as those contacting the testpiece in the second step of measurement, and finally, forces of interaction may stretch or compress the probes by amounts comparable to the distances in Eqn. 16.

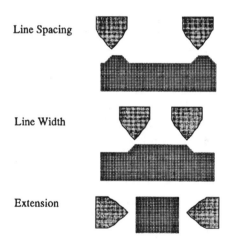

Figure 14. Three Basic Length Measurements of Artifacts.

Table II gives some quantitative estimates of these uncertainties for optical measurements, scanning electron microscopes, and scanning tunneling microscopes for the three types of length measurements just discussed.

The estimates for all three methods assume a response function whose width is comparable to the interaction length of the method. Without more detailed modeling, 1/e and such widths seemed inappropriate. The assumption -- a typical one -- is then made that one can locate the centroid or a defined point on the response function to 1/10th of this response-function width.

For optical methods, a confocal laser scanning microscope was taken as representative. At a light wavelength of 325 nm, the full-width-half-maximum response width is about one wavelength, with a resultant resolution of about 32 nm. For an electron microscope the interaction length is primarily determined by the electron range (ER) of electrons incident and generated by the incident beam in the testpiece. As a best value, a low-voltage SEM source was assumed. The ER for a 1 kV beam of electrons onto silicon is 30 nm, with a resultant resolution of 3 nm.[69]

Table II
Probe Limitations in Nanometrology

	Uncertainties (nm) for Measurements		
	Line Spacing	Line Width	Extension
STM Diameter, d, of probe-specimen interaction = 0.2 nm; Bending, b, of probe shaft = 5×10^{-2} nm; Uncertainty, U, in locat. surf. = 0.1 nm	$\sqrt{2} \times 0.1$ d = 1.4×10^{-2}	2×0.1 d + 2b + ? = 0.15-0.2	2b + 2U + tip dia. uncer. = 1-2
SEM 1 kV onto features on silicon; 30 nm electron range, ER	$\sqrt{2} \times 0.1$ ER = 4.2	2×0.1 ER to $2 \times$ ER = 6-60	---
OPTICAL Wavelength = 325 nm FWHM = 325 nm	$\sqrt{2} \times 0.1$ FWHM = 45	(0.2-2)FWHM = 65-650	$2 \Delta\phi$ = 10-20

For a scanning tunneling microscope, a probe-specimen interaction length of 0.2 nm -- about the nearest-neighbor distance in gold -- was assumed, with a resultant resolution of 0.02 nm. The uncertainty in line spacing measurements was calculated as $\sqrt{2}$ times the stated resolutions, assuming no correlation between the uncertainties at each feature.

For linewidth measurements, one has the same kind of resolution limitations due to probe-specimen interactions. However, for these measurements the asymmetry causes a model-dependent shift, as explained earlier, in addition to the resolution uncertainty. Modeling offset uncertainties are not explicitly indicated in the entries for linewidth measurements. They are assumed to be included within the spread of values shown. Another factor that enters for an STM probe is bending of the probe shaft. A value of 0.05 nm has been estimated for this effect. The analysis given by Griffith[61] shows that for many probe shapes and tip-specimen interactions this is a reasonable value.

In optical and scanning electron microscope measurements, the edge response function can be a strong function of the character of the edge, the types of materials involved, the dimensions of the line, and its thickness-to-width ratio. To account for these variations, the spread in expected uncertainties has been broadened over a range from 0.2 to 20 times the response width to cover values commonly reported.

Extension measurements, as defined here, are not usually attempted with an SEM. In the caliper configuration for conducting these measurements, the entries for the STM assume that a stretching comparable to the bending occurs. An additional uncertainty arises from lack of knowledge of the absolute location of the testpiece surface because of unknown barrier heights and barrier height effects. A similar uncertainty exists for optical calipering measurements in the form of unknown phase changes upon reflection. Uncertainty in the zero exists for optical interferometry because of a lack of knowledge of the spacing of the two reference surfaces with no testpiece present.

Figure 15 is the graphical representation of the data in Table II, to complete the comparisons with the other three summary graphs. The lower range of applicability for the different methods was set at 10 times the respective method's estimated absolute measurement uncertainty. STM extension measurements would, for reasonable uncertainties, then have a lower limit of something like 10 to 20 nanometers. In all but the most highly characterized artifacts, one should be extremely careful in estimating linewidths from electron and optical microscope measurements for widths of fractional micrometer and less.

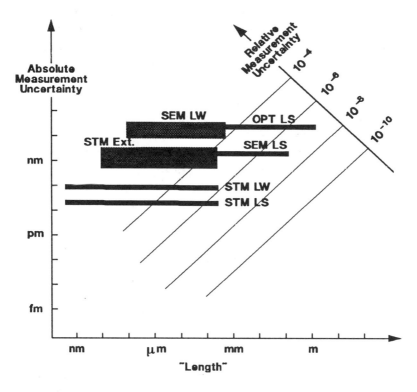

Figure 15. Comparison of probe limitations for SEM (scanning electron microscopy), STM (scanning tunneling microscopy), and OPT (optical methods) in performing LS (line spacing), LW (linewidth), and Ext. (extension) measurements.

Summary

Table III contains the promised estimates of uncertainty for performing measurements on artifacts up to 100 mm in size and for positioning over areas of 100 mm x 100 mm. Estimating overall error budgets for an arbitrary configuration machine is impossible. Even for highly characterized and specified machines the process is complicated because of complex interactions between error sources -- such as those outlined in Figs. 6, 9, 13, and 15 -- and machine properties to produce resultant errors in a measurement. Methods for a well characterized machine are outlined by Donaldson.[45] Entries in Table III are simply sums of values deduced from the cited figures and their associated texts and represent lower bounds on what appear to be possible with current technology. The lower values are sums of best values thought possible for the four tasks. These entries are also for the simplest of measurements, i.e., distance between two points. Contour measurements or measurements on a complicated three-dimensional or two-dimensional part would make estimation of errors more difficult and would certainly increase the entries significantly. The larger entries for each type of measurement depend on the magnitude of the displacements involved and vary with the quantity to be measured.

The uncertainties in realizing the metric, establishing a coordinate reference frame, and generating and measuring translation are the same for all four types of measurements. These uncertainties will determine capabilities for the precise fabrication of components at nanometer scales and tolerances -- regardless of probe type -- since they determine our ability to position and measure the motion of the probe. By the arguments presented in this paper, it will be very difficult to achieve 1 nm positioning accuracy over a reasonable area (cm^2); typical uncertainties could be as bad as 15 nm. Once probe-testpiece interactions are considered, things degrade even more.

Table III

**Overall Uncertainties for Length Measurements and Positioning Involving Displacements Up to 140 mm.
All entries in nanometers.**

	Position to P(xyz)	Line Spac. P1 - P2	Line Width B1 - B2	Extension S1 - S2
Realize Metric	0.1	0.1	0.1	0.1
Establish Coord. Ref.	1.4 - 14	1.4 - 14	1.4 - 14	1.4 - 14
Gen. & Msr. Mot.	0.5 - 5	0.5 - 5	0.5 - 5	0.5 - 5
Probe Linkage	---	0.1 - 0.5	0.1 - 2	1 - 5
Totals	2.0 - 19	2.1 - 20	2.1 - 21	3.0 - 24

There are two clear conclusions from the study presented here. First, we are definitely on target to be able to perform positioning and measuring to nanometer accuracies as far as the tasks of realizing a metric and linking the testpiece to the metrology system, in large part because of the new scanning probes. Second, the tasks for which there are significant problems are in our ability to establish geometrically perfect and stable reference frames and our ability to generate repeatable motion at the nanometer scale.

How the Availability of Large-area, Atomically-flat Surfaces Could Improve Our Ability to Perform the Four Tasks of Metrology.

In addition to providing the capability to image individual atoms and molecules, scanning tunneling microscopy has provided new access to the highly ordered and near-perfect geometry of crystal surfaces. Such access, if the geometry proves as perfect as anticipated, offers much potential for very high accuracy straightness and orthogonality references for use in both positioning and measuring machines. An orthogonality reference of 0.2 nanoradians (a two to three order of magnitude improvement over currently available references) would be possible assuming features may be located to 20 pm over lengths of 100 mm. Equivalent straightness reference accuracies would also be possible and be verifiable over all relevant spatial wavelengths. In combination with the best of conventional techniques for preparing surfaces, scanning probe microscopes acting both as manipulators and sensors have much potential for generating and characterizing large-area, atomically-flat surfaces. Note that atomically-flat and atomically-smooth surfaces are very different requirements. These surfaces could also serve as the reference planes and possibly as substrates for monomolecular lubricant films for sliding bearings. Realization of these ideal specimens will require a thorough determination of the total material properties of the substrate, internal strains, etc.

Dichalcogenide (WSe_2, $NbSe_2$) crystal surfaces are among the most promising materials for these applications. Initial studies of WSe_2 with STM have found that atomically smooth surfaces over large areas, possibly as large as 1 square mm, can be prepared conveniently.[70] These materials appear to be particularly amenable to time-stable atomic site modification of their surfaces, an attribute needed for the preparation of large-area, atomically-smooth surfaces. Preparing specimens with atomic-scale flatness over large areas will require that these smooth surfaces be obtained on strain-free substrates sufficiently thick to deform negligibly under self-induced gravitational loads.

Conclusions

Nanometer accuracy metrology over the ranges discussed in this paper remains a very challenging and relatively unexplored field. It is also a field that will be extremely important for future technologies which are dependent on fabricating objects with these tolerances or of this scale.

Current developments in applying both x-ray interferometry and true angstrom resolution heterodyne interferometry (at reasonable speeds) combined with high-resolution probes provide an excellent basis for developing our ability to do nanometer accuracy metrology. Creative use of proximal probes, both as fabrication and control tools, appear to be a promising means for overcoming the problems of establishing high-accuracy reference frames and generating repeatable motion.

Acknowledgements

The ideas presented here have been shaped by many discussions with my colleagues at NIST and in the precision engineering community. I particularly want to thank Chris Evans, Dennis Swyt, Russell Young, Robert Hocken, Fredric Scire, John Kramar, John Villarrubia, William Penzes, Jay Jun, John Simpson, Tyler Estler, Ted Vorburger, Jack Stone, and Robert Larrabee for their time and their willingness to share ideas. To Christie Poffenbarger, Vita Gagne, Fredric Scire, and Bessmarie Young, I express my many thanks for their highly professional work in preparing the manuscript and the figures.

References

1. E.C. Teague, "The National Institute of Standards and Technology Molecular Measuring Machine Project: Metrology and Precision Engineering Design," *J. Vac. Sci. Technol.* B7, 1898-1902 (1989).
2. L. Curran, "Will It Be Optics or X-rays for Circa-2000 Memories?" *Electronics*, 48-50 (Oct. 91).
3. Certain commercial products are identified in this paper in order to specify experimental procedures adequately. Such identification does not imply recommendation or endorsement by the National Institute of Standards and Technology, nor does it imply that these products are the best available for the purpose.
4. G.F. Strouse and B.W. Mangum, "NIST Measurement Assurance of SPRT Calibrations on the ITS-90: A Quantitative Approach," *Proceedings of 1993 Measurement Science Conference*, Anaheim, California, January 21-22, 1993 and personal communication with B.W. Mangum.
5. No Author Given, "Documents Concerning the New Definition of the Metre," *Metrologia* 19, 163-177 (1984).
6. L.L. Lewis, F.L. Walls, and D.A. Howe, "Prospects for Cesium Primary Standards at the National Bureau of Standards," *Precision Measurement and Fundamental Constants, Proc. of 2nd Int. Conf.*, Gaithersburg, MD, June 8-12, 1981, Eds. B.N. Taylor and W.D. Phillips, Natl. Bur. Stand. Spec. Publ. 617 (1984) pp. 25-29 (U.S. Government Printing Office, Washington, D.C. 20402).

7. W.R.C. Rowley, "Laser Wavelength Measurements and Standards for the Determination of Length," ibid. 6 pp. 57-64.

8. D.A. Jennings, C.R. Pollock, F.R. Petersen, R.E. Drullinger, K.M. Evenson, J.S. Wells, J.L. Hall, and H.P. Layer, "Direct Frequency Measurement of the I_2-Stabilized He-Ne 473 THz (633-nm)Laser," *Optics Lett.* 8, 136-138 (1983), and K.M. Evenson,"Frequency Measurements From the Microwave to the Visible, the Speed of Light, and the Redefinition of the Meter," *Quantum Metrology and Fundamental Physical Constants*, Eds. P.H. Cutler and A.A. Lucas (Plenum Press, New York, NY, 1983), pp. 181-207.

9. G. Hildebrandt, "X-ray Wave Fields in Perfect and Nearly Perfect Crystals - Theoretical Background and Recent Applications," *J. Phys. E: Sci. Instrum.* 15, 1140-1155 (1982).

10. M. Hercher and G. Wyntjes, "Fine Measurements From Coarse Scales," *Proc. 5th Int. Conf. on Prec. Eng.*, Monterey, CA 1989, American Society for Precision Engineering, Raleigh, NC. A simpler optical scale configuration involving only two gratings to generate Moiré fringes is also used. See F.T. Farago, *Handbook of Dimensional Measurement* (Industrial Press, New York, NY, 1968), p. 293 and A. Ernst, *Digital Linear and Angular Metrology; Position Feedback Systems for Machines and Devices*, (Verlag Moderne Industrie; AG&Co; D-8910 Landsberg/Lech., Box 1751; Germany, 1990).

11. U. Bonse and M. Hart, "An X-ray Interferometer," *Appl. Phys. Lett.* 6, 155-156 (1965).

12. M. Hart, "An Angstrom Ruler," *J. Phys. D: Appl. Phys.* 1, 1405-1408 (1968).

13. R.D. Deslattes and A. Henins, "X-ray to Visible Wavelength Ratios," *Phys. Rev. Lett.* 31, 972-975 (1973).

14. P. Becker, K. Dorenwendt, G. Ebeling, R. Lauer, W. Lucas, R. Probst, H.-J. Rademacher, G. Reim, P. Seyfried, and H. Siegert, "Absolute Measurement of the (220) Lattice Plane Spacing in a Silicon Crystal," *Phys. Rev. Letter.* 46, 1540-1543 (1981).

15. M. Hart, "Ten Years of X-ray Interferometry," *Proc. R. Soc. Lond.* A346, 1-22 (1975).

16. M. Tanaka, K. Nakayama, and K. Kuroda, "Experiment on the Absolute Measurement of a Silicon Lattice Spacing at the NRLM," *IEEE Trans. Instrum. Meas.* 38, 206-209 (1989).

17. R.D. Deslattes and E.G. Kessler, Jr., "Status of a Silicon Lattice Measurement and Dissemination Exercise," *Proceedings of CPEM-90 in IEEE Trans. Instrum. Meas.* IM-40, 92-97 (April 1991).

18. D.R. Schwarzenberger, D.G. Chetwynd, and D.K. Bowen, "Phase Measurement X-ray Interferometry," *X-ray Sci. and Tech.* 1, 134-142 (1989).

19. H. Mendlowitz and J.A. Simpson, "On the Theory of Diffraction Grating Interferometers," *J. Opt. Soc.* 52, 520-524 (1962).
20. P. Becker, P. Seyfried, and H. Siegert, "Translation Stage for a Scanning X-ray Optical Interferometer," *Rev. Sci. Instrum.* 58, 207-211 (1987).
21. W.R.C. Rowley, ibid. 7.
22. M. Born and E. Wolf, *Principles of Optics* (Pergamon Press, Elmsford, New York 10523, 1975).
23. W.T. Estler, "High-Accuracy Displacement Interferometry in Air," *Appl. Opt.* 24, 808-815 (1985).
24. K.P. Birch and M.J. Downs, "The Results of a Comparison Between Calculated and Measured Values of the Refractive Index of Air," *J. of Phys. E: Sci. Instrum.* 21, 694-695 (1988).
25. G. Wilkening, *VDI-Berichte 659* (VDI-Verlag, Dusseldorf, Germany 1987).
26. J.M. Vaughn, *The Fabry-Perot Interferometer; History, Theory, Practice, and Applications* (Adam Hilger, IOP Publishing Ltd., Bristol BS1 6NX, England, 1989) Appendix 4.
27. H. Matsumoto, "Recent Interferometric Measurements Using Stabilized Lasers," *Precision Engineering* 6, 87-94 (1984).
28. J. A. Dahlquist, D.G. Peterson, and W. Culshaw, "Zeeman Laser Interferometer," *Appl. Phys. Lett.* 9, 181-183 (1966).
29. J.N. Dukes and G.B. Gordon, "A Two-Hundred Foot Yardstick with Graduations Every Microinch," *Hewlett Packard J.* 21(12), 2-7 (1970).
30. N. Bobroff, "Residual Errors in Laser Interferometry From Air Turbulence and Nonlinearity," *Appl. Opt.* 26, 2676-2682 (1987).
31. W.R.C. Rowley, "Signal Strength in Two-Beam Interferometers with Laser Illumination," *Opt. Acta.* 16, 159-168 (1969) and M.J. Downs, "A Proposed Design for an Optical Interferometer with Sub-nanometer Resolution," *Nanotech.* 1, 27-29 (1990).
32. J.P. Monchalin, M.J. Kelly, J.E. Thomas, N.A. Kurnit, A. Szoke, F. Zernike, P.H. Lee, and A. Javan, "Accurate Laser Wavelength Measurement with a Precision Two-Beam Scanning Michelson Interferometer," *Appl. Opt.* 20, 746-757 (1981).
33. M. Tanaka, Y. Yamagami, and K. Nakayami, "Linear Interpolation of Periodic Error in a Heterodyne Laser Interferometer at Subnanometer Levels," *IEEE Trans. Instrum. & Meas.* 38, 552-554 (1989).
34. F. Reinboth and G. Wilkening, "Optische Phasenschieber fur Zweifrequenz-Laser-Interferometrie," *PTB - Mitteilungert* 93, 168-174 (1983) and W. Hou and G. Wilkening, "Investigation and Compensation of the Non-Linearity of Heterodyne Interferometers," *Progress In Precision Engineering, Proc. 6th Int. Prec. Eng. Seminar (IPES 6)*, Eds. P. Seyfried, H. Kunzmann, P. McKeown, and M.Weck (Springer-Verlag, Berlin, Heidelberg 1991).

35. S.R. Patterson, "Interferometric Measurement of the Dimensional Stability of Superinvar," *Lawrence Livermore National Laboratory Report #UCRL-53787*, Feb. 19, 1988 (NTIS, U.S. Department of Commerce, 5285 Port Royal Road, Springfield, VA 22161).

36. A. Yariv, *Introduction to Optical Electronics* (Holt, Rhinehart and Winston, New York, 1971) pp. 60-66.

37. H. Kunzmann, "Today's Limits of Accuracy in Dimensional Metrology," *Proc. Of 2nd IMEKO TC14 Int. Symposium on Metrology for Quality Control in Production*, Beijing, China, May 9-12, 1989 (International Academic Publishers, A Pergamon - CNPIEC Joint Venture, Beijing, China, 1989).

38. H. Kawakatsu, Y. Hoshi, T. Higuchi, "Crystalline Lattice for Metrological Applications and Positioning Control by a Dual Tunneling-Unit Scanning Tunneling Microscope," *J. Vac. Sci. Tech.* (March/April 1991).

39. C. Evans, *Precision Engineering: An Evolutionary View*, (Cranfield Press, Bedford, MK43 OAL, U.K., 1989).

40. K. Becker and E. Heynacher, "M400 - A Coordinate Measuring Machine with 10 nm Resolution," *Proc. SPIE* 802, 209-216 (1987).

41. Trademark of Schott Glass Company.

42. A. Shimokohbe, H. Aoyama, and I. Watanabe, "A High Precision Straight-Motion System," *Precision Engineering* 8, 151-156 (1986).

43. D.A. Swyt, "Uncertainties in Dimensional Measurements at Nonstandard Temperatures," *J. of Res. NIST*, In Press, 1993.

44. J.B. Bryan, "The Lawrence Livermore National Laboratory 84-inch Diamond Turning Machine," *Precision Engineering* 1, 13-18 (1979).

45. R.R. Donaldson, "Error Budgets," in *Technology of Machine Tools Report # UCRL-52960-5* (Lawrence Livermore National Laboratory, Livermore, California 94550, 1980) Volume 5, Chapt. 9.14.

46. W.R. Moore, *Foundations of Mechanical Accuracy* (The Moore Special Tool Co. Bridgeport, Connecticut 1970) pp. 21-33.

47. F. Reif, *Fundamentals of Statistical and Thermal Physics* (McGraw-Hill Book Company, New York 1965) p. 300.

48. J.C. Maxwell, "General Considerations Concerning Scientific Apparatus" in *The Scientific Papers of James Clerk Maxwell*, Ed. W.D. Niven (Dover Press, 1890). Reprint from *The Handbook to the Special Loan Collection of Scientific Apparatus*, 1876.

49. E.C. Teague and C. Evans, *Patterns for Precision Instrument Design*, Tutorial Notes (American Society for Precision Engineering, Raleigh, North Carolina, 1988).

50. E.G. Loewen, "Metrology Problems in General Engineering: A Comparison with Precision Engineering," *Annals CIRP* 29, 451-453 (1980). Also, E.G. Loewen, "Perceived Limitations on Future Advances in Ultraprecision Machining," *Annals CIRP* 33, 413-414 (1984).

51. P.A. McKeown, "High Precision Manufacturing and the British Economy," *Proc. Institution Mech. Eng.* 200, 1-19 (1986).
52. R.J. Hocken, J. Simpson, B. Borchardt, J. Lazar, C. Reeve, and P. Stein, "3-Dimensional Metrology," *Annals CIRP* 26, 12-15 (1977).
53. M.A. Donmez, C.R. Liu, and M.M. Barash, "A Generalized Mathematical Model For Machine Tool Errors," in *Modeling, Sensing, and Control of Manufacturing Processes - PED Vol. 23/DSC Vol. 4*, Eds. K. Srinivasan, D.E. Hardt, and R. Komanduri (American Society of Mechanical Engineers, 345 47th St. New York, NY 10017) pp. 231-243.
54. E. Abbe, "Messapparate fur Physiker," *Zeits. fur Intrumentenkunde* 10, 446-448 (1890) and J.B. Bryan, "The Abbé Principle Revisited: An Updated Interpretation," *Prec. Eng.* 1, 129-132 (1979).
55. E.C. Teague and C. Evans, ibid. 49.
56. R.V. Jones, *Instruments and Experiences. Papers on Measurement and Instrument Design* (John Wiley and Sons, 1988) Paper XI.
57. J. Strong, *Procedures in Applied Optics* (Marcel Dekker, 1989), Chapters 7 and 8.
58. J.E. Furse, "Kinematic Design of Fine Mechanisms in Instruments," *J. Phys. E.; Sci. Instrum.* 14, 264-271 (1981).
59. J. Strong, "New Johns Hopkins Ruling Engine," *J. Opt. Soc.* 41, 3 (1951).
60. K. Lindsey, S.T. Smith, and C.J. Robbie, "Sub-Nanometre Surface Texture and Profile Measurement with Nanosurf 2," *Annals CIRP* 37, 519-522 (1988).
61. J.E. Griffith, D.A. Grigg, G.P.Kochanski, M.J. Vasile, and P.E. Russell, "Metrology with Scanning Probe Microscopes," This Volume.
62. J.F. Song and T.V. Vorburger, "Stylus Profiling at High Resolution and Low Force," *Appl. Opt.* 30, 42-50 (1991).
63. D. Nyyssonnen, Proc. SPIE 921, (1988).
64. M.G. Rosenfield, "Measurement Techniques for Submicron Resist Images," *J. Vac. Sci. Tech. B, Proc. of the 1988 Symposium on Electron, Ion, and Photon Beams.*
65. M.T. Postek and D.C. Joy, "Microelectronics Dimensional Metrology in the Scanning Electron Microscope," *Sol. State Tech. Pt.1* 29 (11), 145-150 (1986) and *Pt.2* 29 (12), 77-85 (1986).
66. A.J. den Boef, *Scanning Force Microscopy Using Optical Interferometry* (CIP - Gegevens Koninklizke Bibliotheek, Den Haay, Eindhoven, The Netherlands, 1991) pp. 218-232.
67. E.C. Teague, F.E. Scire, S.M. Baker, and S.W. Jensen, "Three-Dimensional Stylus Profilometry," *Wear* 83, 1-12 (1982).
68. M. Stedman, "Mapping the Performance of Surface-Measuring Instruments," *Proc. SPIE* 183, 138-142 (1987).
69. Data provided by Robert Larrabee, NIST.

70. H. Fuchs, R. Laschinski, and Th. Schimmel, "Atomic Resolution of Nanometer Scale Plastic Surface Deformations by Scanning Tunneling Microscopy," *Europhy. Lett.* 13, 307-311 (1990) and private communication with H. Fuchs.

Metrology with Scanning Probe Microscopes

J. E. Griffith, D. A. Grigg +, G. P. Kochanski, M. J. Vasile, and P. E. Russell *

AT&T Bell Laboratories
Murray Hill, New Jersey 07974

+ Digital Instruments
Santa Barbara, California 93117

* North Carolina State University
Department of Material Science and Engineering
Raleigh, North Carolina 27695

ABSTRACT

One of the more demanding requirements of sub-micron lithography is dimensional measurement of the patterned features. Probe microscopes can perform this task non-destructively on most solids in a wide range of ambient conditions, though not without careful attention to several sources of measurement error. The most serious problem arises from the finite size of the probe, which has an intrinsically nonlinear interaction with the surface to be measured.

1. INTRODUCTION

A necessary condition for successful lithography is accurate measurement of the position and the size of the features produced. A widely accepted rule of thumb, the gauge maker's rule, holds that measurement uncertainty should be an order of magnitude smaller than the characteristic size, or critical dimension, of the smallest feature. Critical dimensions in semiconductor lithography are rapidly shrinking below a half micron. There are already structures, such as MOSFET gate oxides, whose thickness must be controlled to within a nanometer. The demand for lateral measurements accurate to better than 50 nm is causing some difficulty for the dominant metrology tools, optical microscopes [1] and and scanning electron microscopes [2]. They suffer from measurement uncertainties arising from the interaction of the incident photon or electron beam with the sample.

Because of their ability to achieve high resolution simultaneously in all three-dimensions, scanning probe microscopes are promising candidates for performing measurements of surface topography [3-5]. (These instruments are sometimes called scanning tip microscopes or proximal probe microscopes.) Fig. 1 exemplifies the advantages of obtaining true three-dimensional data. Cross-sectional and perspective views can be generated, non-destructively, at any location once an image has been acquired. Probe microscopes offer many other practical advantages for measurement. They operate in a wide variety of ambient conditions including ultra-high vacuum , air, and fluids. Probe microscopes can scan insulators as well as conductors. Finally, they are compact so they can be made relatively immune to environmental disturbances such as vibration or fluctuating temperature.

Figure 1. Scanning tunneling microscope image of 0.5 μm tall and 1.5 μm wide polysilicon pillars scanned with a very slender Ir probe.

Probe microscopes are not free, however, of the probe-sample interaction problems that bedevil other metrology tools. Error in a measurement can arise from uncertainties in the position of the probe apex, in the size of the probe, and in the position of the proximal point, which is the point on the probe that is closest to the surface. Since even a very sharp probe is an extended object on the size scale that we are considering, the proximal point does not necessarily coincide with the probe apex. Geometry alone makes the probe-sample interaction strongly nonlinear.

Fig. 2 illustrates two frequently used measures of a surface, pitch and linewidth [6]. Pitch is the spacing between lines. Measurement of pitch is not affected by the size or shape of the probe, but it is susceptible to errors in the position of the probe's apex. In Fig. 2 the distortion caused by the finite size of the probe subtracts out of the pitch calculation. Linewidth measurement is, on the other hand, much more difficult because the width of the stylus is added to the true width of the line being measured. During the measurement shown, the proximal point wanders from one side of the probe to the other, so the width of the line is unknown until the width of the probe is known. Once the probe width is determined, however, the extraction of the linewidth is much simpler than in electron or optical microscopy. The situation shown in Fig. 2 is an idealized case in which a cylindrical probe closely follows vertical sidewalls of trenches wide enough to accommodate the probe. In general, there may be areas of a surface that are completely inaccessible to the probe. Because of the three-dimensional nature of the data, from now on we will use the term line profile instead of linewidth.

In adapting proximal probe microscopes for measurement we have added several features to our probe microscopes, which are illustrated in Fig. 3. Our main concern here is the measurement of features with very steep topography, so we will concentrate on those aspects of the design that affect such measurements. We will not cover issues such as sample stage design, interferometry of the sample stage, temperature stability or mechanical stability. A thorough discussion of these topics can be found in a review by Teague[7]. These problems are being addressed in an ambitious instrument designed to achieve atomic-level metrology at the National Institute of Standards and Technology (NIST)[8]. Its aim is to achieve 1 nm measurements over several centimeters.

The first requirement in scanning probe metrology is to know the position of the probe so pitch measurements can be made. Because of hysteresis and creep in the piezo-ceramic actuators used in these microscopes, the potential applied to the actuator does not in itself determine the position of the probe. A reliable, independent monitor of the probe's position is needed. For this we have developed a capacitance-based position monitor[9]. Fig. 3 illustrates the use of the monitor, which measures displacements along all three axis. Monitoring the tube position is not enough, however, because the end of the tube tilts as it moves. This induces an extra motion in the probe or the sample, which is proportional to the distance from the plane of the position monitor. It is a type of Abbé error, and it can be large[7]. Careful characterization, modeling and calibration of the tube is necessary to achieve the accuracy we desire.

The second requirement in scanning probe metrology is careful measurement and control of the probe shape, especially for line profile and surface roughness measurements. The shape and size of the probe determines the amount of topographic information that is accessible; the probes must be slender enough to reach down into the features of interest. When measuring surface roughness, a blunt probe can cause a surface to appear either smoother or rougher than it really is, depending on the circumstances. Narrow cracks in the surface may not appear in the scan, while protrusions can appear too large. Even worse, the nonlinear nature of the probe-sample interaction can severely distort the spatial frequency spectrum. The probe must not, on the other hand, be too slender. The probe must be stiff enough to resist flexing under the influence of lateral forces generated by attraction to the sample's side walls. This can cause large uncertainties in the position of the probe apex. We have studied the effects of surface forces on these probes and how ambient conditions may affect measurements. We have developed methods to controllably fabricate sharp tungsten and iridium probes with a Ga^+ focussed ion beam (FIB)[10,11]. These probes typically have 5–6 nm radii and cone angles between 12° and 15° over the first two microns from the apex. We have also developed *in-situ* characterization techniques that assure us that the probe had the desired shape during the scan[12,13].

We mention in passing one other feature shown in Fig. 3, the rocking beam force sensor. In most practical situations the surface to be measured has both conducting and insulating areas. For this reason, scanning force microscopes (SFM) are usually required. We have chosen to use the rocking beam force balance sensor developed by Joyce and

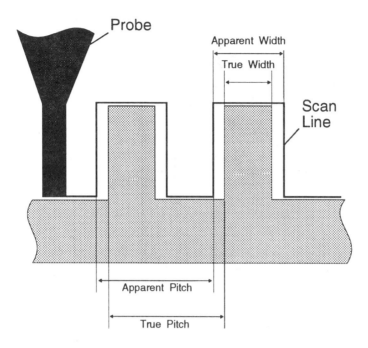

Figure 2. Comparison of pitch and linewidth measurements. The pitch measurement is not affected by the size or shape of the probe. The linewidth measurement requires that the size of the probe be known, because the apparent width is the true width plus the width of the probe.

Houston and Miller et. al [14, 15]. This sensor allows us to measure large, conductive or nonconductive surfaces in air or vacuum. The rocking beam force sensor can operate at a high stiffness without compromising its sensitivity ($\approx 10^{-8}$ N). This technique allows measurement of surface in a dual tunneling-force mode [16], permitting us to use tunneling and force sensing simultaneously.

We will concentrate in this chapter on the techniques necessary for pitch, line profile and surface roughness measurements. We will see that the performance obtained depends on many factors in the microscope, in the sample and in the ambient. Because of this, there is no simple answer to how well these microscopes perform in general. A probe tip capable of resolving atoms may be incapable of profiling a half-micron wide trench. Nevertheless, it is clear that proximal probe microscopes are an important addition to our metrology toolbox.

2. PITCH MEASUREMENT

The position of the probe apex is determined by a piezoceramic actuator, which in most designs is a hollow tube [17, 18]. The ceramic is a lead-zirconate-titanate ($Pb(Zr,Ti)O_3$), which is ferroelectric, so it exhibits hysteresis and creep [19]. Most, though

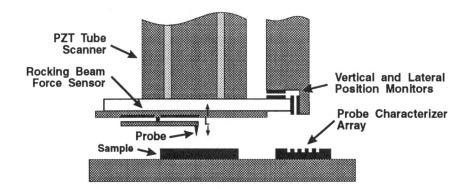

Figure 3. Schematic diagram of some essential elements for scanning probe metrology. The distance L from the probe apex to the measurement plane of the position monitor affects the size of the Abbe offset error.

not all, of the piezoceramic's nonlinear behavior can be eliminated by driving electrode charge rather than electrode potential [20, 21]. Alternatively, the nonlinearities in the tube can be partially canceled through a compensating scan algorithm based on modeling of the piezoceramic material. Unfortunately, the remaining nonlinear component is complicated and poorly understood, so there is no way to reduce the residual error of a few percent. Independently monitoring actuator motion with a reliable and easily modeled transducer is a better way to achieve precise motion. Barrett and Quate [22] used an optical system to measure the actuator's position in two dimensions. Their monitor was incorporated into a servo loop that corrected the tube's position in real time. In the capacitance-based system of Griffith et. al [9], the tube position is simply monitored so a correction can be made after the scan. The configuration of two of the capacitors is schematically illustrated in Fig. 3 by the vertical and lateral position monitors located to the side of the tube scanner. The output of the capacitive sensing circuit is linear to the gap, d, and has a wide-band noise limit of ≈10 nm. With new electronics developed by G. L. Miller and E. R. Wagner [23], we can now monitor actuator motions in all three axis.

A troublesome property of the tube scanner is that the end tilts as part of the lateral displacement. This induces an extra motion in the probe, not seen by the monitors. It is proportional to the distance L from the plane of the position monitor to the probe apex. The effect, illustrated in Fig. 4., is an example of an Abbé offset error [7], and it can be large. Abbé errors arise when the probe tip is not in the plane of the displacement sensors. The calibration of the monitors will change any time L is changed. The tube tilting has been modeled by Carr [24]. The tilt angle, θ, affects both the capacitive sensor output and the true displacement at the probe tip with respect to the capacitor plates. The displacement of the probe, or the sample surface, with respect to the position monitor is a systematic error that can be removed if the calibration takes into account the length of the probe.

To experimentally determine the magnitude of the error induced by the offset of the probe from the plane of the monitors, we have measured the tilt θ shown in Fig. 4 as a function of lateral tube displacement, x_1. We determined θ versus x_1 by comparing the capacitive sensor's output to that of a fiber-optic interferometer[25] located at height L. The capacitive sensor gain was first calibrated to the interferometer by locating the interferometer at the center of the capacitor plate (L = 0 in Fig. 4). The capacitive sensor gain was 3.2 µm/V. Measurements were performed from L = 0 to 7.62 mm above the top of the tube. We tested a tube made from PZT-5H[26] with length, l = 39.5 mm,

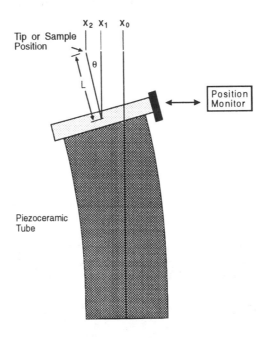

Figure 4. Effect of tube tilting on the lateral displacement of the tip or sample. L is the distance between the probe tip and the measurement plane of the position monitor."

6.35 mm diameter and 0.5 mm wall. The measured variation in θ per x_1 was 2.97 µdeg/nm (51.8 nrad/nm). Alternatively, this can be expressed as $\theta = 2.05(x_1/l)$. Measurements were conducted over a 2.0 µm range in x_1. The angular change was approximately linear over the measured range and consistent for all heights, L. Though an angular variation of 2.97 µdeg/nm seems small, this translates to a 10.0 µm error for a 5 mm probe over a 40.0 µm scan, or 5% error for each millimeter of probe length. Measurements by Barrett and Quate[22] revealed a small quadratic component to the tilt that has also appeared in our tube scanner.

The angular variation also induces a nonlinear term in the output of the capacitive displacement sensor. Imagine square capacitor plates with width W and a tilt θ about an axis parallel to a side of the plates. The area $A = W^2$. For small θ the capacitance, C, obeys[9]:

$$C = \frac{\varepsilon A}{d}\left[1 + \frac{1}{12}\left(\frac{W\theta}{d}\right)^2\right], \qquad (2.1)$$

where ε is the permittivity, and d is the gap at the center. Our capacitors measured 4.5 mm × 4.5 mm and have a mid-range gap, d, of 30 µm. The capacitive output error is 0.45% of x_1 because of the previously mentioned 2.97 µdeg/nm angular variation, so a 40 µm scan has a 180 nm measurement error. This error is small compared to the probe tip displacement error, and it underestimates the true displacement, contributing a minor correction to the former error. It may be possible to design a capacitive sensor system such that θ self corrects the sensor output for a given probe length.

3. LINE PROFILE MEASUREMENT

On lithographically patterned surfaces, a wide variety of surface topographies occur. Deep, narrow trenches and holes are common, and some of them have undercut side walls. How faithfully a probe microscope reveals the surface topography depends strongly on the size and shape of the probe. We will see that the shape of the probe must be tailored to the particular surface features to be measured. An important consideration in choosing the probe shape is the ease with which it allows the true surface shape to be extracted from the scan. There is no universal probe shape that is appropriate for all surfaces.

There are two fundamental sources of error associated with the probe itself. The first simply arises from the geometry of an extended objected moving relative to the surface features. Some features may be completely inaccessible while others will be distorted. The distortion comes from the motion of the proximal point relative to the probe's apex, which is a strongly nonlinear process. If the size and shape of the probe are accurately known, the distortion can sometimes be removed. The second source of error appears if the probe is so slender that it flexes under the lateral forces encountered. Under commonly occurring conditions, the flexing can amount to tens of nanometers, which results in a corresponding loss of measurement accuracy. The elastic properties of the probe must, therefore, be carefully considered in relation to the lateral forces present. The magnitude of these forces are influenced by the ambient conditions.

3.1 The Probe-Sample Geometry

In Fig. 5 we show two of the many ways in which the probe shape affects the scan trace and, thus, the apparent shape of the surface. In Fig. 5a the probe is so blunt that it can not reach the bottom of the trench. The resulting scan reveals the shape of the probe rather than that of the trench. This trace is not a convolution of the two shapes, and the trench shape can not be deconvolved from the trace. Convolution is a linear operation, but the scan formation is not. The cusp at the center of the trench is a characteristic feature of a scan with a probe that is too blunt. Such a cusp can give the impression that the probe is resolving smaller features than is actually the case. Paradoxically, a small cluster of atoms at the probe apex could produce atomic resolution on the flat regions around the trench.

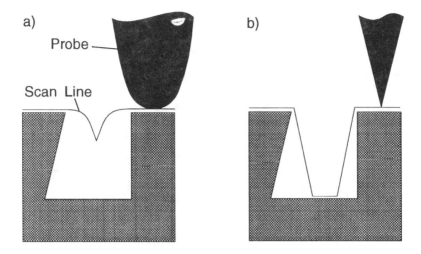

Figure 5. Two ways in which probe shape affects the scan. a) The tip is too blunt to reach the bottom of the trench. The trench features can not be extracted from the scan. b) The cone angle of the tip hides the true angles of the side walls.

One parameter of the trench that definitely can not be extracted in Fig. 5a is the depth of the trench. The only information about depth in this trace is the depth of the cusp in the center, which sets a lower bound on the trench depth. This behavior applies especially to measurement of crack depths. One may not assume that any crack encountered is at its apex wider than the apex of the probe being used. In general, we must expect that the probe, no matter how sharp, will not penetrate to the bottom of a crack. In presenting data on crack depth, the numbers obtained should be quoted as no more than a lower bound on the depth. The problem of mapping those areas inaccessible to the probe has been discussed by Niedermann and Fischer[27] and Reiss, et. al[28].

In Fig. 5b the tip is sharp enough to reach the bottom of the trench, so the scan trace along the bottom represents the shape of the trench. The apparent sidewalls reflect the cone angle of the probe. This probe can not tell the difference between the left and right walls, though it does set a limit on how far the walls slope out into the trench. In the section on probe characterization we will use this effect to measure the shape of the probe with the probe microscope itself.

The probe shape mixes into the scan trace whenever the proximal point wanders away from the apex of the probe. Fig. 5 exhibits extreme cases in which the proximal point hangs at a cusp on the surface. Fig. 6 shows a more subtle shift as the probe climbs a hill. This behavior has been analyzed in detail by Keller[29]. If the probe has the right shape, the true surface can be extracted. The extraction is based on the assumption that the surface and probe have first order contact. In other words, the slope of the surface is equal to the slope of the probe at the proximal point. If the probe has a shape with a one-

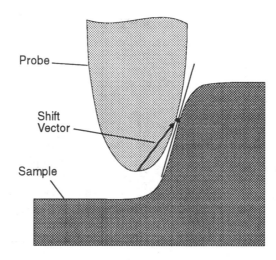

Figure 6. The fundamental source of error is migration of the proximal point away from the probe apex. Under some circumstances, the surface can be reconstructed.

to-one correspondence between slope and shift vector, then the shift vector will be known, allowing the correct position of the surface to be determined. This argument does not work with tips that are conical or cylindrical because there is not a unique probe slope for each shift vector.

In many cases, a special probe shape that is optimized for a particular part of the sample will be necessary. Fig. 7 shows such a shape. The end of the probe is flared so it can reach under the overhang. Nyysonen, Landstein and Coombs[30] have used such a probe in conjunction with a force microscope capable of detecting lateral as well as vertical forces. The scanning algorithm for such a microscope must be considerably more complicated than that usually employed. In most scanning tunneling or scanning force microscopes, the actuator retracts the probe from the surface if the signal becomes too large. In this case, retracting the probe when it is under the overhang would be destructive.

3.2 Probe Fabrication

The nanofabrication techniques being used to pattern surfaces are also having a strong influence on the fabrication of probes for microscopy. We have already seen that a wide variety of probe shapes may be needed to measure samples with steep topography[3, 30]. These shapes may incorporate protrusions with extremely tiny characteristic dimensions. The two basic techniques for forming these shapes are growth and selective erosion. We will discuss growth first.

When an electron or ion beam impinges on a surface along with gas molecules from the ambient, chemical modification of the molecules can cause them to remain on the surface as a permanent deposit. The shape of the deposit can be controlled by manipulating the beam. The use of an electron beam to induce chemical deposition and

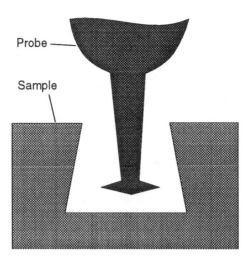

Figure 7. In cases with undercut side walls, a probe with a flared end is appropriate. Special scan algorithms are needed in this case.

growth has been investigated by several groups. In some cases, the molecules present in an imperfect vacuum are sufficient to produce a deposit[31]. Better control of the probes properties can be maintained if a specific gas is introduced into the area around the beam spot. Lee et. al[32] deposited probes from a dimethyl-gold-trifluroacetylacetonate organometallic gas complex using an SEM. These probes are 0.1 µm in diameter and 1–4 µm in length. Nyysonen et. al[30] have used these probes in a two dimensional control scheme to measure trench widths in patterned resist. Ximen and Russell[33] use a combined FIB and electron-beam growth technique that produces probes of similar size, which are stable after hours of scanning. As we will see in the section on probe flexing, it is important to have a high elastic modulus for the probe material. At this time, little is known about the elastic properties of the deposited material.

The most widely used controlled erosion technique is electrochemical etching[34-36]. Electrochemical etching is not likely, however, to achieve the degree of control necessary for metrology applications. We use an FIB technique to sharpen probes, because it can generate a wide variety of shapes in many materials, including refractory metals, with extremely tight tolerances. In contrast to the tip growth techniques, the sharpened material should have elastic moduli close to that of the bulk material. We use polycrystalline W and Ir for our probes. Ir is inert and has a high modulus of elasticity (440 GPa) compared to other metals.

The FIB machining employs a 1.5 nA, 20 kV Ga$^+$ ion beam raster scanned in an annular exposure pattern across the apex of an electrochemical etched probe. The annular exposure pattern is positioned to remove material from around the central axis of the etched probe. This leaves a central pedestal protruding from a larger shank. Fig. 8a and Fig. 8b show scanning electron micrographs of a chemically etched W probe before and after FIB machining, respectively. The beam induced sputtering has removed material from around the central axis down to a point predetermined by the dose of Ga^+ ions. We have found our FIB tips to be 5–6 nm in radius and being no more than 0.5 μm wide at a distance of 4 μm from the apex.

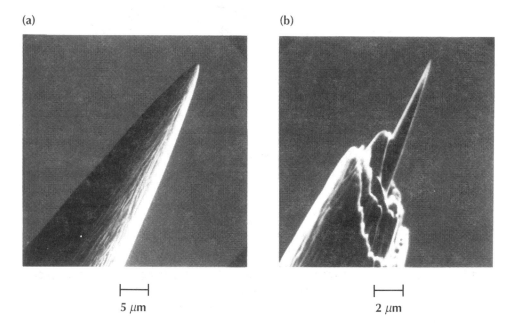

Figure 8. Scanning electron micrographs of a chemically etched W probe a) before and b) after FIB machining. (Note change in scale.)

3.3 Surface Forces and Probe Flexing

It is possible to make a probe tip too slender. We have evidence that extremely sharp tips are flexing in the presence of side walls. For instance, in Fig. 1 the pillars were scanned with a slender FIB-fabricated probe. The cone angle determined from the slope of the side walls is substantially smaller than that determined from SEM observation. Such an effect could only occur if the probes are bending toward the pillars during the scans[12] as shown in Fig. 9.

To support our suggestion that the probe tip is flexing in the proximity of the pillar, we consider the elastic response of a beam the size of the probe with a force F at the apex acting perpendicular to the axis of the beam. For simplicity we first consider a beam having a constant circular cross section with radius R. The moment of inertia is $\frac{1}{4}\pi R^4(x)$. Let L be the beam's length and Y its elastic modulus. For small deflection z perpendicular to the probe axis, the beam obeys[37]

$$z = \frac{4FL^3}{3\pi YR^4}. \tag{3.1}$$

This corresponds to the case $\alpha = 0$ in Fig. 10. We use an elastic modulus of 440 GPa, the value for iridium. Depending on the direction relative to the crystal lattice, the elastic modulus of diamond can be about a factor of two higher. For a given probe material, the elastic modulus is the factor in the equation that we have the least control over. The dominant factors are the geometrical ones, R and L. R is constrained by the narrowness of the trench that we want to explore. L is also constrained by the depth of the features of interest, which is usually no more than 1 µm. If the fabrication procedure allows, the slender part of the tip should be no longer than necessary.

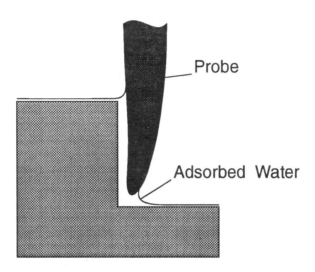

Figure 9. If the surface or probe is coated with an adsorbed fluid, the meniscus between probe and surface can generate strong lateral forces that will flex the probe.

Most tips are actually tapered. The bending calculation for a uniform taper is tedious but straightforward. We consider the elastic response of a frustrum of a right circular cone when the small end, with radius R_1, is subjected to a force F. The geometry is shown in the inset of Fig. 10. The radius of the large end is R_0 and the length is L. Let x be the distance from the large end: $R(x) = R_0 + (R_1 - R_0) x/L$. Let $z(x)$ be the deflection of the cone from its equilibrium position. When the deflection is small, the curvature obeys [37]

$$\frac{d^2 z}{dx^2} = \frac{4F}{\pi Y} \frac{1}{R^4(x)} (L-x). \tag{3.2}$$

Integrating from 0 to x

$$\frac{dz}{dx} = \frac{2FL}{\pi Y (R_1 - R_0)} \left[\frac{x + C_1}{R^3(x)} - \frac{C_1}{R_0^3} \right], \tag{3.3}$$

where $C_1 = \dfrac{L}{3} \left[\dfrac{3R_0 - 2R_1}{R_1 - R_0} \right]$. Integrating again we obtain the deflection

$$z(x) = -\frac{2FL}{\pi Y(R_1-R_0)}\left\{\frac{L}{R_1-R_0}\left[\frac{x+C_2}{R^2(x)} - \frac{C_2}{R_0^2}\right] + \frac{xC_1}{R_0^3}\right\}, \quad (3.4)$$

where $C_2 = \dfrac{L}{3}\left[\dfrac{3R_0-R_1}{R_1-R_0}\right]$. In Fig. 10 we show results from this expression as a function of the radius of the small end for two cases: cone half angles of 5° and 10°. The dominant effect of the moment of inertia of the beam is readily apparent.

Lateral attractive forces between tip and side wall may arise from three sources: electrostatic attraction caused by the tunneling bias, van der Waals attraction, and capillary attraction caused by adsorbed gasses in the gap between tip and sample. These forces may be much stronger here than in the usual scanning force microscope configuration because the tip can run parallel to the side wall for as much as a micron. Repulsive forces caused by collisions with sidewalls have been considered by den Boef[38].

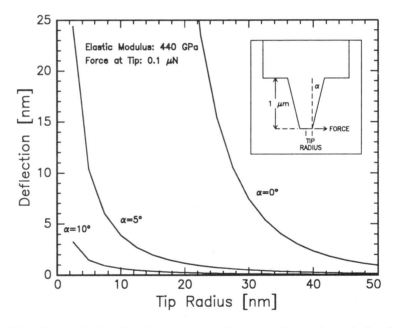

Figure 10. Plot of the calculated deflection perpendicular to the probe axis for three values of the half cone angle. The geometry used is shown in the inset. We assume the slender section to be attached to an infinitely stiff pillar.

We discuss the electrostatic force first. Consider a cylinder with radius R parallel to a plane that is a distance h from the axis of the cylinder. For a potential drop V between the cylinder and plane, the force obeys[39]

$$F = \frac{V^2 L \pi \varepsilon_0}{\text{acosh}^2(h/R) \times \sqrt{h^2 - R^2}}. \tag{3.5}$$

For a bias of 1 V, the electrostatic attractive force on a 0.5 µm long, 50 nm diameter cylinder (neglecting end effects) is 0.05 µN when the gap between cylinder and plane is 1 nm. Since tunneling can in many situations be performed with probe biases less than 0.1 V, electrostatic attraction is not a significant problem.

We discuss the van der Waals force in a similar context. In this case[40]

$$F = \frac{ALR^{1/2}}{8\sqrt{2}(h-R)^{2.5}}, \tag{3.6}$$

where A is the Hamaker constant, $\approx 10^{-19}$ J. Again we choose a length L of 0.5 µm and a gap, h-R, of 1 nm. The force in this case is again ≈ 0.05 µN. It can be reduced by choosing a suitable dielectric medium in which to immerse the system[40].

Capillary forces are much more difficult to estimate. A clean surface in vacuum or immersed in a fluid, it will not be a problem. In air, however, there may be enough water vapor adsorbed on the surface to produce capillary effects. Non-volatile fluids, such as oil, on the surface will also cause capillary attraction. The force generated will depend in detail on a variety of factors including geometry and composition of the fluids and materials. The order of magnitude will be the surface tension times a characteristic length[41]. To set an upper bound we assume a rather high surface tension of 0.1 N/m and a characteristic length of 1 µm, which yields 0.1 µN.

We see that van der Waals and capillary forces can be close to the 0.1 µN value chosen for Fig. 10. In actual scanning the behavior of the tip may be very complicated when it is subject to significant flexing forces. In scanning force microscopy a common problem is that a weak cantilever may allow the tip to become trapped by the deep potential well at the surface. This phenomenon is sometimes called "snap to contact". A flimsy probe may behave similarly in the lateral direction if it gets too close to a side wall.

Several papers have reported large forces between the probe and the surface under some conditions[12,42]. In addition, as the tip radius diminishes, repulsive forces in STM[16,42,43,44] and SFM cause increased stresses due to the small contact area between probe and sample[45,46]. Repulsive forces[16,42,43,44,47] and frictional forces have been measured in STM on a variety of surfaces and under various ambient conditions[48-50]. We have measured repulsive forces from 10^{-7} to 10^{-6} N while tunneling with an STM in air[16].

For soft samples, surface deflection can severely limit the resolution that can be achieved without permanent deformation of the sample. The peak strain ε_{max} under a hard spherical tip of radius R, in a thick sample with modulus Y is[51]

$$\varepsilon_{max} = \left[\frac{.236\, F}{YR^2}\right]^{1/3}. \tag{3.7}$$

All materials will exhibit damage or plastic deformation as ε_{max} approaches unity, typically at $\varepsilon_{max} = 10^{-1}$ to 10^{-3}, so we can calculate a maximum force (or minimum tip radius) for a given material. At a force of 10^{-8} N and at $\varepsilon_{max} = 10^{-1}$, the minimum allowable radius varies from 3 nm for steel (Y = 200 GPa) to 30 nm for polymethyl methacrylate (PMMA, Y = 3 GPa), and even larger for softer surfaces or larger forces. The probe contact area is a disk of radius $0.69 F^{1/2} Y^{-1/2} \varepsilon_{max}^{-1/2}$, which ranges from 0.5 nm for steel to 4 nm for PMMA at 10^{-8} N. This deflection sets an intrinsic resolution for a force microscope in repulsive mode, which is sufficiently small if the contact pressure is small. If capillary forces of 10^{-7} N pull the tip toward the surface, or if the microscope applies excessive force, the deformations can become appreciable. A probe applying 10^{-6} N to PMMA will have a contact radius on the order of 40 nm. Sharp corners on the sample will have less support, and will deform correspondingly more.

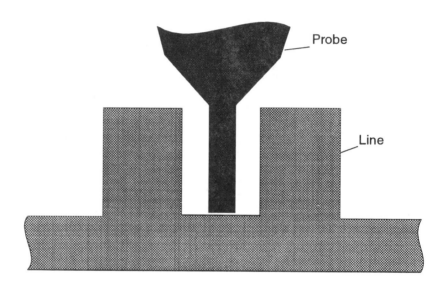

Figure 11. The preferred probe shape for the surface depicted is a relatively thick cylinder that will resist flexing. The probe should not be thinner than that is necessary to reach the bottom of the trench.

The resistance to flexing sets a limit on the accuracy of position measurement. There are several actions that can be taken to reduce the problem. The most effective is to keep the thickness of the probe as large as possible because of the R^{-4} dependence of the deflection. In addition, the thin part of the probe should be kept as short as possible. Such a probe is shown in Fig. 11. Though the dependence on force is linear, it is possible that the force can be suppressed by over an order of magnitude by removing any adsorbed layers through control of the ambient conditions. The parameter that we have the least control over is the elastic modulus of the probe material. Diamond, because of its high stiffness, would be the most desirable material. Probes formed from grown deposits may not attain the bulk elastic constants.

4. SURFACE ROUGHNESS MEASUREMENT

Like their forerunner, the contact-stylus profilometer, probe microscopes are effective tools for measuring surface roughness [28, 27, 52, 53]. Probe microscopes offer higher performance than traditional stylus instruments mainly because significantly sharper probe tips are employed. A sharp probe alleviates, but does not eliminate, distortions caused by the interaction of an extended tip with the surface. These distortions, long ago recognized in stylus profilometers [54, 55], were analyzed by Stedman [56-58] for probe microscopes. In an analytical treatment of sinusoidal surfaces, she mapped regions of wavelength-amplitude space that were accessible to the tips. We extend Stedman's argument with numerical modeling of parabolic probes scanning sinusoidal surfaces. Simulating the image trace allows us to directly model the distortions caused by the probe. In addition, by showing the results from a few representative examples, we hope to convey an intuitive feel for the tip effects, which are more severe than generally assumed.

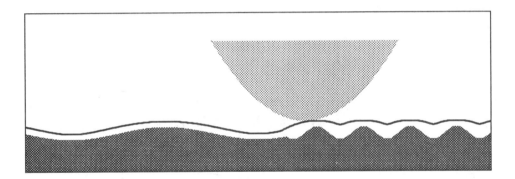

Figure 12. Plot of a simulated scan of a 40 nm radius parabolic probe over sinusoidal surfaces with 10 nm amplitude. On the left the wavelength is 200 nm; on the right the wavelength is 50 nm.

We begin with a strictly geometrical interaction arising from contact of the probe with the surface. In the simulation we move an imaginary probe along a straight line over the surface at a constant, and arbitrary, height h. At each position along the line, the vertical distances of the points of the probe from the surface are calculated. The minimum of these distances is subtracted from h to give the height of the probe when in contact with the surface at that position. The calculation also yields the position relative to the apex of the contact point on the probe. In Fig. 12 we show the results from such a calculation with a parabolic tip having a radius R of 40 nm: $y(\Delta x) = (\Delta x)^2/(2R)$. The surface consists of two sinusoidal segments having an amplitude of 10 nm. On the left the probe produces a good representation of a surface with 200 nm wavelength even though the contact point wanders as much as 13 nm from the apex. On the right, where the surface wavelength is 50 nm, we see that the average height of the probe trace has risen, and the amplitude of the trace is much less than that of the surface. In addition, the lowest points of the trace are cusps, which have high frequency spatial wavelengths. This directly demonstrates that the probe-sample interaction is nonlinear, because the scan contains spatial frequencies not present in the surface. As we will see, a Fourier analysis of a scan can be misleading. The nonlinear behavior of profilometers has been theoretically studied by Church, et al.[59-61] through Taylor series expansions of the path followed by the probe.

In Fig. 13 we summarize results from modeling a 40 nm parabolic tip over a range of sinusoidal surfaces as a function of wavelengths for amplitudes of 10 nm, 5 nm and 1 nm. We plot the RMS roughness, or R_q, of the trace. The true R_q of the sinusoidal surfaces is simply the amplitude divided by $\sqrt{2}$, so the decrease in measured roughness at short wavelengths is a result of errors generated by the probe. As the amplitude increases, the wavelength at which the probe fails to reproduce the surface also increases. This is a manifestation of the boundaries in wavelength-amplitude space mapped by Stedman. The decrease in apparent roughness at short wavelengths is a measure of the error in the scan. Unfortunately, these curves can not be used to estimate an upper bound for the surface roughness. A surface with deep, narrow cracks, for instance, could have an arbitrarily large roughness while a blunt probe would show it to be smooth. The 40 nm radius chosen is typical of a good commercial probe. Such a probe may exhibit better performance than in Fig. 13 if there happens to be a sharp protrusion hanging from it, but in general the user may not assume that this is the case. The FIB-sharpened probes will, of course, produce better results than the 40 nm probe assumed in Figs. 12 and 13.

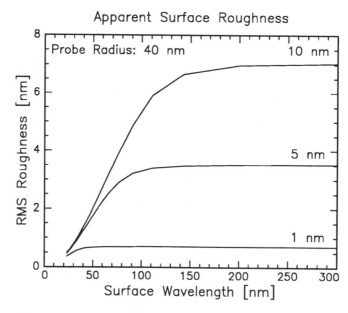

Figure 13. Plot of apparent surface roughness calculated from simulated scans with a 40 nm radius parabolic probe and pure sinusoidal surfaces. The three curves correspond to surface amplitudes of 10 nm, 5 nm and 1 nm.

On the atomic scale, approximating the tip and sample as rigid, contacting objects clearly breaks down. We will now derive a simple microscopic calculation of the image that allows calculation of STM and SFM images down to atomic length scales. We assume an exponential interaction between the tip and the sample, which could be either tunneling current or a perpendicular force. The tip surface can be written as $z_{tip} = t(\mathbf{x} - \mathbf{x}_{tip}) + T(\mathbf{x}_{tip})$, where $\mathbf{x}$ are the coordinates parallel to the surface and $\mathbf{x}_{tip}$ is the position of the end of the tip. The function $t()$ is then the tip shape and $T()$ is the image of the surface. If we assume that the tunneling (or force) is local and primarily perpendicular to the surface, we can write the interaction in terms of a current density $J(\mathbf{x}) = C\exp(-(z_{tip} - s(\mathbf{x}))\kappa)$. In the above expression, $s(\mathbf{x})$ is the sample surface and κ and C describe the tip-sample interaction. The total interaction is then the integral of J over the surface:

$$I(\mathbf{x}_{tip}) = C\int e^{(s(\mathbf{x}) - t(\mathbf{x} - \mathbf{x}_{tip}) - T(\mathbf{x}_{tip}))\kappa} d\mathbf{x}. \tag{4.1}$$

If the microscope operates at constant current or force, we set $I(\mathbf{x}_{tip}) = I_0$, and solve for $T(\mathbf{x}_{tip})$ to get the image:

$$T(\mathbf{x}_{tip}) = \frac{-1}{\kappa} \ln\left[\left[e^{-\kappa t(-\mathbf{x})}\right] * \left[e^{-\kappa s(\mathbf{x})}\right]\right] + C'. \tag{4.2}$$

As written, the asterisk denotes convolution, so the calculation can be efficiently implemented with Fourier transform techniques so long as the terrain is not so rough that rounding error or numerical overflow interferes.

For terrain large compared to κ^{-1}, the results are dominated by the nonlinearity of the exponentials, and the macroscopic results are duplicated. At smaller sizes, the characteristic cusps obtained on macroscopic images become rounded. It is only for terrain small compared to κ^{-1}, that the image approaches a simple convolution of the tip and sample. However, κ^{-1} is on the order of an atomic radius, so the linear regime can only be relevant for atomically flat surfaces.

Fig. 14 shows simulated surface roughness, similar to Fig. 13, but calculated with the microscopic model. Overall, the curves are similar to those of the macroscopic model, but short wavelength roughness is further suppressed. Fig. 14 shows values of κ often seen while tunneling or in repulsive SFM.

The strong nonlinearity of the imaging can cause other difficulties beyond a simple reduction in the image roughness compared to the sample roughness. Fig. 15 displays the Fourier power spectrum of a simulated 1-dimensional STM image, with $\kappa = 10$ nm^{-1}. The surface is the superposition of two sinusoids, but the image shows the characteristic cusps that appear with an insufficiently sharp tip. As can be seen, the image power spectrum contains all intermodulation products of the two wavelengths present in the sample. For the parameters chosen, the image has no response at the spatial frequencies actually present in the sample. In general, however, the sample spatial frequencies will also appear in the image with amplitudes that are strongly tip dependent. There is no

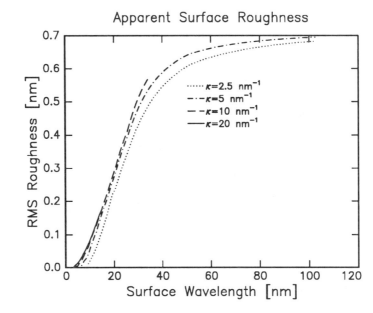

Figure 14. RMS roughness of the image calculated using the microscopic model. The tip was a parabola with 40 nm radius, and the sample was a sinusoid with 1 nm amplitude. The four curves, from left to right, are calculations for $\kappa = 20$ nm^{-1}, 10 nm^{-1}, 5 nm^{-1}, and 2.5 nm^{-1}, respectively.

simple relationship between the spectral power density of the sample and the image. The nonlinearities of the imaging will generally make such a display of a real surface an uninterpretable, meaningless exercise in computer graphics unless the sample is smooth and nearly flat.

5. CHARACTERIZATION AND CALIBRATION

Clearly, the accuracy of a topographic measurement depends on the probe shape. If the shape of the probe is known then it may be possible to remove certain probe-induced distortions from the measured topography. This extraction has been studied by several groups [28, 29]. Unfortunately, the shape of the probe may change during a scan, and it is impractical to frequently remove the probe for inspection in an SEM. It is possible to use the probe microscope itself to determine the shape of the probe if an appropriate structure is scanned [12]. Fraundorf and Tentschert used the tiny holes formed by nuclear particle tracks [62], while Song and Vorburger employed the cutting edge of a razor blade [63]. We use an array of pillars with undercut sidewalls for the characterizer. Fig. 16 illustrates the process of obtaining the probe tip shape with the probe microscope. In a scan, the apparent sidewalls of the pillar are actually an extended image of the probe, inverted and reflected about the vertical axis. A lithographically defined pillar having a slight undercut ensures that the trace reflects the probe shape as it travels along the side of the pillar. The pillars are polysilicon ≈530 nm tall and 1.5 μm on each side having slightly undercut walls. The undercut serves two purposes. First it produces a cusp at the upper edge where the proximal point will hang as the probe scans over it. In addition, the wall of the pillar

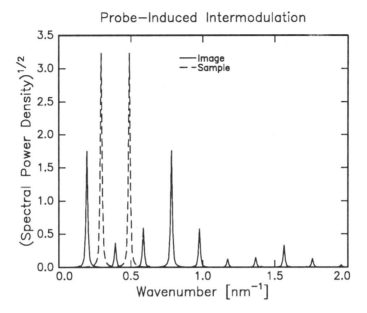

Figure 15. Fourier transform power spectra of the surface (dashed curve) and its image (solid curve) defined by s(x) = (0.5 nm)sin($2\pi x$/(12.8 nm))+(0.5 nm)sin($2\pi x$/(20.48 nm)). A 40 nm radius parabolic tip was used in the calculation. Individually, both of the two sinusoids are imaged without cusps; when superposed, the nonlinearities dominate the image.

is kept away from the probe so attractive forces between the probe and sidewall are kept to a minimum. Though pillars were used here, holes would also work if they were round. To measure the total width of the probe, the width of the pillar or hole scanned must be independently determined.

The accuracy of a calibration can be no better than the uncertainty in the size of the reference artifact used. In Fig. 1 the roughness of the tops of the polysilicon pillars is a significant source of error in the probe shape measurement. This is a common and important problem in calibration of metrology microscopes. For instance, in fabricating reference standards for optical linewidth measurement, a major source of uncertainty is the roughness of the line edges[64]. As with our pillars, the cause of the problem is the polycrystalline structure of the chromium reference lines.

The gauge maker's rule that we mentioned at the beginning of this chapter, was established at a time when instrument makers could employ fabrication techniques that were substantially ahead of the state of the art in manufacturing. In modern manufacturing, the techniques employed are often the most precise known, so it is becoming more and more difficult to develop reference artifacts that exhibit tolerances an order of magnitude better than the objects to be measured. Proximal probe lithography may find an important application in the fabrication of reference artifacts because it may be necessary to literally control the number of atoms that make up reference standards in the future. Another approach is to find natural artifacts, that is objects such as C_{60}, or fullerene tubes, that naturally adopt a given size and shape[65].

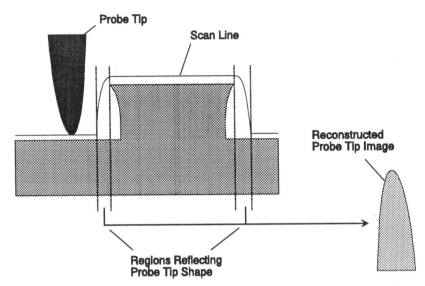

Figure 16. Illustration of the method for characterizing scanning probe tips with the probe microscope.

6. THE FUTURE

In the last ten years probe microscopes have evolved at a rapid, and accelerating, pace. While optical and electron microscope metrology tools have benefited from decades of engineering refinement, that process has just begun for scanning probe microscopes. Many of the advances in probe metrology will come quickly because they merely require straightforward application of already existing technology. Much of the vigor of this field has, after all, arisen from the ongoing revolutions in computers and electronics. Improving computer hardware and software will make the operation of these instruments more automatic, more reliable, and more precise. The ability to manipulate large amounts of information and present it in an easily comprehended format for the operator will continue to grow. There will be real-time modeling of the behavior of the instrument to correct for systematic errors. More sophisticated feedback algorithms will allow faster scan speed and scanning of features under ledges. Two-dimensional force sensing has already been achieved, and three-dimensional sensing will eventually be available. Probe microscopes will have greater range and operate over ever larger samples, while the sensors themselves will be ever smaller. They may incorporate myriads of tips scanning simultaneously. Probe microscopes will combine several types of sensors to exploit different contrast mechanisms, and they will be incorporated into other metrology tools.

The most important, and most difficult, advances will involve the probe-sample interaction. Probe fabrication will become ever more sophisticated as our ability to manipulate materials advances. Exotic probe shapes that are stable, stiff and tiny will be available, perhaps as the result of techniques discussed in other chapters of this volume. There may even be probes that dynamically alter their shapes in response to the surface topography. Ambient conditions will be controlled to reduce forces between tip and sample.

Though atomic resolution has barely been mentioned in this chapter, the ability of scanning probe microscopes to resolve atoms will have a profound influence on the future of metrology in two ways. The first effect is psychological. Though a handful of visionaries at the National Bureau of Standards understood the possibilities two decades ago [66], it took images of atoms to spark the explosion of invention that made it possible to push profilometry to these levels [67]. The lure of atomic-scale fabrication is a significant motivating force. The second effect will be in calibration. The ultimate reference artifact will be a crystal lattice of some sort. A typical bulk terminated lattice has a period of $0.2-0.3$ nm , which is inconveniently small for currently available techniques at atmospheric pressure. But either our technique will improve or lattices with larger periods will be found. Our measurements will then be founded on direct comparison with reliable natural structures.

7. ACKNOWLEDGEMENTS

We thank G. L. Miller and E. R. Wagner for their generous assistance, and we thank Robert Larrabee, Michael Postek and Clayton Teague of NIST for helpful discussions. D. A. Grigg and P. E. Russell acknowledge the support of the National Science Foundation Presidential Young Investigator Award Program (DMR8657813) and of AT&T Bell Laboratories.

REFERENCES

1. D. Nyysonen and R. D. Larrabee, "Submicrometer Linewidth Metrology In the Optical Microscope", J. Res. Nat. Bur. Stand. **92**, pp. 187-204, 1987.

2. M. T. Postek and D. C. Joy, "Submicrometer Microelectronics Dimensional Metrology: Scanning Electron Microscopy", J. Res. Nat. Bur. Stand. **92**, pp. 205-228, 1987.

3. J.E. Griffith, D. A. Grigg, M. J. Vasile, P. E. Russell, and E. A. Fitzgerald, "Scanning Probe Metrology", J. Vac. Sci. Technol. A, **10**, pp. 674-679, 1992.

4. D. A. Grigg, J. E. Griffith, G. P. Kochanski, M. J. Vasile and P. E. Russell, "Scanning Probe Metrology", Proc. SPIE, **1673**, pp. 557-567, 1992.

5. J. E. Griffith, M. J. Vasile, G. L. Miller, E. R. Wagner, E. A. Fitzgerald, D. A. Grigg, and P. E. Russell, "Pitch and Linewidth Measurements in Scanning Probe Metrology", *1991 ULSI Science and Technology*, J. Andrews and G. Celler, eds., Electrochem. Soc. Proc., 1991.

6. J. M. Jerke, ed., "Accurate Linewidth Measurements on Integrated-Circuit Photomasks", Natl. Bur. Stand. Spec. Publ. 400-43, 1980.

7. E. C. Teague, "Nanometrology", in *Scanned Probe Microscopy: STM and Beyond*, Engineering Foundation Conference Proceedings, Santa Barbara, CA, 1991.

8. E. C. Teague, "The National Institute of Standards and Technology molecular measuring machine project: Metrology and precision engineering design", J. Vac. Sci. Technol. B **7**, pp. 1898-1902, 1989.

9. J. E. Griffith, G. L. Miller, C. A. Green, D. A. Grigg, and P. E. Russell, "A Scanning Tunneling Microscope with a Capacitance-Based Position Monitor", J. Vac. Sci. Technol. B **8**, pp. 2023-2027, 1990.

10. M. J. Vasile, D. A. Grigg, J. E. Griffith, E. A. Fitzgerald, and P. E. Russell, "Scanning Probe Tips Formed by Focused Ion Beams", Rev. Sci. Instrum, **62**, pp. 2167-2171, 1991.

11. M. J. Vasile, D. A. Grigg, J. E. Griffith, E. A. Fitzgerald, and P. E. Russell, "Scanning Probe Tip Geometry Optimized for Metrology by Focused Ion Beam Ion Milling", J. Vac. Sci. Technol., **B** 9, pp. 3569-3572, 1991.

12. J.E. Griffith, D. A. Grigg, M. J. Vasile, P. E. Russell, and E. A. Fitzgerald, "Characterization of Scanning Probe Microscope Tips for Linewidth Measurement", J. Vac. Sci. Technol. **B** 9, pp. 3586-3589, 1991.

13. D. A. Grigg, P. E. Russell, J. E. Griffith, M. J. Vasile, and E. A. Fitzgerald, "Probe Characterization for Scanning Probe Metrology", Ultramicroscopy, **42-44**, pp. 1616-1620, 1992.

14. S. A. Joyce, and J. E. Houston, "A New Force Sensor Incorporating Force-Feedback Control for Interfacial Force Microscopy", Rev. Sci. Instrum., **62**, pp. 710-715, 1991.

15. G. L. Miller, J. E. Griffith, E. R. Wagner, and D. A. Grigg, "A Rocking Beam Electrostatic Balance for the Measurement of Small Forces", Rev. Sci. Instrum., **62**, pp. 705-709, 1991.

16. D. A. Grigg, P. E. Russell, and J. E. Griffith, "Tip-Sample Forces in Scanning Probe Microscopy in Air and Vacuum", J. Vac. Sci. Technol. A, **10**, pp. 680-683, 1992.

17. C. P. Germano, "A Study of a Two-channel Cylindrical PZT Ceramic Transducer for Use in Sterco Phonograph Cartridges", IRE Transactions on Audio, pp. 96-100, July/August 1959.

18. G. Binnig, and D. P. E. Smith, "Single-tube Three-dimensional Scanner for Scanning Tunneling Microscopy", Rev. Sci. Instrum., **57**, pp. 1688-1689, 1986.

19. O. Nishikawa, M. Tomitori and A. Minakuchi, "Piezoelectric and Electrostrictive Ceramics for STM", Surf. Sci. **181**, pp. 210-215, 1987.

20. C. V. Newcomb and I. Flinn, "Improving the Linearity of Piezoelectric Ceramic Actuators", Electron. Lett., **18**, pp. 442-444, 1982.

21. H. Kaizuka, "Application of Capacitor Insertion Method to Scanning Tunneling Microscopes", Rev. Sci. Instrum., **60**, pp. 3119-3122, 1989.

22. R. C. Barrett and C. F. Quate, "Optical Scan-Correction System Applied to Atomic Force Microscopy", Rev. Sci. Instrum., **62**, pp. 1393-1399, 1991.

23. G. L. Miller and E. R. Wagner, private communication.

24. R. G. Carr, "Finite Element Analysis of PZT Tube Scanner Motion for Scanning Tunneling Microscopy", J. Microscopy, **152** Pt. 2, p. 379, 1988.

25. R. Cook, Opto Acoustic Sensors Inc., Raleigh, N.C. 27607.

26. PZT-5H is a trademark of the Vernitron Corp.

27. Ph. Niedermann and O. Fischer, "Imaging of Granular high-T_c Thin Films Using a Scanning Tunneling Microscope with Large Scan Range", J. of Microscopy, **152**, pp. 93-101, 1988.

28. G. Reiss, F. Schneider, J. Vancea, and H. Hoffmann, "Scanning Tunneling Microscopy on Rough Surfaces: Deconvolution of Constant Current Images", Appl. Phys. Lett. **57**, pp. 867-869, 1990.

29. D. Keller, "Reconstruction of STM and AFM Images Distorted by Finite-size Tips", Surf. Sci. **253**, pp. 353-364, 1991.

30. D. Nyysonen, L. Landstein, and E. Coombs, "Two-dimensional Atomic Force Microprobe Trench Metrology System", J. Vac. Sci. Technol., **B** 9, pp. 3612-3616, 1991.

31. Y. Akama, E. Nishimura, A. Sakai, and H. Murakami, "New Scanning Tunneling Microscopy Tip for Measuring Surface Topography", J. Vac. Sci. Technol. A, **8**, pp. 429-433, 1990.

32. K. L. Lee, D. W. Abraham, F. Secord, and L. Landstein, "Submicron Si Trench Profiling with an Electron-beam Fabricated Atomic Force Microscope Tip", J. Vac. Sci. Technol., **B** 9, pp. 3562-3568, 1991.

33. H. Ximen and P. E. Russell, "Atomic Force Microscope Observation of Microfabricated Patterns with High Aspect Ratios with Electron Beam Contamination Microtips", Ultramicroscopy, **42-44**, pp. 1526-1532, 1992.

34. I. H. Musselman and P. E. Russell, "Platinum/iridium tips with controlled geometry for scanning tunneling microscopy", J. Vac. Sci. Technol A **8**, pp. 3558-3562, 1990.

35. J. P. Ibe, P. P. Bey, Jr., S. L. Brandow, R. A. Brizzolara, N. A. Burnham, D. P. DiLella, K. P. Lewe, C. R. K. Marrian, and R. J. Colton, "On the electrochemical etching of tips for scanning tunneling microscopy", J. Vac. Sci. Technol. A **8**, pp. 3570-3575, 1990.

36. M. Fotino, "Nanotips by Reverse Electrochemical Etching", Appl. Phys. Lett. **60**, pp. 2935-2937, 1992.

37. R. P. Feynman, R. B. Leighton, and M. Sands, *The Feynman Lectures on Physics*, Addison-Wesley, Reading, MA, 1964.

38. A. J. den Boef, "The influence of lateral forces in scanning force microscopy", Rev. Sci. Instrum., **62**, pp. 88-92, 1991.

39. W. R. Smythe, *Static and Dynamic Electricity*, Hemisphere Publishing Corporation, New York, 1989.

40. J. N. Israelachvili, *Intermolecular and Surface Forces*, Academic Press, London, 1985.

41. A. W. Adamson, *Physical Chemistry of Surfaces*, John Wiley & Sons, New York, 1967.

42. M. Salmeron, D. F. Ogletree, C. Ocal, H. C. Wang, G. Neubauer, and W. Kolbe, "Tip-Surface Forces During Imaging by Scanning Tunneling Microscopy", J. Vac. Sci. Technol. B 9, pp. 1347-1352, 1991.

43. H. Yamada, T. Fujii, and K. Nakayama, "Experimental Study of Forces Between a Tunneling Tip and the Graphite Surface", J. Vac. Sci. Technol., A 6, pp. 293-295, 1988.

44. U. Durig, J. K. Gimzewski, and D. W. Pohl, "Experimental Observation of Forces Acting during Scanning Tunneling Microscopy", Phys. Rev. Lett., **57**, pp. 2403-2406, 1986.

45. U. Landman, W. D. Luedtke, N. A. Burnham, and R. J. Colton, "Atomistic Mechanisms and Dynamics of Adhesion, Nanoindentation, and Fracture", Science, **248**, pp. 454-461, 1990.

46. N. A. Burnham, and R. J. Colton, "Measuring the Nanomechanical Properties and Surface Forces of Materials Using an Atomic Force Microscope", J. Vac. Sci. Technol., A 7, pp. 2906-2913, 1989.

47. S. C. Meepagala, F. Real, and C. B. Reyes, "Tip-Sample Interaction Forces in Scanning Tunneling Microscopy: Effects of Contaminants", J. Vac. Sci. Technol., B 9, pp. 1340-1342, 1991.

48. S. R. Cohen, G. Neubauer, and G. M. McClelland, "Nanomechanics of a Au-Ir Contact Using a Bidirectional Atomic Force Microscope", J. Vac. Sci. Technol., A 8, pp. 3449-3454, 1990.

49. O. Marti, J. Colchero, and J. Mlynak, "Combined Scanning Force and Friction Microscopy of Mica", Nanotechnol., **1** 2, pp. 141-144, 1990.

50. C. M. Mate, G. M. McClelland, R. Erlandson, and S. Chiang, "Atomic-Scale Friction of a Tungsten Tip on a Graphite Surface", Phys. Rev. Lett., **59**, pp. 1942-1945, 1987.

51. *Marks' Standard Handbook for Mechanical Engineers*, 8th Ed., T. Baumeister, E. Avallone, and T. Baumeister III, eds., pp. 5-51, McGraw-Hill, New York, 1978.

52. M. Green, M. Richter, J. Kortright, T. Barbee, R. Carr and I. Lindau, "Scanning Tunneling Microscopy of X-ray Optics", J. Vac. Sci. Technol., A **6**, pp. 428-431, 1988.

53. K. Nakajima, S. Aoki, T. Koyano, E. Kita, A. Tasaki and S. Fujiwara, "Surface Roughness Evaluation of Multilayer Coated X-ray Mirrors by Scanning Tunneling Microscope", Japan. J. of Appl. Phys., **28**, pp. L854-L857, 1989.

54. H. Dagnall, in: *Exploring Surface Texture*, Rank Taylor Hobson, Leichester, England, 1980.

55. J. Bennett and L. Mattson, in: *Introduction to Surface Roughness and Scattering*, Optical Society of America, Washington, D.C., 1989).

56. M. Stedman, "Limits of Topographic Measurement by the Scanning Tunneling and Atomic Force Microscopes", J. of Microscopy, **152**, pp. 611-618, 1988.

57. M. Stedman and K. Lindsey, "Limits of Surface Measurements by Stylus Instruments", Proc. SPIE, **1009**, pp. 56-61, 1988.

58. M. Stedman, "Mapping the Performance of Surface Measuring Instruments", Proc. SPIE, **83**, pp. 138-142, 1987.

59. E. L. Church and P. Z. Takacs, Proc. SPIE **1332**, 504 (1990).

60. E. L. Church and P. Z. Takacs, Proc. SPIE **1530**, 71 (1991).

61. E. L. Church, J. C. Dainty, D. M. Gale and P. Z. Takacs, Proc. SPIE **1531**, 234 (1991).

62. P. Fraundorf and J. Tentschert, "The Instrument Response Function in Air-Based Scanning Tunneling Microscopy", Ultramicroscopy, **73**, pp. 125-129, 1991.

63. J. F. Song and T. V. Vorburger, "Stylus Profiling at High Resolution and Low Force", Appl. Optics **30**, pp. 42-50, 1991.

64. C. F. Vezzetti, R. N. Varner and J. E. Potzick, "Bright-Chromium Linewidth Standard, SRM 476, for Calibration of Optical Microscope Linewidth Measuring Systems", Natl. Bur. Stand. Spec. Publ. 260-114, 1985.

65. A. F. Hebard, "Buckminsterfullerene", Annu. Rev. Mater. Sci., **23**, 1993.

66. R. Young, J. Ward, and F. Scire, "The Topografiner: An Instrument for Measuring Surface Microtopography", Rev. Sci. Instrum. **43**, pp. 999-1011, 1972.

67. G. Binnig, H. Rohrer, C. Gerber, and E. Weibel, "Surface Studies by Scanning Tunneling Microscopy", Phys. Rev. Lett. **49**, pp. 57-61, 1982.

METROLOGY APPLICATIONS OF SCANNING PROBE MICROSCOPES

by
Leigh Ann Files-Sesler
John Randall
Francis Celii

Texas Instruments Incorporated

ABSTRACT

Metrology, broadly defined as the science of dimensional measures, encompasses a wide range of applications in the semiconductor industry. This article presents examples of scanning probe microscopy (SPM) [specifically scanning tunneling microscopy (STM) and atomic force microscopy (AFM)], demonstrating the unique capabilities of these techniques to meet many of the increasingly stringent requirements of semiconductor analysis. We also address a number of problems encountered with these techniques, as well as our goals and plans for future developments and applications.

INTRODUCTION

Device geometries continue to shrink, thereby requiring techniques to perform line width and defect analysis on ever smaller scales. Conventional metrology tools such as optical microscopy and scanning electron microscopy are reaching their limit, and researchers are exploring new techniques to extend the range. This review summarizes our efforts and results in the application of STM and AFM to semiconductor metrology, and lends some insight into our thoughts regarding future areas for growth and development.

INSTRUMENTATION

The scanning tunneling microscope used in this study was commercially produced (TopoMetrix, Inc.) and operates in air. The system allows scan ranges between atomic scale and 75×75 µm^2 for both low- and high-resolution scanning (i.e., 50×50 to 500×500 data points per image). Many different combinations of scanners and sample holders allow a wide range of versatility and resolution. Scans are obtained by computer control of voltages applied to either a piezoelectric tripod arrangement (for large scans) or a tube scanner (for higher resolution) to which the tip is attached. All images reproduced here were obtained in constant-current mode by adjusting the z-voltage to maintain the desired current and recording the voltage applied as a function of x and y position to define three-dimensional (3-D) views of the surface. Coarse lateral positioning is accomplished by using piezoelectric inchworms to translate the baseplate to which the sample is attached and allows access to a 1×1 cm^2 area. The sample is attached to a plate that sits on two threaded rods connected to the base of the microscope and a motor-driven third rod that allows coarse vertical positioning and tilting of the sample plane relative to the tip. The complete microscope assembly is suspended by rubber o-rings for vibration isolation. An optical microscope is positioned to view the STM tip and its reflection at the sample surface. Sample positioning relative to the STM tip is accomplished while viewing the output of the optical microscope on a video monitor. The positional accuracy using the optical microscope is approximately 5 µm, and the optical microscope remains active during STM scanning. We use electrochemically etched tungsten tips (commercially produced by TopoMetrix, Inc., and Materials Analytical Services, Inc.). These tips are imaged with a scanning electron microscope to characterize the tip shape.

The atomic force microscope used in the studies described in this article is commercially produced (TopoMetrix, Inc.) and also operates in air. The primary difference between the STM and the AFM is the ability to image insulating surfaces with the AFM without the need for gold coating. In the AFM, a probe tip that senses the sample surface is held on a spring cantilever. Probe motions of less than 0.1 nanometer and forces as small as 1 nanonewton are measurable as the probe scans across the sample. A laser beam is reflected from the back of the cantilever onto a two- or four-section photodetector. When the probe moves up or down, the ratio of light intensity on the upper versus lower detectors changes, providing a signal to the system. Typical images are obtained in constant-force mode (i.e., the sample height is adjusted to maintain constant force). As with the STM, the voltage applied to the z piezo as a function of x and y position is recorded to define a 3-D representation of surface topography, and an optical microscope is positioned to view the cantilever and sample. (Note: Lateral force measurements can also be made with this system by determining the ratio of the light intensity between right and left detectors in the four-section detector scheme. However, this area of study is not reviewed in this article.) Commercially available cantilevers (TopoMetrix, Inc.) were used in this work; in some applications, modifications (described later) were made to the standard pyramidal tips to enhance resolution.

QUANTUM EFFECT DEVICES

The exponential progress in semiconductor integrated circuits that has stimulated growth of electronics and associated industries faces serious roadblocks in the near future.[1] Among other troubling effects brought on by continued down-scaling, the role of quantum size effects and quantum transport becomes increasingly important as semiconductor devices scale down to the sub-0.1-µm regime. Many current physics experiments have been made possible with semiconductor fabrication technology.[2] There are also attempts to exploit quantum effects to develop novel semiconductor devices. Of the many quantum effects available, resonant tunneling offers the best opportunity to develop practical devices.[1] We believe that quantum-effect devices will eventually replace conventional VLSI devices and carry the exponential growth of semiconductor computing power well into the next century. We refer to this quatum-effect device technology as nanoelectronics.

At present, nanoelectronics is a field of study dedicated to the development of methods for using the speed and discrete nature of quantum transitions in the design of electronic devices. Extensive work has been done to develop theoretical models to describe the properties of such devices as a function of size, composition, and layer thickness.[3-5] Nevertheless, devices are often fabricated only to discover they do not exhibit the expected properties. At this point, new issues arise: Is the model incorrect? Did the fabricated device actually conform to design criteria? (In other words, people start pointing fingers...). In the past, scanning electron microscopy (SEM) and optical microscopy (OM) have been extensively used for critical dimension measurements. Unfortunately, at the scale of the devices described here, these methods are reaching their limit, forcing researchers to explore new techniques to extend the range. Although many researchers have compared and contrasted SPM and SEM with varying degrees of disparity among techniques, we believe these two techniques are complementary, and we use both extensively in our work.

STM

Figure 1 shows a 3-D image of an array of gold metal contacts to gallium arsenide defined by electron-beam (e-beam) lithography and gold liftoff. This scan was obtained with STM; because of the nonconductive nature of the substrate, the sample was gold-coated before imaging. In addition to the classical 3-D and color-shaded height images, valuable quantitative information can be obtained. Figure 1(b) shows a combination of line scans that indicates the ability to choose lines of interest and then position the cursor to determine lengths, widths, and heights on the image. The widths of the dots can be used to optimize and calibrate the e-beam system while the periodicity of the pattern (which is very accurate in e-beam lithography) provides an internal standard. We attribute the rounded appearance of the dots primarily to the finite tip size, which we made no effort to deconvolute from the images, and in part to smoothing by the gold coating.

Three different arrays such as the one imaged in Figure 1 were printed on the same wafer. One problem we encountered in this study was sample damage caused by high field strengths associated with the STM tip. A scanning electron micrograph (taken at normal incidence), Figure 2(a), shows direct correlation to an STM image, Figure 2(b), and confirms the damage caused by the STM. (Correlation of the two images is easier if one locates the pair of small dots in the center of each micrograph.) The problem with sample damage manifested itself on only one array of contacts. We hypothesize that the ability to "move" the contacts results because of contamination on the GaAs surface before gold deposition. We suspect that the ability to obtain images on other arrays without damage compared to this array is caused by variation in adhesion and conduction between the gold contacts and the GaAs substrate. (Scanning electron micrographs of regions outside the STM scan area showed no damage.) Based on these results and further experiments, we found it best to operate at the lowest currents and bias voltages at which the system will stabilize. It is interesting to note that perhaps the STM is a good measure of the adhesive strength of contacts and could be used as a quality control test.[6]

Sample Preparation

In our efforts to determine the errors associated with STM imaging of samples, we discovered a problem related to sample preparation. As mentioned in the preceding discussion, it was necessary to coat the sample with gold for imaging. It is typically assumed that the gold will conformally coat and allow accurate imaging of the surface of interest. This assumption is true for many samples, especially when structural features of the surface are of micrometer dimension. However, transmission electron microscopy (TEM) cross-section images of coated samples indicate that it is possible for the coating to form islands or to assume a morphology completely independent of the underlying surface[6] (see Figure 3). The film of interest was titanium nitride and the coatings were sputtered gold palladium alloy and evaporated pure gold.

AFM

After obtaining these results and realizing the limitations of the STM, we pursued AFM for metrology applications. Figure 4(a) shows an AFM image of a new e-beam defined quantum dot array. Although we appreciated the ability to image without gold coating and we appear to have eliminated damage problems with the AFM, we discovered that the resolution was limited by the finite size of the cantilever tips, as indicated by the line scan. Through

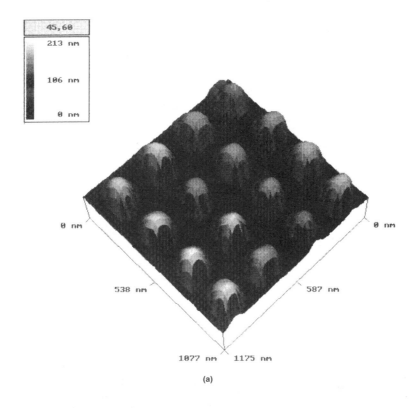

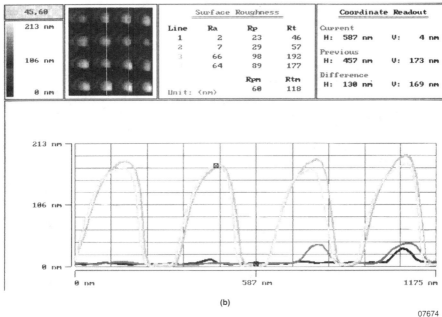

Figure 1.
(a) STM image of quantum-dot array on gallium arsenide;
(b) line scans through image. [Ref. 6]

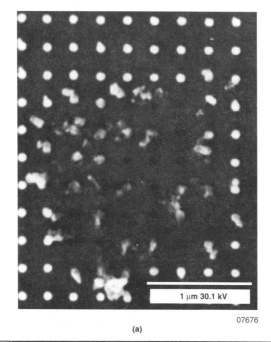

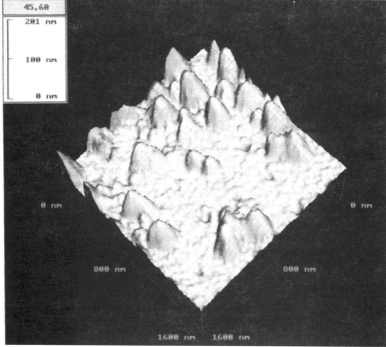

Figure 2.
(a) Electron micrograph showing damage caused by STM tip;
(b) STM image directly corresponding to electron micrograph. [Ref. 6]

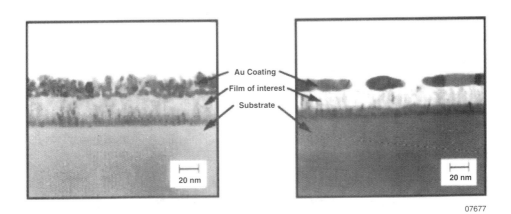

Figure 3.
Transmission electron micrographs showing nonconformity of gold coating. [Ref.6]

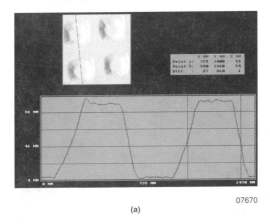

(a)

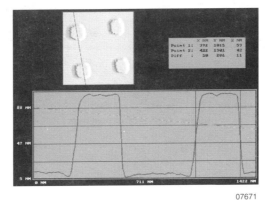

(b)

Figure 4.
AFM image and line scan of quantum dot array with
(a) standard cantilever and (b) enhanced cantilever. [Ref. 8]

interactions with other researchers,[7] we found a new technique that allows the formation of sharper tips for higher resolution. Figure 5 shows such a tip, which was formed by focusing an electron beam on the apex of the cantilever in an SEM and leaving it in spot mode. The tip "grows" as a function of time and is composed of hydrocarbon contamination in the vacuum, which is cracked and deposited onto the cantilever. Figure 4(b) shows an image obtained on the same array of quantum dots with the new, sharper tips, and the associated line scan shows the increase in resolution.[8] This enhancement in resolution is especially critical for quantum-effect devices, and we hope to see even better developments in the future. Standardization and commercial availability of high-performance tips are critical needs for routine SPM applications in this area.

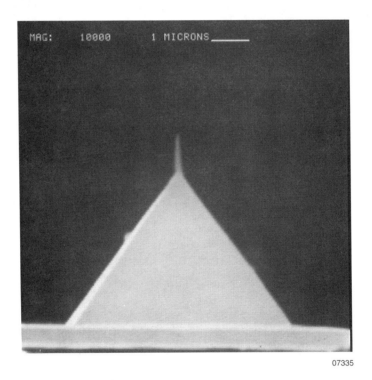

**Figure 5.
Enchanced tip "grown" on a standard cantilever in a scanning electron microscope. [Ref. 8]**

MBE FILM CHARACTERIZATION

Many advanced devices, such as pseudomorphic high electron mobility transistors (PHEMTs) and resonant tunneling diodes (RTDs),[9] contain materials that are strained because of lattice mismatch. Strained layers (e.g., InGaAs/GaAs and Si/SiGe) can be deposited defect-free only as long as the layer thickness remains below a critical value, h_c, after which misfit dislocations relieve mismatch strain, but degrade electrical performance. While a theoretical value of h_c can be calculated from the lattice mismatch,[10] the critical thickness will also depend on variables such as the deposition temperature and the initial concentration of surface defects. Therefore, it is desirable to have an *in situ* monitor of strained-layer relaxation.

We recently demonstrated the utility of laser light scattering (LLS) for determining critical thickness and monitoring surface morphology during strained epitaxial layer growth in molecular-beam epitaxy (MBE).[11-13] Because light scattering does not directly measure surface quantities, such as RMS roughness or surface step height,[14] it is important to determine the structures that give rise to the scattering. Here, we describe the correlation between *in situ* LLS observations during InGaAs/GaAs MBE growth and *ex situ* characterization with AFM, TEM, and x-ray diffraction (XRD).

MBE Experimental

Samples were prepared on epitaxy-ready, 50-mm on-orientation (100) GaAs wafers using a Perkin-Elmer 425B MBE system. For LLS measurements, a 543-nm HeNe laser (0.5 mW, randomly polarized), optically chopped at 1 kHz, irradiated the wafers at normal incidence during growth. Collection optics (telescope, interference filter, and GaAs photomultiplier tube) were placed outside one of the eight symmetric effusion cell ports, 23 degrees from the surface normal, while a UTI 100 C mass spectrometer occupied a second port. The mass spectrometer provided a sharp desorption peak of the native oxide (m/e 156, Ga_2O) for pyrometer calibration at 638°C.[15] For AFM analysis, samples were positioned in a known orientation with respect to the scan direction. TEM was performed using a Philips EM430, operating at 300 kV, on cross-sectioned or plan-view samples prepared using bromine/methanol etching.

The typical growth sequence for an InGaAs sample was the following: desorption of the GaAs native oxide by temperature ramping to ~660°C; deposition of a GaAs buffer layer at 600°C with a rate of 170 nm/min; InGaAs layer deposition at 515°C with rates from 43 to 53 nm/min; and (if used) deposition of a GaAs cap layer at 515°C. Wafer rotation speed was 1 RPM during GaAs buffer growth. InGaAs was often deposited on stationary wafers, with the azimuthal angle aligned with respect to the detection geometry for high sensitivity to oriented misfit dislocations. The InGaAs composition, varied by changing the In flux while maintaining a constant Ga flux, was strain-corrected using rocking curve measurements of the symmetric (004) and asymmetric (224) reflections.[13] The InGaAs growth rates were interpolated, using the In fraction determined by XRD between those of five calibration samples, with thicknesses determined by either cross-sectional TEM (of single layers) or XRD (of superlattices). To prepare stop-growth samples for *ex situ* characterization, the typical growth sequence was halted at various stages.

MBE Results

We first summarize the LLS observations during desorption of the GaAs native oxide and growth of GaAs buffer layers. During the initial temperature ramping of the GaAs wafer, an increase in LLS intensity was observed simultaneously with the occurrence of a sharp peak in the m/e 156 (Ga_2O) mass spectrum.[13] Desorption of the native oxide, under $\sim 1 \times 10^{-6}$ torr As ambient, causes an increase in surface roughness. *Ex situ* AFM analysis showed the presence of 3- to 10-nm-deep surface pits, with lateral size of 30 to 100 nm and a density of 10^9 to 10^{10} cm^{-2}. The RMS roughness following oxide desorption was ~10 nm, compared with 0.13 nm from a nascent GaAs wafer. These findings[13] are in agreement with a previous AFM/LLS study,[16] although different oxide thicknesses or preparation can affect the LLS signal[17,18] and, presumably, the surface morphology.

Deposition of GaAs buffer layers produced a complex LLS signature.[13] The LLS intensity decreased during the initial 20 nm of GaAs deposition, presumably because of filling of the oxide desorption pits. The LLS signal subsequently rose to a maximum with about 100 nm GaAs growth, then decreased. During the ensuing decrease, the LLS signal anisotropy (i.e., the change in signal intensity during sample rotation) also slightly increased. AFM scans of a completed buffer layer showed a smooth morphology (0.4 to 0.6 nm RMS roughness) with undulations (~0.8 to 1.2 nm high with period of 1 to 1.5 having slight alignment toward $(01\bar{1})$.[12] Also observed on some samples were oval defects (0.2×0.5 to 1.0 mm, 5 to 100 nm deep, ~10^9 cm^{-2} density), which were preferentially oriented with major axes parallel to (011).

At first glance, the LLS and AFM results for GaAs buffer growth appear to be inconsistent: a large change in surface roughness following oxide desorption gave only a small change in LLS intensity, while smoothing of the surface following GaAs buffer growth resulted in a higher LLS level. We rationalized these results by considering that our fixed-geometry detection system is most sensitive to scattering features with correlation length of 0.7 mm.[14,16,19] Thus, the LLS signal is more sensitive to the wavy GaAs morphology than the oxide desorption pits. The observed maximum in the buffer growth LLS signature may result because the correlation length of the surface features matches that of our detection system. Anisotropic GaAs buffer growth has been previously detected by LLS[16,20] and STM.[21]

We now discuss the LLS results characteristic of InGaAs layer growth. For InGaAs layers with In content below 25%, the LLS intensity typically remains constant or decreases slightly during the first few minutes of InGaAs deposition. After the critical layer thickness is deposited, misfit dislocations are generated in perpendicular planes, aligned with the (011) and $(01\bar{1})$ crystal axes, which scatter light in preferential directions. This anisotropic LLS distribution appears as a set of two or four peaks during one period of wafer rotation about the (100) axis, shown in Figure 6(a).[12,13] The more intense LLS peaks are caused by the α dislocations, aligned along $(01\bar{1})$, which have a higher density than the perpendicular β dislocations running along (011). The LLS azimuthal rotation pattern is quite different for growth of InGaAs layers with >25% In content, also seen in Figure 6(a). For example, $In_{0.25}Ga_{0.75}As$ deposition results in mostly isotropic light scattering, with weak intensity maxima that do not correspond to the major crystal axes. These observations indicate that island (3-D) growth occurs for films with In content of 25% or higher, while films with In content below 25% deposit in layer-by-layer (2-D) growth mode, and relaxation occurs by generation of misfit dislocations running in two perpendicular directions. AFM scans, shown in Figures 6(b) and 6(c), confirmed this interpretation. Cross-sectional TEM also showed randomly oriented dislocations in the 25% In sample. The onset behavior of the LLS signal for $In_xGa_{1-x}As$ films (with x = 0.25 and 0.35) suggests, however, that initial layer growth is still 2-D in these films.

Studies were also conducted with the wafer stationary and aligned specifically for detection of misfit dislocations.[11-13] While monitoring the LLS signal indicative of primary α $(01\bar{1})$ dislocations, the intensity sharply increased following attainment of the critical layer thickness, continued to increase until InGaAs deposition was halted, and remained constant during wafer cooldown. During GaAs overlayer growth, the scattering from β (011) dislocations increased proportionally to that of α dislocations. With thick InGaAs layers (up to 200 nm), the density of β dislocations, assumed proportional to the LLS intensity, always remained less than the α dislocations.

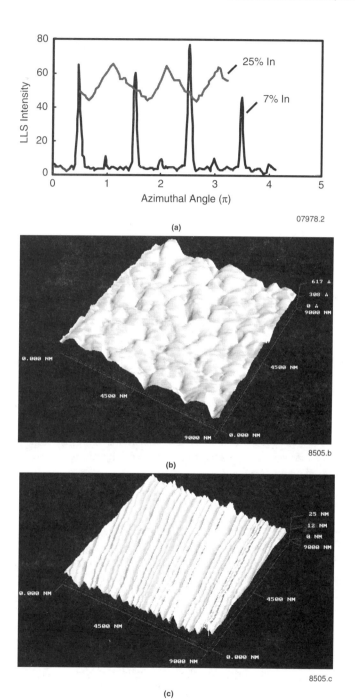

Figure 6.
(a) LLS azimuthal scans of $In_{0.07}Ga_{0.93}As$ and $In_{0.25}Ga_{0.75}As$ samples, showing highly anisotropic and isotropic scattering, respectively; (b) AFM scan of $In_{0.35}Ga_{0.65}As$ surface showing effects of 3-D growth; (c) AFM image of 25 nm GaAs/110 nm $In_{0.20}Ga_{0.80}As$ surface, showing the surface morphology caused by α dislocations. [Ref. 13]

Figure 7 shows AFM images of InGaAs samples with different layer thickness. For thicknesses close to the onset value, h_c, only surface steps running parallel to $(01\bar{1})$ were observed, in agreement with plan-view TEM, which showed only α misfit dislocations.[11] The LLS signal probably arises because of the surface steps, which are formed as misfit dislocations glide along the surface, although interface contributions cannot be discounted. The density of surface steps was about an order of magnitude lower than the TEM defect density, but the step heights were 2 to 4 monolayers (0.5 to 1 nm), and grew larger (up to ~15 nm) as the film thickness increased (up to ~200 nm). Surface steps running parallel to (011) are only weakly observed, although an AFM artifact (sensitivity to scan direction) may partly explain this result.[13]

We found that the onset time period increased as the In content of the film decreased.[12] We interpret this onset to represent the critical thickness, h_c, and found the LLS-determined h_c values to lie above the Matthews-Blakeslee model predictions.[10] Correlation with the TEM-determined misfit dislocation density showed that our current LLS detection limit is ~10^4 cm^{-1}.[12] The use of higher laser power should increase this sensitivity.

In conclusion, we found laser light scattering to be a useful *in situ* monitor of strain relaxation, especially in conjunction with *ex situ* analysis by AFM and TEM for interpretation and calibration of the LLS signatures. We used LLS to determine critical layer thickness for InGaAs/GaAs films under growth conditions, as well as to monitor surface morphology during other MBE growth processes (i.e., oxide desorption and GaAs buffer growth). Relaxation of InGaAs layers by formation of misfit dislocations caused an increase in LLS signal, probably because of surface step formation. Dislocations were generated primarily in the α direction [i.e., running parallel to $(01\bar{1})$], with secondary relaxation forming β dislocations along (011) directions. The azimuthal dependence of the LLS signal, obtained by wafer rotation, distinguished between 2-D and 3-D growth. In the future, we hope to correlate the LLS signatures with device performance to demonstrate the utility of this technique for *in situ* or *ex situ* relaxed wafer screening.

DEVICE ANALYSIS

A nanoelectronic device of particular interest is the lateral resonant tunneling transistor (LRTT).[22-24] LRTTs are essentially dual-gate high electron mobility transistors (HEMTs) where both gates are ultrashort and closely spaced. The depletion regions under each gate create tunnel barriers in the plane of the two-dimensional electron gas (2DEG). For gate lengths and spacing below 100 nm, the quantization in the region between gates becomes detectable through resonant tunneling. Because the tunnel barriers are created by depletion regions, the barrier height and width are a function of the bias applied to the gates. The resulting transport includes negative differential resistance peaks and negative transconductance. These complex characteristics may be exploited through compressed logic functions where a single device performs the same functions as many conventional transistors and is also a candidate to perform multivalued logic.[25]

However, devices such as the LRTT require sub-100-nm features to achieve the desired functionality and are consequently difficult to fabricate and characterize. The AFM can be used in a more classical failure analysis mode to solve fabrication problems. Figure 8 indicates a scan of the dual-gate area of a lateral resonant tunneling device where the gate

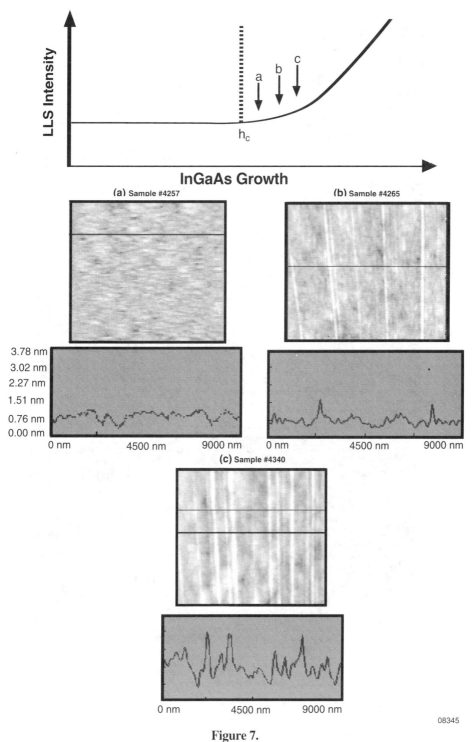

Figure 7.
Schematic of LLS signal observed during deposition of InGaAs on GaAs, and corresponding AFM images of stop-growth samples. (Ref. 13)

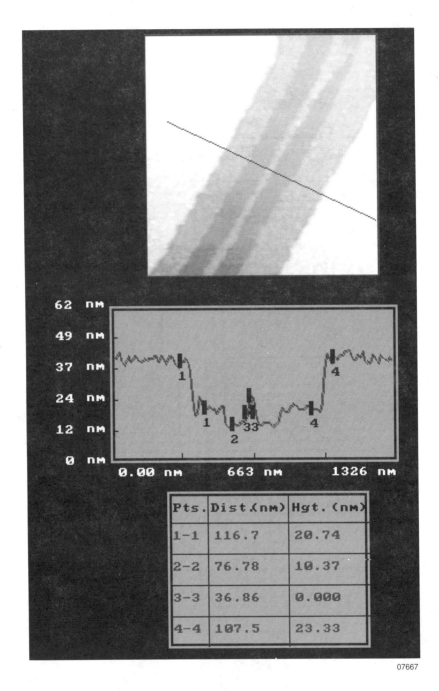

Figure 8.
AFM image and line scan of a dual-gate resonant tunneling diode device with metallization failure. [Ref. 8]

metallization failed. Line scans of this area indicated that, not only did the gold not adhere properly, but it appeared to have pulled away a portion of the substrate in the liftoff process. Scanning Auger analysis of depressed areas such as this showed a lower oxygen concentration than the surrounding areas, perhaps indicating the gold adhered to an oxide layer and then pulled the oxide layer away.

Our primary work has been in the area of topographic analysis. However, we have also explored the possibility of using STM for electrical probing of samples.[8] Figure 9 indicates the ability to distinguish between p- and n-doped silicon. These results were obtained on a cross section of an n-type substrate with a p-type film grown by liquid-phase epitaxy. The STM in constant-current mode scanned the cross section and located the edge, then the tip was moved different distances from the film edge toward the substrate. At various distances from the edge, the system was removed from constant-current feedback mode and current-versus-voltage (I/V) curves were obtained. The I/V curves seem to indicate a diode-type response. The current appears to flow strongly over the p-type film when the sample is biased positively (i.e., forward bias); at distances beyond the known thickness of the LPE film, the I/V curves reverse and the current appears to flow more readily when the sample is biased negatively. We find these results encouraging and hope to more fully explore the implications of this analysis as well as pursue more detailed electrical probing of devices in the future.

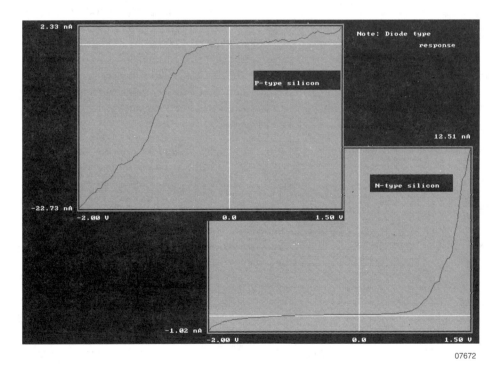

Figure 9.
I/V curves obtained on a cross section over (a) p-type region, and (b) n-type region. [Ref. 8]

SURFACE ROUGHNESS

Surface roughness is a significant figure of merit for characterization of samples. Many processes are recognized to be a function of topography of the surfaces involved.[26-28] Greater pattern resolution requires photoresist processes involving very thin layers, and particulate contamination can act as a mask or generate pinholes and limit the ultimate resolution of a process.[29] As device geometries shrink and pattern densities increase, the size of particles that can become "killer" defects decreases.[29] Therefore, one concern about choosing stainless-steel tubing for installation in advanced wafer fabrication facilities involves the possibility that the tubing will generate and/or entrap particles and provide a source of contamination.

Stainless-Steel Tubing

Figure 10 presents a comparison among samples that were (a) annealed, (b) electropolished, and (c) oxygen-passivated.[6] The software available with the SPM system allows us to calculate traditional surface roughness parameters such as R_a, R_p, and R_t for individual line scans. (R_a is the arithmetic mean of the departures of the roughness profile from the mean line, R_p is the maximum height of the profile above the mean line for a defined assessment length, and R_t is the maximum peak-to-valley height of the profile in a given assessment length.) We also developed a number of programs that allow us to determine additional statistical descriptions of the surfaces. One program calculates the increase in surface area resulting from surface morphology by finding the area of a best-fit plane (ideal surface), subtracting this value from the true area calculated for the surface, and then dividing by the ideal surface area. The surface area increase multiplied by 1,000 gives a figure of merit we call the surface index. Another program calculates the standard deviation of the complete data file (equivalent to RMS, root-mean-square roughness, for the complete image) and the average deviation from the arithmetic mean (equivalent to R_a, average roughness, for the complete image). Table 1 summarizes these surface-roughness parameters for the samples pictured in Figure 10.

For comparative roughness analysis of different samples, it is imperative that the statistical parameters for each sample be obtained under the same conditions (i.e., identical scan speeds, tip shapes, and scan ranges). Also, it is important to obtain scans over a range that is truly representative of the surface topography of all samples. In some cases, this may involve a number of scans at different ranges to identify roughness on different scales. For example, metallization is often found to exhibit a bimodal roughness nature (e.g., annealed aluminum will often demonstrate a well-defined short-range roughness and large hillocks that require long-range scans for statistically valid representation).[30]

Another important aspect of surface topography analysis is evaluation of the statistical method by which the roughness values are obtained to determine their usefulness and applicability to the problem being investigated. For example, since RMS calculations involve squaring the deviations from the average, this value is much more sensitive to large deviations in the topology of a sample (such as one finds with hillock or crack formation) than R_a. The surface index, which quantifies increases in surface area, is expected to correlate well with parameters such as moisture contamination in vacuum systems and surface

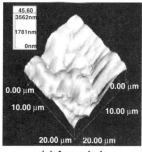
(a) Annealed

(b) Electropolished

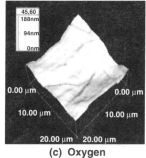
(c) Oxygen

Figure 10.
Stainless-steel samples (a) annealed, (b) electropolished, and (c) oxygen passivated. [Ref. 6]

Table 1.
Surface Roughness Parameters for Stainless-Steel Samples

Stainless-Steel Sample	SA Index (nm)	Z_{max} (nm)	Z_{avg} (nm)	RMS (nm)	RA (nm)
Bright annealed pipe	192.60	3562	1798.00	627.90	520.60
Electropolished tubing	11.07	469	318.70	50.05	38.19
Oxygen passivated tubing	0.19	188	91.99	39	32.74

SA Index = $[(\text{True surface area} - \text{Ideal surface area})/\text{Ideal surface area}]10^3$

Ideal surface area = area of least squares fitted plane

RMS (root mean square) = $\left[\left(\frac{1}{N}\right)\sum_{i=1}^{N}(z_i - \overline{Z})^2\right]^{1/2}$

RA (average roughness) = $\left(\frac{1}{N}\right)\sum_{i=1}^{N}|z_i - \overline{Z}|$

where:

$Z_{avg} = \left(\frac{1}{N}\sum_{i=1}^{N} z_i\right)$

z_i = height at point (x, y)

i = number of measurement

N = number of data points

reactivity of catalysts. Limitations to these calculations should also be recognized. For example, profiles defined by a function such as a sine wave or a triangular wave with different periodicities but identical amplitudes have equal R_a and RMS for the two profiles.[31]

FRACTALS

The roughness parameters described previously are widely accepted, but are often insufficient to correlate surface topography with other surface-related phenomena such as bondability, reflectivity, and particle removal. One method that has been applied by a number of groups[32-34] as a means of characterizing metal surfaces and their corresponding treatments is fractal analysis. However, methods for determining fractal dimensions are not yet standardized or well understood. Here, we briefly summarize the relevant background and calculation factors regarding fractal calculations. References 35 through 37 provide more complete explanations.

Euclidian dimensions of 0, 1, 2, and 3 correspond to a dot, a line, a surface, and "three-dimensional" objects, respectively. Mandlebrot introduced the idea of fractal objects to describe geometries that do not quite fit into one of these categories. For example, a dashed line might be described as having a dimension between 0 and 1, while a line profile with increasing oscillations might be described as having a dimension between 1 and 2 (approaching 2 with increasing oscillation density). For our work, the line profile case is of particular interest. Richardson[38] is responsible for demonstrating that the length estimate of certain natural boundaries is a function of the stride used to measure that length (e.g., the infamous coast of Norway). Richardson found that the measured perimeter is proportional to the stride length raised to a power that depends on the object measured, and is not an integer. If we plot the perimeter as a function of stride length versus normalized stride length on a

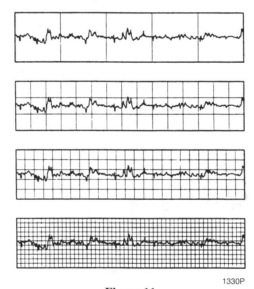

Figure 11.
"Box counting" methods for determining fractal dimension of surface profile. As the box size used to tile over a profile decreases, the number required to do so increases.
[Courtesy of Chesters *et al., Applied Surface Science* 40 (1989)].

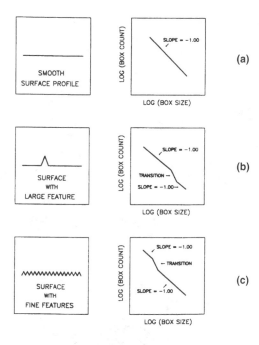

Figure 12.
(a) A flat surface yields a slope of –1. (b) Large features produce a break in the linearity corresponding to the feature size. (c) Smaller features produce a break at a smaller feature size. The number of these features affects only the slope.
[Courtesy of Chesters *et al.*, *Applied Surface Science* **40** (1989)].

log-log scale, we produce a straight line tending to infinity. Mandlebrot showed that the slope of these plots is related to the fractal dimension of the profiles.[35]

Following the example of Chesters *et al.*,[39] we apply a similar method using squares to tile over individual line scans from our STM images (Figure 11). Figure 12 demonstrates examples of plots for model surfaces. The Richardson plot for a straight line yields a slope of –1, while plots for curves with repetitive features of a particular size yield a break in the plot corresponding to the size of the features. A Richardson plot for a line-scan curve that exhibits similar roughness on magnification (termed "infinitely self-similar") would exhibit a single linear range. However, as reported by Kaye,[40] treatment processes will be evident in the resulting surface morphology, and we have seen corresponding details in Richardson plots. For typical plots, the smallest box size corresponds to the resolution limit of our microscope (or the smallest pixel size within a given scan, whichever is larger), while the largest size is chosen to extend beyond the z range. Once the boxes are large enough to contain the largest feature in the profile, the slope of the plot beyond this box size returns to –1.

Tungsten Film Analysis

Figure 13 shows tungsten films deposited under three different proprietary processes. Figure 14 shows Richardson plots for these three scans. These plots show that the break points at

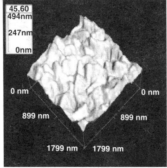

**Figure 13.
Tungsten samples with three different sets of processing parameters.**

which the slopes return to −1 with increasing box size correspond to the maximum z range, and the variations in slope can be attributed to the range and frequency of feature sizes.

Reflectivity is an important property in semiconductor processing. Low reflective films can cause registration mark misalignment in lithographic processing.[41] Also, setup of automated bonding systems is difficult if the reflectivity of samples is not consistent.[42] We have been able to correlate the results of Richardson plots with the reflectivity of samples, such as in Figure 13. Sample 1, which exhibits the highest slope and multiple break points, also has the lowest reflectivity (29% versus silicon at 480 nm). Samples 2 and 3, which have lower slopes and break points at the smallest box size, show higher reflectivity (>70%), in keeping with the general assumption that smoother samples yield higher reflectivity (assuming equivalent surface chemical composition). These preliminary results are encouraging but are not yet complete. As mentioned in the MBE section, reflectivity and light scattering measurements are known to be a function of the geometry and wavelength used in the analysis setup. The correlation length[19] defines the size and range of surface features to which a system is most sensitive. Therefore, further study into reflectivity as a function of correlation length (e.g., various wavelengths) versus fractal analysis would be particularly interesting.

We are also investigating the utility of Chesters' extension to fractal analysis, which involves calculating the slope of the log-log Richardson plot versus box size. These plots give information about feature size distributions caused by processing or growth parameters. Model fractal plots are shown in Figure 15. Plot (a) is a perfectly smooth surface—probably ideal for reducing contamination and increasing reflectivity; plot (b) is a possible surface with

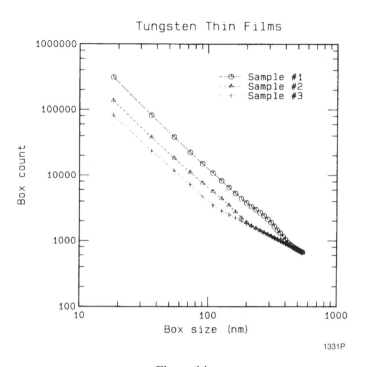

Figure 14.
Richardson plots for the three tungsten film images shown in Figure 13.

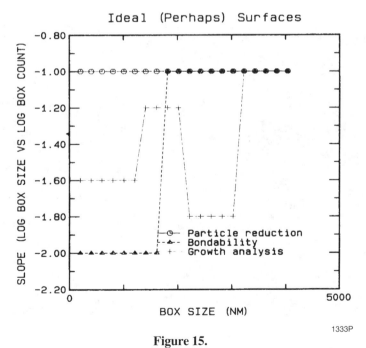

Figure 15.
Model fractal plots proposed for (a) contamination reduction/maximum reflectivity, (b) enhanced bondability, and (c) surface with multiple growth mechanisms.

good adhesion properties (i.e., a great deal of short-range roughness but no large-scale features to cause voiding and stress); and plot (c) is a surface of varying feature size related to growth mechanisms. We anticipate ultimately tuning the desired surface by correlating fractal results with processing parameters.

As with the standard roughness measurements described, fractal analysis must be performed in a consistent manner on various samples to achieve comparable results. (Note: Through collaboration with the vendor of our commercial system, many of the extended surface-roughness analysis programs we have written, including fractal dimension measurements, are being incorporated into their system. The ready availability of these programs is expected to greatly enhance research into the areas described.)

PLANARIZATION

Figure 16 demonstrates another important application of AFM. These scans indicate the ability to determine not only contact via dimensions, but also deviations from planarity as a function of process levels.[8] The device pictured in Figure 16 showed a definite conformation of top metal levels to underlying layers, yielding depressions up to ~100 nm. For devices such as capacitors, deviations from planarity can result in dramatic changes in performance, if not complete failures.[43] For other devices, lack of planarity can result in electrical shorts between multilevel-metal ayers, increased levels of contamination,[44] and ill-defined device characteristics.

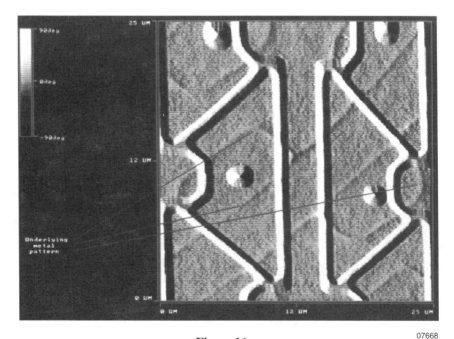

Figure 16.
Large-scale scan of device demonstrating the ability to determine metal planarity. [Ref. 8]

PROBLEMS

One of the problems unique to SPMs is the inherent nonlinearity and hysteresis associated with piezo-electric materials used in these systems. However, methods for minimizing these problems continue to be developed and incorporated rapidly into these systems.[45] Another issue is the fact that these systems involve a tip of finite size, which becomes convoluted with the sample dimensions measured.[46,47] This is particularly difficult since the shape of the tip required is a function of the sample geometry. For example, for extremely flat samples, a blunt tip with a tiny protrusion appears to provide the highest stability and resolution, while high-aspect ratio features, such as quantum dot arrays, need a long slim tip. Still another issue is the problem of understanding interactions between the tip and sample that will be critical in routine use of SPMs in metrology. The field also currently lacks standards covering the full range of metrology levels scanned by SPMs, especially z ranges.

FUTURE

Recently, Marchman et al. reported I/V measurements of quantum effect device structures.[48] Also, dopant profiling on a 100-nm scale has been demonstrated.[49] We would like to extend applications in the area of electrical probing of samples with correlation to device geometries. For example, scanning quantum effect devices (both lateral and vertical structures composed of various materials) in topography mode and then obtaining I/V curves *in situ*.

In the future, we plan to continue to enhance the resolution and accuracy of our topography measurements, and to further explore fractal analysis with correlation to device and material properties. Preliminary work using SPMs to determine tribological information is also very promising. Measurements of adhesion forces,[50,51] solvation forces,[52] and frictional forces[53,54] have been demonstrated, and we anticipate rapid growth and exciting developments in these areas. Other areas receiving a great deal of interest include measurements of temperature[55-57] and magnetic forces[58,59] in the submicrometer range.

We recently obtained a new AFM stage that allows analysis of large samples (e.g., whole wafers), and we consider developments such as this crucial to eventual *in situ* process monitoring. However, work must still be done to make these systems compatible with cleanroom technology (i.e., reduced particle generation and contamination control). Automation of systems for particular applications will also be necessary. Most critical of all will be improvements in the precision and accuracy of these systems. Credibility of process monitoring equipment is of utmost importance. Million-dollar decisions are based on the output of such equipment, and SPMs introduced too early into a wafer fabrication facility could cause serious errors, leading to loss of credibility for the metrology equipment (and the fabrication facility).

SUMMARY

We summarized a wide range of applications of scanning probe microscopes to analysis of nanostructure devices, including determination of line widths, substrate roughness, and lithography development failure analysis. We presented results of surface topography analysis of aluminum metallization to determine planarity and roughness measurements of stainless-steel supply tubing where particle formation and contamination were the primary concerns. We demonstrated correlation between fractal analysis and surface reflectivity of thin films of tungsten. We also summarized an in-depth study correlating *in situ* laser light scattering results with *ex situ* AFM analysis. We indicated problem areas such as tip-sample interactions and loss in resolution because of finite tip size. However, we are very encouraged by our results to date and the rapid progress in improving these techniques and resolving the associated problems. We plan to continue our present efforts in topographical analysis as well as explore electrical and tribological analysis with these techniques.

ACKNOWLEDGEMENTS

The authors would like to thank Rocky Krueger for preparing the enhanced tips and the scanning electron micrographs and Mike Coviello for providing the transmission electron micrographs. We acknowledge the technical assistance of Faye Phillips, Kathy Rice, and C. David Smith in the MBE and LLS work. Also, Steven Chesters of American Air Liquide was especially helpful in our efforts to understand and implement fractal analysis programs

REFERENCES

1. R.T. Bate, *Nanotechnology* 1 (1990).
2. Examples of this work are found in: *Nanostructure Physics and Fabrication*, edited by Mark Reed and Wiley Kirk (Academic Press, Inc., New York, 1989).
3. J.H. Luscombe, J.N. Randall, and A.M. Bouchard, *Proceedings of the IEEE* 79:8 (1991).
4. W.R. Frensley, *Rev. Mod. Phys.* 62 (1990).
5. C.S. Lent, D.J. Kirkner, *J. Appl. Phys.* 67:10 (1990).
6. L.A. Files-Sesler, J.N. Randall, and D. Harkness, *J. Vac. Sci. Technol.* B 9:2 (1991).
7. D. Keller, Univ. of New Mexico (private communications).
8. L.A. Files-Sesler, *Texas Instruments Technical Journal* 10:3 (1993).
9. A.C. Seabaugh, Y.-C. Kao, and H.-T. Yuan, *IEEE Electron. Dev. Lett.*, 13 (1992).
10. J.W. Matthews and A.E. Blakeslee, *J. Cryst. Growth* 27 (1974).
11. F.G. Celii, E.A. Beam III, L.A. Files-Sesler, H.-Y. Liu, and Y.-C. Kao, *Appl. Phys. Lett.*, 62: 21 (1993).
12. F.G. Celii, Y.-C. Kao, H.-Y. Liu, L.A. Files-Sesler, and E.A. Beam III, *J. Vac. Sci. Technol.* B 11:3 (1993).
13. F.G. Celii, L.A. Files-Sesler, E.A. Beam III, and H.-Y. Liu, *J. Vac. Sci. Technol.*, A 11:4 (1993).
14. E.L. Church, H.A. Jenkinson, and J.M. Zavada, *Opt. Eng. 16* (1977).
15. A.J. SpringThorpe and P. Mandeville, *J. Vac. Sci. Technol.* B6 (1988).
16. G.W. Smith, A.J. Pidduck, C.R. Whitehouse, J.L. Glasper, A.M. Keir, and C. Pickering, *Appl. Phys. Lett.* 59 (1991).
17. C. Lavoie, S.R. Johnson, J.A. Mackenzie, T. Tiedje, and T. van Buuren, *J. Vac. Sci. Technol.* A10 (1992).
18. M.K. Weilmeier, K.M. Colbow, T. Tiedje, T. van Buuren, and L. Xu, *Can. J. Phys.* 69 (1991).
19. D.J. Robbins, A.J. Pidduck, A.G. Cullis, N.G. Chew, R.W. Hardeman, D.B. Gasson, C. Pickering, A.C. Daw, M. Johnson, and R. Jones, *J. Crys. Growth* 81 (1987).
20. F. Briones, D. Golmayo, L. Gonzalez, and J.L. De Miguel, *Jpn. J. Appl. Phys.* 24: L478 (1985).
21. J. Sudijono, M.D. Johnson, C.W. Snyder, M.B. Elowitz, and B.G. Orr, *Phys. Rev. Lett.* 69:19 (1992).
22. K.E. Ismail, P.G. Bagwell, T.P. Orlando, D.A. Antoniadis, and H.I. Smith, *Proc. IEEE* 79 (1991).
23. S. Chou, D.R. Allee, R.F. Pease, J.S. Harris, Jr., *Proc. IEEE* 79 (1991).
24. A.C. Seabaugh, J.N. Randall, Y.-C. Kao, J.H. Luscombe, and A.M. Bouchard, *Electronics Letters* 27 (1991).

25. A.H. Taddiken, A.C. Seabaugh, G.A. Frazier, and J.N. Randall, GOMAC, November 10, 1992, Las Vegas, NV.
26. G. Kasper, and H.Y. Wen, in *Handbook of Contamination Control in Microelectronics*, D.L. Tolliver, ed. (Noyes, Park Ridge, NJ, 1988).
27. D.A. Toy, *Semiconductor International* (1990).
28. S.C. Langford, M.A. Zhenyik, L.C. Jensen, and J.T. Dickinson, *J. Vac. Sci. Technol. A* 8:4 (1990).
29. W.G. Fisher, in *Particle Control for Semiconductor Manufacturing*, R.P. Donovan, ed. (Marcel Dekker, Inc., New York, NY, 1990).
30. L.A. Files-Sesler, T. Hogan, and T. Taguchi, *J. Vac. Sci. Technol. A* 10:4 (1992).
31. S. Chesters, H.C. Wang, and G. Kasper, *Proc. Inst. Env. Sci.* (1990).
32. E.E. Underwood, and K. Banerji, *Matl. Sci. & Eng.* 80 (1986).
33. H. Schwarz and H.E. Exner, *Powder Tech.* 27 (1980).
34. P. Pfeifer, *Appl. of Surf. Sci.* 18 (1984).
35. B.B. Mandelbrot, *The Fractal Geometry of Nature* (Freeman, New York, 1983).
36. J. Feder, *Fractals* (Plenum Press, New York, 1988).
37. M. Barnsley, *Fractals Everywhere* (Academic Press, California, 1988).
38. L.F. Richardson, *General Systems Yearbook* 6 (1961).
39. S. Chesters et al., *Appl. Surf. Sci.*, 40 (1989).
40. B.H. Kaye, in *Particle Characterization and Technology* Vol. 1, J.K. Beddow, ed. (CRC, Boca Raton, FL, 1984) Ch. 5.
41. R.S. Nowicki, *J. Vac. Sci. Tech.* 17 (1980).
42. G. Harmon, in *Reliability and Yield Problems in Wire Bonding in Microelectronics* (International Society for Hybrid Microelectronics, 1989).
43. S.M. Sze, in *VLSI Technology* (McGraw-Hill Book Co., New York, 1983) Ch. 9.
44. H.P. Klein, U. Dermutz, S. Pauthner, and H. Rohrich, *IPFA Symposium Proceedings* (1989), unpublished.
45. D.A. Grigg, J.E. Griffith, G.P. Kochanski, M.J. Vasile, and P.E. Russell, *Proceedings SPIE Symposium on Microlithography* 1673 (1992).
46. S. Akamine, R.C. Barrett, and C.F. Quate, *Appl. Phys. Lett.* (1990).
47. J.E. Griffith, D.A. Grigg, M.J. Vasile, P.E. Russell, and E.A. Fitzgerald, *J. Vac. Sci. Technol. A* 10:4 (1992).
48. H.M. Marchman, G.C. Wetzel, and J.N. Randall, unpublished work.
49. C.C. Williams, J. Slinkman, W.P. Hough, and H.K. Wickramasinghe, *J. Vac. Sci. Technol. A* 8:2 (1990).
50. H.A. Mizes, K.-G. Loh, R.J.D. Miller, S.K. Ahuja, and E.F. Grabowski, *Appl. Phys. Lett.* 59:22 (1991).
51. U. Landman, W.D. Luedtke, N.A. Burnham, and R.J. Colton, *Science* 248:2954 (1990).
52. S.J. O'Shea, M.E. Welland, and T. Rayment, *Appl. Phys. Lett.* 60:10 (1992).
53. R.M. Overney, E. Meyer, J. Frommer, D. Brodbeck, R. Luthi, L. Howald, H.J. Guntherodt, M. Fujihira, H. Takano, and Y. Gotoh, *Nature* 359 (1992).
54. J. Krim and R. Chiarello, *J. Vac. Sci. Technol.* B 9:2 (1991).
55. M. Nonenmacher, M. O'Boyle, and H.K. Wickramasinghe, *Ultramicroscopy Part A* 42-44 (1992).
56. M. Nonenmacher and H.K. Wickramasinghe, *Appl. Phys. Lett.* 61:2 (1992).
57. A. Majumdar, J. Lai, A. Padmanabhan, and J.P. Carrejo, AVS International Symposium (November 1992).
58. J.H. Wandass, J.S. Murday, and R.J. Colton, *Sensors and Actuators* 19:3 (1989).
59. P.N. First, J.A. Stroscio, D.T. Pierce, R.A. Dragoset, and R.J. Celotta, *J. Vac. Sci. Technol. B* 9:2 (1991).